Withdrawn
University of Waterloo

Current Topics in Microbiology and Immunology

268

Editors

R.W. Compans, Atlanta/Georgia
M.D. Cooper, Birmingham/Alabama · Y. Ito, Kyoto
H. Koprowski, Philadelphia/Pennsylvania · F. Melchers, Basel
M.B.A. Oldstone, La Jolla/California · S. Olsnes, Oslo
M. Potter, Bethesda/Maryland
P.K. Vogt, La Jolla/California · H. Wagner, Munich

Springer
Berlin
Heidelberg
New York
Barcelona
Hong Kong
London
Milan
Paris
Tokyo

The Proteasome – Ubiquitin Protein Degradation Pathway

Edited by P. Zwickl and W. Baumeister

With 17 Figures and 6 Tables

Springer

Dr. Peter Zwickl
Professor Dr. Wolfgang Baumeister
Department of Molecular Structural Biology
Max-Planck-Institute for Biochemistry
Am Klopferspitz 18 a
82152 Martinsried
Germany
e-mail: zwickl@biochem.mpg.de
  baumeist@biochem.mpg.de

Cover Illustration: Schematic representation of the MHC class I-antigen processing and presentation pathway (G. Niedermann, this volume).

ISSN 0070-217X
ISBN 3-540-43096-2 Springer-Verlag Berlin Heidelberg New York

This work is subject to copyright. All rights are reserved, whether the whole or part of the material is concerned, specifically the rights of translation, reprinting, reuse of illustrations, recitation, broadcasting, reproduction on microfilm or in any other way, and storage in data banks. Duplication of this publication or parts thereof is permitted only under the provisions of the German Copyright Law of September 9, 1965, in its current version, and permission for use must always be obtained from Springer-Verlag. Violations are liable for prosecution under the German Copyright Law.

Springer-Verlag is a company in the specialist publishing group BertelsmannSpringer

http://www.springer.de

© Springer-Verlag Berlin Heidelberg 2002
Library of Congress Catalog Card Number 15-12910
Printed in Germany

The use of general descriptive names, registered names, trademarks, etc. in this publication does not imply, even in the absence of a specific statement, that such names are exempt from the relevant protective laws and regulations and therefore free for general use.

Product liability: The publishers cannot guarantee the accuracy of any information about dosage and application contained in this book. In every individual case the user must check such information by consulting other relevant literature.

Production Editor: Christiane Messerschmidt, Rheinau
Cover Design: *design & production GmbH*, Heidelberg
Typesetting: Scientific Publishing Services (P) Ltd, Madras
Printed on acid-free paper   SPIN: 10718401   27/3020 5 4 3 2 1 0

Preface

This volume gives an overview of proteasome-mediated protein degradation and the regulatory role of the ubiquitin system in cellular proteolysis.

The first chapter describes the molecular evolution of the proteasome and its associated activators, i.e., the 20S core, the base and the lid of the 19S cap, and the 11S regulator. The ensuing chapter gives an overview of the structure and assembly of the 20S proteasome and the regulation of the archaeal proteasome by PAN. The third contribution summarizes our knowledge on the eukaryotic 26S proteasome and its regulation by the 19S regulator, followed by a chapter devoted to the 11S regulator, which elucidates the structural basis for the 11S-mediated activation of the 20S proteasome. The fifth chapter reviews in detail the role of the proteasome in the immune response. The subsequent chapter gives a comprehensive description of the natural substrates of the proteasome and their recognition by the enzymes of the ubiquitination machinery. The penultimate chapter rounds up the information on intracellular distribution of proteasomes in yeast and mammalian cells, while the last contribution highlights proteasome inhibitors, tools which proved to be very valuable for dissecting the cellular roles of the proteasome and which might turn out to be of pharmacological importance. In summary, this volume gives a complete overview of the structure and function of the proteasome, its regulators and inhibitors, and its cellular importance, in conjunction with the ubiquitin system, for the degradation of regulatory proteins and the generation of immunogenic peptides.

January 2002

PETER ZWICKL
WOLFGANG BAUMEISTER

List of Contents

C. VOLKER and A.N. LUPAS
Molecular Evolution of Proteasomes   1

P. ZWICKL
The 20S Proteasome. 23

M.H. GLICKMAN and V. MAYTAL
Regulating the 26S Proteasome 43

C.P. HILL, E.I. MASTERS, and F.G. WHITBY
The 11S Regulators of 20S Proteasome Activity 73

G. NIEDERMANN
Immunological Functions of the Proteasome 91

H.D. ULRICH
Natural Substrates of the Proteasome
and Their Recognition by the Ubiquitin System 137

C. GORDON
The Intracellular Localization of the Proteasome 175

M. BOGYO and E.W. WANG
Proteasome Inhibitors:
Complex Tools for a Complex Enzyme 185

Subject Index. 209

List of Contributors

(Their addresses can be found at the beginning of their respective chapters.)

BOGYO, M.   185

GLICKMAN, M.H.   43

GORDON, C.   175

HILL, C.P.   73

LUPAS, A.N.   1

MASTERS, E.I.   73

MAYTAL, V.   43

NIEDERMANN, G.   91

ULRICH, H.D.   137

VOLKER, C.   1

WANG, E.W.   185

WHITBY, F.G.   73

ZWICKL, P.   23

Molecular Evolution of Proteasomes

C. VOLKER and A.N. LUPAS*

1	Introduction	1
2	Molecular Evolution of the 20S Proteolytic Core Complex	2
2.1	The Ntn Hydrolase Family	2
2.2	Evolution of Self-Compartmentalization	5
2.3	Subunit Differentiation in Eukaryotes	7
2.4	Problems in the Phylogenetic Reconstruction of Proteasome Evolution	9
3	Molecular Evolution of the 11S Regulator (PA28)	10
4	Molecular Evolution of the 19S Cap Complex	11
4.1	The Base Subcomplex: ATPase Subunits	12
4.2	The Base Subcomplex: Non-ATPase Subunits	15
4.3	The Lid Subcomplex and Its Relationship to COP9 and eIF3	17
5	Conclusion	18
6	Summary	18
	References	19

1 Introduction

Because of the potential damage to the cell, intracellular protein degradation is one of the most tightly regulated processes in living systems. As a consequence, all protein breakdown in the cell is energy dependent, even though proteolysis is fundamentally exergonic. The basic regulatory strategy has been the confinement of proteolytic active sites to internal compartments with tightly controlled access. In eukaryotes this has eventually led to the evolution of a dedicated organelle, the lysosome, in which the primary energy-dependent step is the translocation of substrates across the membrane. The oldest solution, however (which still handles the bulk of intracellular proteolysis even in eukaryotes) is a set of proteases that form barrel-shaped complexes through self-association, enclosing a central

---

Bioinformatics, SmithKline Beecham Pharmaceuticals, UP 1345, 1250 South Collegeville Road, Collegeville, PA 19426-0989, USA
* *Present address*: Department of Protein Evolution, Max Planck Institute for Developmental Biology, Spemannstr. 35, 72076 Tuebingen, Germany

proteolytic cavity (LUPAS et al. 1997b). Access to this cavity is provided by polar pores guarded by ring-shaped ATPases, which unfold and translocate substrate proteins in an energy-dependent manner. Prokaryotes contain several such proteases, including Lon, ClpAP, ClpXP, FtsH, and proteasomes, but the only one present outside of organelles in eukaryotes is the proteasome.

Proteasomes are, in one form or another, ubiquitous to life (for reviews see COUX et al. 1996; BAUMEISTER et al. 1998; PETERS et al. 1998). The simplest form is found in bacteria and consists of two rings of core subunits (HslV), which associate with ATPase rings of the Clp/HSP100 subfamily (HslU). This complex is induced during heat shock and is not essential under normal growth conditions. Archaea contain a more elaborate version, formed by four rings of core subunits (the 20S proteasome) that interact with ATPase rings of the AAA subfamily. This form is also found in one branch of the bacteria, the actinomycetes, where it was presumably acquired by horizontal transfer. In the archaeon *Thermoplasma acidophilum* the proteasome is essential under stress conditions, but not during normal growth, whereas in the actinomycete *Mycobacterium smegmatis* its knock-out does not have a phenotype under any of the conditions tested. The most complicated form occurs in eukaryotes, where a 20S proteasome complex forms the core of a much larger protease, the 26S proteasome. In this form, the 20S core complex is capped by two 19S complexes containing ATPases of the AAA subfamily. The 26S proteasome is the central protease of ubiquitin-dependent protein degradation and is essential for cell survival. It is found free in the cytosol, attached to the endoplasmic reticulum (ER), and in the nucleus, and is involved in the removal of abnormal proteins, antigen processing, signal transduction, transcription, cell cycle progression, and apoptosis.

In this chapter we discuss the evolution of proteasomes from simple monomeric precursors to structures of increasing complexity (Fig. 1), leading up to the highly complicated 26S proteasome, a molecular machine of 32 subunits and 2.2MDa.

2 Molecular Evolution of the 20S Proteolytic Core Complex

2.1 The Ntn Hydrolase Family

The protease subunits of energy-dependent proteases appear to have originated from a number of unrelated hydrolase families, most of which are not typically associated with proteolysis (LUPAS et al. 1997b). Only FtsH, a Zn-dependent metalloprotease, belongs to one of the classic protease families (clan MB, family M41 in the MEROPS classification; http://merops.iapc.bbsrc.ac.uk). The proteolytic subunits of proteasomes belong to the N-terminal nucleophile (Ntn) hydrolases, a group of proteins that also encompasses penicillin acylases, class II glutamine amidotransferases, and glycosylasparaginase (BRANNIGAN et al. 1995;

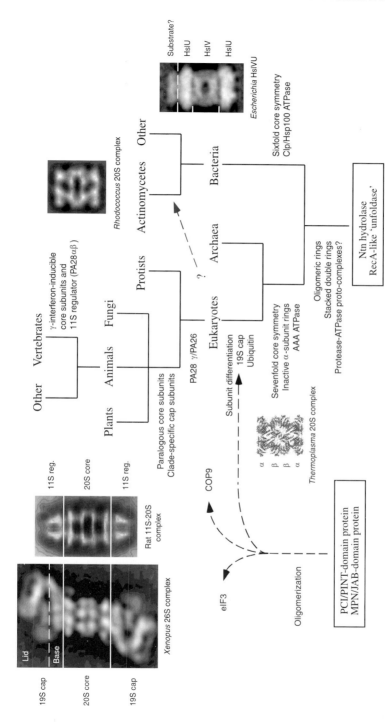

Fig. 1. Schematic representation of proteasome evolution. The branchpoints represent the most likely sequence of events, as discussed in the text. (Images courtesy of the Baumeister laboratory)

see the SCOP classification; http://scop.mrc-lmb.cam.ac.uk/scop/index.html). All Ntn hydrolases cleave amide bonds, suggesting that this is an ancestral property of the family; however, their substrates differ substantially and only the proteasome cleaves proteins and peptides. Also, each Ntn hydrolase has a different quaternary structure, suggesting that the ancestor was not an oligomeric protein.

Ntn hydrolases are synthesized as inactive precursors and are activated either by an internal cleavage, which generates two subunits, or by removal of a propeptide. The cleavage exposes an internal threonine, serine, or cysteine residue at the newly created N terminus (hence the family name). This forms a 'single-residue active site' by providing the catalytic nucleophile in its side-chain and the general base in the free backbone amino group (DUGGLEBY et al. 1995). Although the catalytic activity depends primarily on the N-terminal residue, other residues appear to facilitate its task. These differ between Ntn hydrolases and presumably arose after the separation from the common ancestor. In the proteasome, a lysine, a glutamate, and an aspartate residue (Lys33, Glu17, and Asp166 in the *Thermoplasma* β subunit) appear to delocalize the charge of the free N terminus (SEEMULLER et al. 1995). This system of interacting charges seems to lie halfway between true single-residue catalysis and a full charge-relay system. Specialization of the catalytic mechanism after divergence from the common ancestor is also seen in the fact that proteasomes require threonine as a nucleophile, although serine and cysteine still provide some functionality (SEEMULLER et al. 1995; AKOPIAN et al. 1997).

The cleavage required for activation is known to be autocatalytic in proteasomes (CHEN and HOCHSTRASSER 1996; SEEMULLER et al. 1996), penicillin acylases (LEE et al. 2000), and glycosylasparaginase (GUAN et al. 1996), suggesting that this is a common property of Ntn hydrolases. The reaction appears to be intramolecular both in glycosylasparaginase and proteasomes; however, efficient intermolecular processing is also seen by incubation or coexpression of processing-competent and -incompetent *Thermoplasma* proteasome subunits. Processing is generally dependent on the presence of the residues involved in forming the mature active site; however, in glycosylasparaginase, processing requires a histidine–aspartate pair (residues 204 and 205 in the human enzyme), which is subsequently removed in a trimming reaction. Overall, it would seem that autocatalytic processing by an intramolecular mechanism is an ancestral feature of Ntn hydrolases, as proposed in the initial description of the superfamily (BRANNIGAN et al. 1995), but that variations on this reaction evolved separately in the different enzyme families.

How may an inactive ancestor have evolved into an autocatalytically processed Ntn proto-hydrolase? An essential step must have been the ability to undergo an N–O acyl shift at the catalytic nucleophile, in which the main-chain carbonyl group is transferred from the main-chain amino group to the side-chain hydroxyl (or sulfhydryl) group. This reaction is probably easy to attain in proteins, as it has evolved independently in many enzyme classes (such as Ntn hydrolases, inteins, and pyruvoyl-dependent enzymes) and even within classes (pyruvoyl-dependent enzymes belonging to different folds). In Ntn hydrolases, the acyl-shifted intermediate is resolved by hydrolysis, yielding a free N terminus; in pyruvoyl-dependent enzymes, it is resolved by beta elimination, resulting in an N-terminal pyruvate group (van Poelje and Snell 1988; Xiong and Pegg 1999); and in inteins it is resolved by transfer to a second internal nucleophile, leading to a branched-chain intermediate and eventually to excision of the intein (Xu et al. 1994).

A second essential step must have been the acquisition of catalytic activity. In Ntn hydrolases (but not in pyruvoyl-dependent enzymes or in inteins), catalysis and processing use the same nucleophile and are closely related reactions, suggesting that one activity evolved from the other. But which came first? In an autoprocessing-first scenario, the propensity for internal cleavage would have resulted from the unusual Ntn hydrolase fold, subsequently leading to the evolution of an active site. In agreement with this scenario, another very similar fold (albeit with substantial differences in the connectivity and orientation of secondary structure elements) appears to have evolved convergently, resulting in a similar catalytic site and processing mechanism (Bompard-Gilles et al. 2000). However, in this scenario the cleavage site would be expected to occur in an integral part of the fold and not, as in Ntn hydrolases, at a domain boundary. This location is more compatible with a catalysis-first scenario, in which the nucleophile was initially located in the second position of the polypeptide chain and the only processing required for activation was the removal of the N-terminal methionine by Met aminopeptidase (as observed in propeptide deletion mutants of the *Thermoplasma* proteasome expressed in *Escherichia coli*; Seemuller et al. 1995). Subsequent evolution would have added autocatalytic processing as a regulatory step, by addition of a propeptide (as in the proteasome), fusion with a second protein (as in aspartylglucosaminidase), or circular permutation.

2.2 Evolution of Self-Compartmentalization

The starting point of proteasome evolution was most probably a monomeric hydrolase; the end points, observable in organisms today, are barrel-shaped structures of stacked subunit rings associated with ATPase complexes. The intermediate steps on this path would have been ring formation, ring stacking, and association with ATPases. Ring formation must have preceded ring stacking, but it is unclear at what point the first association with an ATPase occurred.

If protease–ATPase proto-complexes had formed first in proteasome evolution, assembly could have proceeded on a template provided by the ATPases. As will be discussed later, proteasome ATPases belong to an ancient superfamily of ring-shaped proteins, which had almost certainly assumed their oligomeric structure by the time proteasomes evolved. However, ring-shaped complexes appear to have arisen independently in many families, without template structures and driven primarily by functional requirements (translocation along DNA in helicases, substrate segregation in chaperones, protection of intracellular proteins from random degradation in proteases). Therefore, we favor a scenario in which proteases and ATPases evolved independently towards mutually compatible ring-shaped structures. Interactions between them formed subsequently and, at least in part, as a result of convergence.

Complexes formed by single rings of protease and ATPase subunits may have represented an evolutionarily stable solution; such an arrangement is still seen in the present-day proteases FtsH and Lon. However, in Clp and proteasomes oligomerization proceeded further, producing complex dimers by head-to-head stacking of the protease rings. The assembly of 20S complexes from 'half-proteasomes' may represent a remnant of this event (CHEN and HOCHSTRASSER 1996; SCHMIDTKE et al. 1997; ZÜHL et al. 1997). Stacking was probably driven by the need to maintain an internal cavity that is sequestered from the cytosol independently of protease–ATPase association. In FtsH and Lon, where protease and ATPase are part of the same polypeptide, the cavity enclosed by the two rings never becomes exposed as the two rings cannot dissociate.

In ClpP and HslV, oligomerization stopped at the two-ring stage, but in 20S proteasomes, further stacking occurred to produce a four-ringed structure, in which the two inner rings are stacked head-to-head (as in HslV) and the two outer rings are stacked head-to-tail. In a homo-oligomeric complex, such an arrangement seems to be evolutionarily unstable (because it would promote filament formation), requiring duplication and divergence of the subunits forming the inner and outer rings. In 20S proteasomes this led to the origin of α-subunits.

It is unclear what evolutionary incentive could have driven the formation of four-ring complexes. One rationale comes from the observation that in the *Rhodococcus* proteasome, αβ dimers represent an assembly intermediate (ZÜHL et al. 1997). If α-subunits arose from β-subunits as an assembly aid, they would have had to interact in a head-to-tail orientation in order not to block the head-to-head assembly sites of the β-subunits, leading directly to the complex observed today. A second rationale could come from a need for more specific protease–ATPase interactions. α-Subunits contain a highly conserved N-terminal extension that forms a docking surface at each end of the barrel. This extension is absent from β-subunit rings or HslV and is incompatible with catalytic activity (by precluding the processing required for activation). Therefore, this solution requires the duplication and specialization of subunits, yielding inner catalytic and outer docking rings. Prior to the origin of α-subunits, proteasome–ATPase interactions may have been more promiscuous (as still seen with ClpP), providing a rationale for the observation that HslV associates with a different ATPase subfamily than other proteasomes.

In conclusion, the most likely scenario for proteasome evolution involves ring formation, followed by ATPase association, followed by stacking into progressively more elaborate complexes. The steps preceding the origin of α-subunits would have taken place prior to the divergence of bacteria from other organisms. Subsequent steps up to the formation of a specific interaction with an AAA ATPase would have occurred in the common ancestor of archaea and eukaryotes. The endogenous bacterial complex (HslV) has a subunit stoichiometry different from that of other proteasomes, raising the possibility that proteasomes had not evolved to an oligomeric structure by the time of bacterial divergence. However, we believe that this difference is superficial. Packing interactions within and between subunit rings are so similar in HslV and 20S β-subunits as to essentially preclude convergent evolution (LÖWE et al. 1995; BOCHTLER et al. 1997). Stoichiometric heterogeneity is observed in other ring-shaped complexes, for example in the HslV-associated ATPase HslU (ROHRWILD et al. 1997).

2.3 Subunit Differentiation in Eukaryotes

The common ancestor of archaeal and eukaryotic proteasomes was a barrel-shaped structure of four homo-oligomeric rings, consisting of two inner active β-rings and two inactive outer α-rings; it interacted with homo-oligomeric AAA ATPase rings. This organization is still observable in the present-day proteasomes of archaea. In the proteasomes of all eukaryotes, however, the rings are each formed by different but related subunits (seven in the case of α and β rings, six in the case of the ATPase). A fully diversified set of subunits is observed even in the most primitive eukaryotes, for example the amitochondriate diplomonad *Giardia* (see Table 1, BOUZAT et al. 2000), suggesting that eukaryotes evolved proteasomes of maximal complexity in an evolutionarily brief period following immediately upon their divergence from archaea. Similarly rapid evolution to a state of maximal complexity is seen in other large oligomeric complexes, such as class II chaperonins (WALDMANN et al. 1995). Presumably, the differentiation of protease and chaperone subunits was driven by the need to act specifically on multiple related proteins, which had themselves arisen by multiplication and differentiation in response to the large increase in cellular complexity during the prokaryote–eukaryote transition. In the case of proteasomes, differentiation may have been accelerated further by the fact that in a ring with a prime number of subunits (seven), states of intermediate complexity are probably evolutionarily unstable.

It is not clear why eukaryotes did not use the bacterial solution to multiple specificities, namely multiple different complexes (FtsH, Clp, Lon, proteasomes), but it has been noted in other contexts – for example transport processes (SAIER 1999) – that eukaryotes have shown a comparative lack of evolutionary inventiveness at the level of proteins, generally proceeding by multiplication and differentiation of existing systems. The last common ancestor, however, probably contained only a proteasome precursor and (possibly) a Lon-like protease, limiting the evolutionary options for the proto-eukaryotic cell.

A peculiar (and not understood) aspect of differentiation is the origin of four inactive β-subunits (β3, β4, β6 and β7), which reduced the number of active site specificities to three (β1, peptidylglutamyl-hydrolyzing; β2, tryptic; β5, chymotryptic), unevenly spaced in the inner proteolytic cavity. As noted above, inactive β-subunits are seen in the deepest-branching eukaryotes and must be considered an integral part of the differentiation process.

After the origin of multi-cellularity, proteasomes further increased in complexity through the origin of paralogous core subunits (i.e., which can substitute for each other in active complexes). In *Arabidopsis*, for example, only four subunits are encoded by single genes; all others are encoded by families of at least two genes (23 genes in total; FU et al. 1998). The reasons for such extensive paralogy are, as yet,

Table 1. Subunits of the 19S cap complex, separated into lid and base subcomplexes. The table lists the human and yeast subunits, their sequence motifs, and their phylogenetic spectra. The subunits of the lid are related to subunits of the COP9 and eIF3 complexes. Subunits S1 and S2 of the base are related to subunit APC1 of the cyclosome/APC complex. Subunit S5a, which appears to mediate the association of the lid and the base, contains an N-terminal VWA domain resembling that of subunit p44 of basic transcription factor TFIIH

	19S Human	19S Yeast	Motif	Phylogenetic spectrum	COP9	Phylogenetic spectrum	eIF3	Phylogenetic spectrum	Other complexes	Phylogenetic spectrum
Lid	S3	RPN3	PCI/PINT	C, G	Sgn3 (S3)	C, D	S8? (p110)	C		
		RPN4a	Zn finger ARM	fungi Coelomata					Karyopherin A	C, D
	p55	RPN5	PCI/PINT	C, D, T	Sgn4 (S4)	C, D				
	S9	RPN6	PCI/PINT	E	Sgn2 (S2)	C, D, T	S10? (θ, p170)	C		
	S10a	RPN7	PCI/PINT		Sgn1 (S1)	C	S6 (p48)	C		
	S12	RPN8	MPN/JAB	C, D, G	Sgn6 (S6)	C, D, T				
	S11	RPN9	PCI/PINT	C, D, G	Sgn7 (S7)	C, D	S5 (ε, p47)	C, D		
	S13	RPN11	MPN/JAB	C, D, G	Sgn5 (S5)	C, D				
	S14	RPN12	PCI/PINT	C, D, T	Sgn8 (S8)	C, D	S3 (γ, p40)	C, D		
	S5a	RPN10	VWA	C, G					TFIIH p44	C, G
Base	S2	RPN1	SIS2	E					Cyclosome APC1	C, T
	S1	RPN2	SIS2	E					Cyclosome APC1	C, T
	S7	RPT1	AAA	A, E						
	S4	RPT2	AAA	A, E						
	S6′	RPT3	AAA	A, E						
	S10b	RPT4	AAA	A, E						
	S6′	RPT5	AAA	A, E						
	S8	RPT6	AAA	A, E						

Subunit RPN4 is not present in humans and subunit S5b is not present in yeast.

Motif abbreviations: PCI/PINT, proteasome/COP9/IF3 domain; MPN/JAB, Mpr1p and Pad1p N-terminal domain; VWA, von Willebrand factor type A domain; ARM, armadillo repeats; SIS2, repeats specific to proteasome S1 and S2 subunits and cyclosome/APC subunit APC1; AAA, AAA ATPase domain.

Phylogenetic spectrum abbreviations: A, archaea; E, eukaryotes; C, eukaryote crown group; D, *Dictyostelium* (Dictyosteliida); T, *Trypanosoma* (Euglenozoa); G, *Giardia* (Diplomonadida). Sequences found in all eukaryotic groups (C, D, T, and G) are labelled as E. However, the genomes of *Dictyostelium*, *Trypanosoma*, and *Giardia* are not completely sequenced, so the absence of a particular sequence is not evidence that the respective protein is absent from that organism. Searches for the sequences were made in the non-redundant and EST databases at NCBI (www.ncbi.nlm.nih.gov) and on the *Giardia* genome web site (evol3.mbl.edu/Giardia-HTML/giardia_data.html). Sequences for which the Blast P-values were greater than e-10 were analyzed for their correct assignment by reciprocal Blast searches or by motif identification. Preliminary sequence data from the *Giardia lamblia* Genome Project were obtained from The Josephine Bay Paul Center WEB site at the Marine Biological Laboratory (www.mbl.edu/LABS/JBPC/). Sequencing of *Giardia lamblia* is supported by the National Institute of Allergy and Infectious Diseases using equipment from LI-COR Biotechnology.

unknown. In vertebrates, subunit paralogy is linked to the development of the immune system. The cytokine γ-interferon induces the synthesis of three β-type subunits (LMP2, MECL-1 and LMP7) that replace the three active β-subunits (β1, β2 and β5, respectively) to alter the catalytic properties of the proteasome. These 'immunoproteasomes' improve the yield of peptides that are transported to the ER by the Tap transporter, leading to major histocompatibility (MHC) class I display (for reviews see TANAKA 1994; GOLDBERG et al. 1995; YORK et al. 1999). LMP2 and LMP7 are clustered with the subunits of the Tap transporter in the MHC class II locus on chromosome 6 (6p21.3); MECL-1 is located on chromosome 16 (16q22.1).

2.4 Problems in the Phylogenetic Reconstruction of Proteasome Evolution

Various molecular phylogenies have been published for the proteasome core subunits (LUPAS et al. 1993; COUX et al. 1994; PÜHLER et al. 1994; MONACO and NANDI 1995; HUGHES 1997; BOUZAT et al. 2000). Based on the considerations we have presented in Sections 2.2 and 2.3, the deepest division would be expected between HslV and 20S proteasome subunits, followed by the division between α- and β-type subunits. Within the latter two branches, the archaeal subunits should form an outgroup to the eukaryotic subunits. This, however, is not what is observed in phylogenetic reconstructions from sequence data. These agree instead on placing the deepest branchpoint between α- and β-type subunits, followed by the branchpoint between active and inactive β-type subunits [a disagreeing phylogeny by HILT and WOLF (1996) in which α-type subunits do not constitute a monophyletic group, is most probably the result of arbitrary rooting of an unrooted tree]. Although this branch topology is reproduced by different methods, it is unlikely to be correct. The most likely explanation for the discrepancy lies in the constrained evolution of active subunits, which leads to the incorrect grouping of HslV and archaeal β subunits with the three active eukaryotic subunits. It is, however, unclear why the archaeal α subunits are never found at the root of α-type subunits.

Another oddity of molecular phylogeny is the position of actinomycete subunits at or near the root of the α- and β-subunit branches (LUPAS et al. 1997c). Proteasomes of the 20S type, resembling those of archaea and eukaryotes, are found in all actinomycetes (*Rhodococcus*, TAMURA et al. 1995; *Mycobacterium*, KNIPFER and SHRADER 1997; *Streptomyces*, NAGY et al. 1998; *Frankia*, POUCH et al. 2000), but in no other bacterium. With more than 50 bacterial genomes determined and found to lack 20S proteasomes, it seems fairly clear that an early actinomycete acquired genes for 20S subunits by horizontal transfer. The alternative, that 20S genes are ancestral and were lost in every single bacterial lineage except actinomycetes, requires such a large number of independent genetic events as to be excluded with near certainty. Horizontal transfer, however, implies a considerably more recent origin than that suggested by the phylogenies. We propose that the odd position of actinomycete subunits is a branch-length artifact. Such artifacts arise from the tendency of phylogenetic methods to reduce the length of very long (i.e., very divergent) branches by placing them near the root. Why would the molecular clock have run so much faster in actinomycetes than in other branches? Probably because proteasomes were acquired on top of a fully differentiated machinery of energy-dependent proteases and are therefore dispensable under most conditions. Experiments in *Mycobacterium smegmatis* have shown that proteasome mutants are essentially indistinguishable from wild-type under all conditions tested (KNIPFER et al. 1999). Divergence in the α subunits may also have been a consequence of the fact that actinomycetes lack AAA ATPases of the proteasome subfamily and probably use a divergent AAA ATPase, ARC, for proteasome activation (WOLF et al. 1998). Because of the artefactual position of actinomycete subunits in molecular phylogenies, it is impossible to decide with any confidence whether the 20S genes were acquired from an archaeon or a eukaryote.

A final oddity of proteasome phylogenies is the poor reproducibility of branch order within individual eukaryotic subfamilies and the unusual position of some organisms. For example, nematode subunits are repeatedly found near the root of their respective branches, more distant from vertebrates than fungi, plants, and occasionally even slime molds. The grouping of vertebrates and arthropods is also not always monophyletic with respect to fungi and plants. This poor reproducibility of branch order within individual subfamilies contrasts with the ease with which sequences from even the most divergent eukaryotes (such as diplomonads and microsporidians; BOUZAT et al. 2000) can be assigned to their respective subfamily. The large number of discrepancies between proteasome phylogenies and the probable sequence of evolutionary events raises questions about the extent to which details of proteasome evolution can be deduced from their sequences.

3 Molecular Evolution of the 11S Regulator (PA28)

In vertebrates, the 20S core may associate with 11S complexes (variously referred to as 11S regulator or PA28 activator), which accelerate the degradation of peptides, but not of proteins or ubiquitin–protein conjugates, in an energy-independent manner (CHU-PING et al. 1992; DUBIEL et al. 1992; KUEHN and DAHLMANN 1997; BAUMEISTER et al. 1998). The 11S regulator is a ring-shaped complex containing two related subunit types, α and β, which are induced by γ-interferon and do not occur in other organisms. Both map to 14q11.2 in the human genome and have very similar gene structures, in line with a comparatively recent origin by gene duplication. The 11S complex has been shown to assist in the generation of dominant T-cell epitopes (DICK et al. 1996) and enhance the efficiency of viral antigen processing (USTRELL et al. 1995), suggesting a specific function in the vertebrate immune response. Surprisingly for a complex formed of two subunit types, 11S appears to be heptameric (three α, four β; ZHANG et al. 1999), mirroring the stoichiometry of the 20S core.

Both 11S subunits appear to have originated from a third protein, referred to as PA28γ or Ki antigen (located at 17q12–21 in the human genome), whose function is unknown. This is also up-regulated by γ-interferon, but is much more broadly distributed than PA28α and PA28β (Fig. 2), with clear homologs in other metazoans, such as arthropods, nematodes, and flatworms. Jointly, α, β, and γ sequences have clear relatives in many deeply branching eukaryotes, such as the well-characterized PA26 protein of *Trypanosoma brucei*, which can activate the peptidase activity of trypanosome and rat proteasomes (To and WANG 1997; YAO et al. 1999) and whose structure has been solved in complex with the yeast 20S proteasome (WHITBY et al. 2000). Surprisingly, no sequence relative of 11S subunits is detectable in plants or fungi. Given the broad representation of 11S/PA28-like sequences in the most primitive eukaryotes, it seems likely that they were lost from these two major lineages early in eukaryotic evolution.

A phylogenetic analysis of 11S/PA28-like sequences yields a tree in which the α- and β-subunits appear to form an outgroup to the γ-subunits (Fig. 2). The phylogenetic spectrum of individual subunits, however, suggests that this is not correct and that the α- and β-subunits arose by gene duplication from the γ subunit during the evolution of the vertebrate immune system. We propose that the discrepancy is due to a branch-length artifact, resulting from the much more rapid rate of evolution in the α and β branches (witness the large difference in branchpoint depth between zebrafish and mammals in the three branches).

The degree of conservation in most 11S/PA28-like sequences is astonishing, particularly considering that their structure, a coiled-coil-like bundle of four helices, is amenable to rapid sequence divergence in the solvent-exposed positions. The sequence identity between the slime mold and mammalian proteins, for example, is 35% (ungapped!), over the 146 residues of the slime mold expressed sequence tag (EST) fragment. This suggests that 11S/PA28-like proteins have an important role in proteasome function, making their apparent absence from plants and fungi all the more puzzling.

4 Molecular Evolution of the 19S Cap Complex

In eukaryotes, and only eukaryotes, the 20S core is capped at one or both poles by a 19S complex, to form the 26S proteasome (RECHSTEINER 1998; VOGES et al. 1999). Association with the 19S cap complex enhances the proteolytic activity of the core in an energy-dependent manner. In addition, it mediates the ability to degrade ubiquitinated substrates and adds an additional layer of regulation not found in prokaryotes. The 26S proteasome also degrades specific non-ubiquitinated proteins, a function of the ancestral proteasome complex that has not been completely supplanted.

The 19S cap can be dissociated in vitro into defined subcomplexes, called the base and the lid (GLICKMAN et al. 1998). The base contains six ATPases of the AAA family that probably unfold and translocate substrate proteins into the 20S

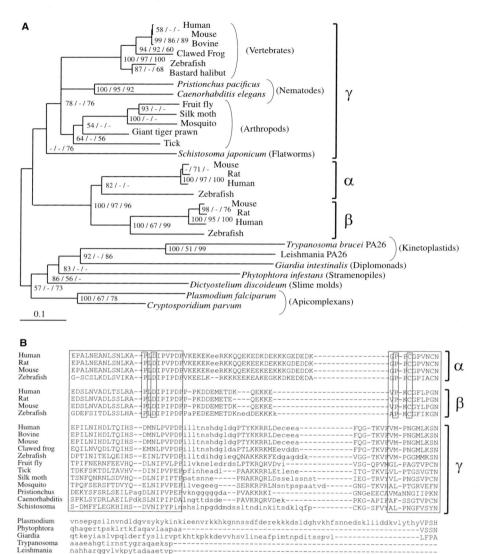

core (BRAUN et al. 1999). In addition, the base contains three and the lid nine non-ATPase subunits (Table 1).

4.1 The Base Subcomplex: ATPase Subunits

The oldest part of the 19S cap complex is probably the set of six AAA ATPases, which arose by duplication and differentiation from an ancestral proteasome-associated ATPase after divergence from the archaea. A single descendent of this

Fig. 2. A Phylogeny of 11S subunits and related proteins. The tree was obtained by distance-based reconstruction, using the Phylip 3.572 package (Protdist/Neighbor). Percent support in 1,000 bootstrap replicates by distance (Clustal), maximum likelihood (Puzzle 3.1), and maximum parsimony (Paup 3.1.1) is shown in this order for key branchpoints. The alignment from which the phylogeny was produced can be obtained by request from A. Lupas (andrei@lupas.org). Gapped positions and ambiguously aligned regions, particularly the long loop connecting helices 1 and 2 (see **B**) were excluded from computations. The database accession codes are as follows. α-Subunits: human, AAA16521; mouse, AAC53295; rat, BAA08206; zebrafish, AAF17492; β-subunits: human, AAF02218; mouse, AAC53296; rat, BAA08207; zebrafish, AAF05817; γ-subunits: human, AAB60335; bovine, B60537; mouse, BAA22041; zebrafish, AAF05816; fruit fly, AAF19529; tick, AAB41818; other: *Trypanosoma brucei*, AAD50581. All other sequences were assembled from ESTs. **B** Sequence alignment of the loop connecting helices 1 and 2, as defined by the crystal structure of the human 11S α-complex (PDB: 1AVO). Sequences have been divided into four blocks: α, β, γ, and other; residues aligned within each block are in *capital letters*. Positions of the alignment that distinguish αβγ from other and αβ from γ are *boxed*. These positions provide independent support for the main branchpoints observed in **A**

ATPase is still found in most archaea (ZWICKL et al. 1999), though ironically not in *Thermoplasma* (RUEPP et al. 2000), where so much of proteasome biology has been elucidated. In agreement with this scenario, phylogenies of the AAA family (BEYER 1997; SWAFFIELD and PURUGGANAN 1997; WOLF et al. 1998; AAA web site at http://yeamob.pci.chemie.uni-tuebingen.de/AAA/Tree.html) show proteasome ATPases as a monophyletic group; however, only the phylogenetic reconstruction of WOLF et al. 1998 places the archaeal representatives at the root.

Although the eukaryotic proteasome ATPases have not been imaged yet as a stand-alone complex, it seems clear that they form a ring, like their archaeal homologues (ZWICKL et al. 1999) and other AAA proteins (LENZEN et al. 1998; WOLF et al. 1998; YU et al. 1998; ROCKEL et al. 1999), and are located at the base of the 19S cap. Why did proteasome ATPases differentiate fully in eukaryotes, when many of their AAA homologues remained homo-oligomeric (like Cdc48/p97 and Sec18/NSF)? One reason, which has already been fielded in the context of core subunit differentiation, is the need to discriminate between a broad range of proteins. In contrast to homo-oligomeric AAA proteins, which typically act only on a restricted set of substrates (for reviews see BEYER 1997; LANGER 2000), proteasome ATPases need to interact with a substantial fraction of cellular proteins, as well as with foreign proteins introduced during infection. ATPase differentiation in response to substrate diversity may have in turn accelerated the differentiation of the underlying α-subunit rings. Clearly, the two processes happened in close temporal proximity, as subunit differentiation in both rings was complete prior to the origin of the deepest-branching eukaryotes observable today (Table 1). Recently WOLLENBERG and SWAFFIELD (2001) have presented an intriguing model for ATPase differentiation through successive duplications of an ancestral, homo-oligomeric subunit. Such a model is supported by the presence of coiled-coil segments in proteasomal ATPases, which suggest that the hexameric ring is in fact a trimer of dimers.

As noted, the protease subunits of energy-dependent proteases have quite different evolutionary origins and folds. The ATPase subunits, however, belong to the same protein family (LUPAS et al. 1997b), which is very large and contains many proteins not involved in proteolysis (Fig. 3). This family has been named 'AAA+'

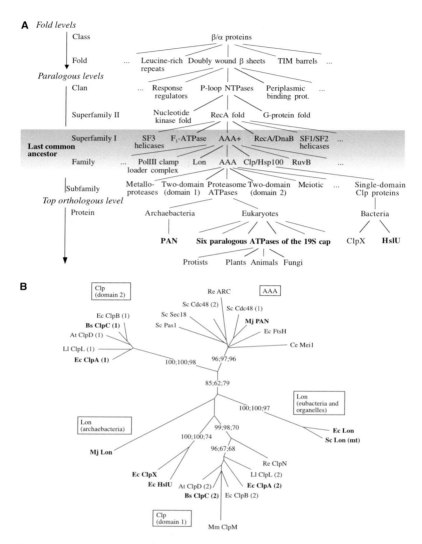

Fig. 3A,B. Phylogenetic relationship of protease-associated ATPases. **A** Schematic representation of a systematic classification for proteasome-associated ATPases (*bold*). **B** Phylogeny of protease-associated ATPases (AAA, Clp/Hsp100, Lon). The tree was obtained by distance-based reconstruction, using the Phylip 3.572 package (Protdist/Neighbor). Percent support in 1,000 bootstrap replicates by distance (Phylip), maximum parsimony (Paup 3.1.1), and maximum likelihood (Puzzle 3.1) is shown in this order for key branchpoints. Protease-associated ATPases are shown in *bold*. The alignment from which the phylogeny was produced was published in LUPAS et al. 1997b. The *two letters* that precede each gene name in the tree indicate the organism: At, *Arabidopsis thaliana*; Bs, *Bacillus subtilis*; Ce, *Caenorhabditis elegans*; Ec, *Escherichia coli*; Ll, *Lactococcus lactis*; Mj, *Methanococcus jannaschii*; Mm, *Mus musculus*; Re, *Rhodococcus erythropolis*; Sc, *Saccharomyces cerevisiae*. The database accession codes for the sequences are: Sc Pas1,P24004; Sc Sec18, P18759; Sc Cdc48, P25694; Re ARC, AAC68690; Mj PAN, Q58576; Ec FtsH, S35109; Ce Mei1, P34808; Ec ClpA, P15716; Ll ClpL, Q06719; At ClpD, P42762; Bs ClpC, P37571; Ec ClpB, P03815; Mm ClpM, Q60649; Re ClpN, Q01357; Ec ClpX, P33138; Ec HslU (ClpY), P32168; Ec Lon, P08177; Sc Lon (mt), P36775; Mj Lon, AAB99427

(NEUWALD et al. 1999), because many of its most prominent members are AAA proteins. Recent crystal structures (GUENTHER et al. 1997; LENZEN et al. 1998; YU et al. 1998) show that AAA+ proteins have a RecA fold and are related to F- and V-type ATPases, SF1, SF2, and SF3 helicases, DnaB, and RecA. All proteins with a RecA fold for which structural information is available form oligomeric (typically hexameric) rings, except for SF1 and SF2 helicases. However, their structure reveals that they contain two domains (KOROLEV et al. 1998), which mimic the interactions of two adjacent subunits in an AAA ATPase ring. It therefore seems likely that all proteins with a RecA fold evolved from a ring-forming ancestor. In addition, most proteins of this fold are helicases, chaperones, and proteases involved in the unfolding of macromolecules, suggesting that the first protein with a RecA fold already had some sort of 'unfoldase' activity.

A peculiarity of the RecA fold, which sets it apart from other P-loop NTPases, is an inserted $\beta\alpha$ element in the active site region. This element separates the P-loop (Walker A motif) from the Walker B motif and inserts an additional residue into the active site. This residue, called sensor-1, is located at the C-terminal end of the inserted β-strand and makes contact with the γ-phosphate of the bound nucleotide. Its role, however, has remained unclear. Based on an analysis of the sequences and structures of proteins with a RecA fold, we propose the following explanation for the inserted $\beta\alpha$ element: All catalytically active AAA+ domains (though not some inactive domains, such as the NSF D2) contain an arginine residue at the C-terminal end of the inserted α helix (LUPAS et al. 1997b). The same is true for F-type ATPases, SF1, SF2, and SF3 helicases, DnaB, and RecA, suggesting that this is a conserved feature of the RecA fold. In crystal structures, it can be seen that this residue forms an 'arginine finger', positioned to interact with the β- and γ-phosphates of the nucleotide bound to an adjacent subunit. The role of arginine fingers in stimulating nucleotide hydrolysis was first discovered in G proteins (AHMADIAN et al. 1997; SCHEFFZEK et al. 1997) and has since been recognized in the active sites of many other P-loop NTPases (KOROLEV et al. 1998; NADANACIVA et al. 1999). In the RecA fold, the sensor-1 residue and the arginine finger are located at opposite ends of the inserted α-helix and make contact with the γ-phosphates of nucleotides in two adjacent binding sites. This suggests a mechanism for concerted hydrolysis, in which loss of the γ-phosphate in one binding site is conveyed to the next active site in the ring via rigid-body motion of the helix connecting the sensor-1 and arginine finger residues (SCHMIDT et al. 1999). This model implies that the peculiar fold of RecA-like proteins, and thus also of the proteasome ATPases, was caused by the need of an ancestral ring-shaped NTPase to undergo concerted hydrolysis at its active sites. As discussed, the ring shape probably arose as an adaptation to the functional requirements of macromolecule disassembly and unfolding.

4.2 The Base Subcomplex: Non-ATPase Subunits

In addition to the ATPases, the base subcomplex contains three non-ATPase subunits, S1/Rpn2, S2/Rpn1 and S5a/Rpn10. Whereas the individual ATPases of the base are approximately 75% identical between humans and yeast, the non-ATPase subunits are only 30%–45% identical.

S1/Rpn2 and S2/Rpn1 are homologous proteins of more than 900 residues each, making them the largest proteins in the 19S cap complex, and contain two characteristic patterns. One is a 'KEKE motif', defined as a string of at least 13 amino acids that starts and ends with a lysine and/or glutamate and in which more than 60% of the residues are lysine or glutamate/aspartate, there are no more than four consecutive residues with identical positive or negative charge, and there are no cysteines, prolines, or aromatic residues in the sequence (REALINI 1994). KEKE motifs have also been identified in S5a/Rpn10, as well as in two of the lid subunits, the C-terminal extensions of two of the core α-subunits, a subunit of 11S/PA28, chaperonins, calcium-binding proteins, and others. In fact, KEKE motifs, as defined, are found in 6% of the entries in an edited library of human proteins (RECHSTEINER 1998). An examination of sequences shows that among nearest neighbors it is not unusual for one to contain a KEKE motif where the other does not, and KEKE motifs are found among proteins that do not appear to otherwise be related. Further, there does not appear to be any conserved pattern among residues within KEKE motifs. These observations indicate that KEKE motifs have arisen independently and probably do not represent a sequence motif in the stricter sense.

The second characteristic pattern found in the S1/Rpn2 and S2/Rpn1 subunits is a stretch of nine 35–40 residue repeats that generally contain a predicted β-strand of alternating large and small aliphatic residues, followed by a predicted amphipathic α-helix. These repeats have been suggested to fold into successive β/α elements in an analogous manner to the leucine-rich repeats of ribonuclease inhibitor (LUPAS et al. 1997a). A string of repeats homologous to those of S1/Rpn2 and S2/Rpn1 is also observed in the largest subunit of the 20S anaphase-promoting complex (APC)/cyclosome (Apc1/BimE/Tsg24/Cut4). The APC/cyclosome is a complex that ubiquitinates cyclin B in a cell cycle-dependent manner, targeting it for degradation by the 26S proteasome, but outside their largest subunit(s), the two complexes do not appear to be related. This suggests that the common ancestor of S1, S2, and APC1 may have been a ubiquitin-binding protein. This ancestor is not recognizable in bacteria or archaea and differentiated into the three proteins before the divergence of any eukaryote observable today.

The third non-ATPase subunit is S5a/Rpn10. This subunit appears to be the 'glue' that links the lid to the base. Electron micrographs of proteasomes purified from Δ*rpn*10 yeast indicate that the interaction of lid and base is destabilized in the absence of Rpn10 (GLICKMAN et al. 1998). The S5a/Rpn10 subunit has a von Willebrand factor type A (VWA) domain and appears to be most similar to subunit p44 of the RNA polymerase II basal transcription factor TFIIH (ARAVIND and PONTING 1998). VWA domains were first identified as "adhesive" domains important to the blood coagulation properties of von Willebrand factor, suggesting that it might play the same role in the proteasome.

Significant amounts of S5a/Rpn10 are also found unassociated with the 26S proteasome (HARACSKA and UDVARDY 1995). Its role outside of the 26S proteasome is unclear, but as it is the only subunit that has been shown to bind ubiquitin, it has been suggested that S5a/Rpn10 may shuttle ubiquitinated proteins to the proteasome for degradation (DEVERAUX et al. 1994). This view supports an accretion model for the 19S complex, in which the ATPase ring formed a complex with a ubiquitin recognition system consisting of the S1 and S2 subunits and a soluble shuttle, S5a. S5a subsequently became fixed when the lid was added to the 19S cap. Based on homology between S1, S2 and APC1, and on the fact that RPN10 is not essential, we conclude that S5a may have been a peripheral element of this putative ancestral ubiquitin recognition system.

4.3 The Lid Subcomplex and Its Relationship to COP9 and eIF3

The lid forms the most distal portion of the 19S cap. Although certain lid subunits have been shown to interact with one another, their arrangement is unknown. Proteasomes missing the lid subcomplex seem to have activity comparable to that of wild-type in the degradation of non-ubiquitinated substrates (GLICKMAN et al. 1998). However, the lid is required for degradation of ubiquitinated proteins. This is intriguing given that none of the lid subunits is known to bind ubiquitin and that the lid is related to two other cellular complexes that are not ubiquitin-dependent.

Eight lid subunits are present in all eukaryotes (Table 1). Although their sequences are not as well conserved as those of the ATPase subunits, they can be readily identified even in the most divergent eukaryotes. In addition, clade-specific subunits have evolved in various lineages, for example Rpn4 in fungi, which contains a zinc finger domain, or S5b in vertebrates and arthropods, which contains armadillo repeat motifs (ARM). The ARM repeats of S5b appear to be most similar to those of α-importin/karyopherins, cytosolic signal peptide-binding proteins that mediate nuclear transport.

The eight basic subunits of the lid have homologs in two other eukaryotic multi-subunit complexes, the COP9 signalosome and eIF3 (GLICKMAN et al. 1998; WEI et al. 1998). This relationship is puzzling as the three complexes do not seem to have related functions. COP9 is an approximately 450kDa complex that was originally identified in *Arabidopsis* as a repressor of photomorphogenesis, but is now believed to play a role in a number of cell-signaling pathways in eukaryotes. eIF3 plays a role in the initiation of protein synthesis and is the largest of the eukaryotic initiation factors. It is a multi-subunit complex of at least 10 non-identical subunits and has a molecular mass of approximately 600kDa. It appears that the 19S lid is the most ancient of the three complexes, as no COP9 subunits were found in *Giardia*, and none of the eIF3 subunits were found in either *Trypanosoma* or *Giardia* (Table 1). However, it cannot be ruled out that the complexes were lost in these species, as COP9 appears to have been lost in yeast (WEI and DENG 1999).

COP9 and the lid subcomplex are substantially more similar to one another than either is to eIF3; their subunits can be mapped one-to-one by sequence comparisons (Table 1) and have similar sizes. In eIF3, only five subunits correspond to subunits of the lid, and these have low sequence similarity and poorly matching sizes. In addition, eIF3 contains multiple unrelated subunits, although we can't exclude that some of these are in fact homologous but have diverged beyond recognition. Two conserved domains are found in the homologous subunits of lid, COP9, and eIF3: MPN/JAB and PCI/PINT (Table 1; ARAVIND and PONTING 1998; HOFMANN and BUCHER 1998). Both were probably once single-domain proteins, but only MPN/JAB is still found as a stand-alone protein in prokaryotes (SMART web site: http://smart.embl-heidelberg.de/). PCI/PINT domains are not found outside the three eukaryotic complexes. Each of the eight lid and COP9 subunits contains either an MPN/JAB or a PCI/PINT domain, suggesting that they evolved by duplication and differentiation from a much simpler precursor. However, no internal symmetry is observable in these particles by electron microscopy (WALZ et al. 1998; KAPELARI et al. 2000). As is true with the base subunits, lid subunits are fully differentiated throughout eukaryotes (Table 1). In the absence of any functional knowledge, the significance of these observations remains unclear.

5 Conclusion

In this chapter, we have attempted to provide an outline of proteasome evolution. Such an endeavor is by necessity speculative, as proteins do not have a fossil record and all reconstructions work by extrapolation of current properties to a hypothetical ancestral state. In some areas, for example in the lid complex, too few data are available as yet to allow meaningful hypotheses. In others, for example the evolution of the ATPases, powerful hypotheses with substantial implications for experimental approaches have become possible. Overall, proteasomes provide an excellent model system for many problems in evolutionary biology.

6 Summary

Proteasomes are large, multisubunit proteases that are found, in one form or another, in all domains of life and play a critical role in intracellular protein degradation. Although they have substantial structural similarity, the proteasomes of bacteria, archaea, and eukaryotes show many differences in architecture and subunit composition. This article discusses possible paths by which proteasomes may have evolved from simple precursors to the highly complicated and diverse complexes observed today.

References

Ahmadian MR, Stege P, Scheffzek K, Wittinghofer A (1997) Confirmation of the arginine-finger hypothesis for the GAP-stimulated GTP-hydrolysis reaction of Ras. Nature Struct Biol 4:686–689

Akopian TN, Kisselev AF, Goldberg AL (1997) Processive degradation of proteins and other catalytic properties of the proteasome from *Thermoplasma acidophilum*. J Biol Chem 272:1791–1798

Aravind L, Ponting CP (1998) Homologues of 26S proteasome subunits are regulators of transcription and translation. Protein Sci 7:1250–1254

Baumeister W, Walz J, Zühl F, Seemuller E (1998) The proteasome: paradigm of a self-compartmentalizing protease. Cell 92:367–380

Beyer A (1997) Sequence analysis of the AAA protein family. Protein Sci 6:2043–2058

Bochtler M, Ditzel L, Groll M, Huber R (1997) Crystal structure of heat shock locus V (HslV) from Escherichia coli. Proc Natl Acad Sci USA 94:6070–6074

Bompard-Gilles C, Villeret V, Davies GJ, Fanuel L, Joris B, Frere JM, Van Beeumen J (2000) A new variant of the Ntn hydrolase fold revealed by the crystal structure of L-aminopeptidase D-ala-esterase/amidase from *Ochrobactrum anthropi*. Structure Fold Des 8:153–162

Bouzat JL, McNeil LK, Robertson HM, Solter LF, Nixon JE, Beever JE, Gaskins HR, Olsen G, Subramaniam S, Sogin ML, Lewin HA (2000) Phylogenomic analysis of the alpha proteasome gene family from early diverging eukaryotes. J Mol Evol 51:532–543

Brannigan JA, Dodson G, Duggleby HJ, Moody PC, Smith JL, Tomchick DR, Murzin AG (1995) A protein catalytic framework with an N-terminal nucleophile is capable of self-activation. Nature 378:416–419, 644

Braun BC, Glickman M, Kraft R, Dahlmann B, Kloetzel PM, Finley D, Schmidt M (1999) The base of the proteasome regulatory particle exhibits chaperone-like activity. Nature Cell Biol 1:221–226

Chen P, Hochstrasser M (1996) Autocatalytic subunit processing couples active site formation in the 20S proteasome to completion of assembly. Cell 86:961–972

Chu-Ping M, Slaughter CA, DeMartino GN (1992) Identification, purification, and characterization of a protein activator (PA28) of the 20S proteasome (macropain). J Biol Chem 267:10515–10523

Coux O, Nothwang HG, Silva Pereira I, Recillas Targa F, Bey F, Scherrer K (1994) Phylogenic relationships of the amino acid sequences of prosome (proteasome, MCP) subunits. Mol Gen Genet 245:769–780

Coux O, Tanaka K, Goldberg AL (1996) Structure and functions of the 20S and 26S proteasomes. Annu Rev Biochem 65:801–847

Deveraux Q, Ustrell V, Pickart C, Rechsteiner M (1994) A 26S protease subunit that binds ubiquitin conjugates. J Biol Chem 269:7059–7061

Dick TP, Ruppert T, Groettrup M, Kloetzel PM, Kuehn L, Koszinowski UH, Stevanovic S, Schild H, Rammensee HG (1996) Coordinated dual cleavages induced by the proteasome regulator PA28 lead to dominant MHC ligands. Cell 86:253–262

Dubiel W, Pratt G, Ferrell K, Rechsteiner M (1992) Purification of an 11S regulator of the multicatalytic protease. J Biol Chem 267:22369–22377

Duggleby HJ, Tolley SP, Hill CP, Dodson EJ, Dodson G, Moody PC (1995) Penicillin acylase has a single-amino-acid catalytic centre. Nature 373:264–268

Fu H, Doelling JH, Arendt CS, Hochstrasser M, Vierstra RD (1998) Molecular organization of the 20S proteasome gene family from *Arabidopsis thaliana*. Genetics 149:677–692

Glickman MH, Rubin DM, Coux O, Wefes I, Pfeifer G, Cjeka Z, Baumeister W, Fried VA, Finley D (1998) A subcomplex of the proteasome regulatory particle required for ubiquitin-conjugate degradation and related to the COP9-signalosome and eIF3. Cell 94:615–623

Goldberg AL, Gaczynska M, Grant E, Michalek M, Rock KL (1995) Functions of the proteasome in antigen presentation. Cold Spring Harb Symp Quant Biol 60:479–490

Guan C, Cui T, Rao V, Liao W, Benner J, Lin CL, Comb D (1996) Activation of glycosylasparaginase. Formation of active N-terminal threonine by intramolecular autoproteolysis. J Biol Chem 271:1732–1737

Guenther B, Onrust R, Sali A, O'Donnell M, Kuriyan J (1997) Crystal structure of the delta' subunit of the clamp-loader complex of *E. coli* DNA polymerase III. Cell 91:335–345

Haracska L, Udvardy A (1995) Cloning and sequencing a non-ATPase subunit of the regulatory complex of the Drosophila 26S protease. Eur J Biochem 231:720–725

Hilt W, Wolf DH (1996) Proteasomes: destruction as a programme. Trends Biochem Sci 21:96–102

Hofmann K, Bucher P (1998) The PCI domain: a common theme in three multiprotein complexes. Trends Biochem Sci 23:204–205

Hughes AL (1997) Evolution of the proteasome components. Immunogenetics 46(2):82–92

Kapelari B, Bech-Otschir D, Hegerl R, Schade R, Dumdey R, Dubiel W (2000) Electron Microscopy and Subunit–Subunit Interaction Studies Reveal a First Architecture of COP9 Signalosome. J Mol Biol 300:1169–1178

Knipfer N, Shrader TE (1997) Inactivation of the 20S proteasome in *Mycobacterium smegmatis*. Mol Microbiol 25:375–383

Knipfer N, Seth A, Roudiak SG, Shrader TE (1999) Species variation in ATP-dependent protein degradation: protease profiles differ between mycobacteria and protease functions differ between *Mycobacterium smegmatis* and *Escherichia coli*. Gene 231:95–104

Korolev S, Yao N, Lohman TM, Weber PC, Waksman G (1998) Comparisons between the structures of HCV and Rep helicases reveal structural similarities between SF1 and SF2 super-families of helicases. Protein Sci 7:605–610

Kuehn L, Dahlmann B (1997) Structural and functional properties of proteasome activator PA28. Mol Biol Rep 24:89–93

Langer T (2000) AAA proteases: cellular machines for degrading membrane proteins. Trends Biochem Sci 25:247–251

Lee H, Park OK, Kang HS (2000) Identification of a new active site for autocatalytic processing of penicillin acylase precursor in *Escherichia coli* ATCC11105. Biochem Biophys Res Commun 272:199–204

Lenzen CU, Steinmann D, Whiteheart SW, Weis WI (1998) Crystal structure of the hexamerization domain of N-ethylmaleimide-sensitive fusion protein. Cell 94:525–536, 95:289

Löwe J, Stock D, Jap B, Zwickl P, Baumeister W, Huber R (1995) Crystal structure of the 20S proteasome from the archaeon *T. acidophilum* at 3.4 Å resolution. Science 268:533–539

Lupas A, Koster AJ, Baumeister W (1993) Structural features of 26S and 20S proteasomes. Enzyme Protein 47:252–273

Lupas A, Baumeister W, Hofmann K (1997a) A repetitive sequence in subunits of the 26S proteasome and 20S cyclosome (anaphase-promoting complex). Trends Biochem Sci 22:195–196

Lupas A, Flanagan JM, Tamura T, Baumeister W (1997b) Self-compartmentalizing proteases. Trends Biochem Sci 22:399–404

Lupas A, Zühl F, Tamura T, Wolf S, Nagy I, De Mot R, Baumeister W (1997c) Eubacterial proteasomes. Mol Biol Rep 24:125–131

Monaco JJ, Nandi D (1995) The genetics of proteasomes and antigen processing. Annu Rev Genet 29:729–754

Nadanaciva S, Weber J, Wilke-Mounts S, Senior AE (1999) Importance of F1-ATPase residue alpha-Arg-376 for catalytic transition state stabilization. Biochemistry 38:15493–15499

Nagy I, Tamura T, Vanderleyden J, Baumeister W, De Mot R (1998) The 20S proteasome of *Streptomyces coelicolor*. J Bacteriol 180:5448–5453

Neuwald AF, Aravind L, Spouge JL, Koonin EV (1999) AAA+: A class of chaperone-like ATPases associated with the assembly, operation, and disassembly of protein complexes. Genome Res 9:27–43

Peters J-M, Harris JR, Finley D (eds) (1998) Ubiquitin and the biology of the cell. Plenum Press, New York

Pouch MN, Cournoyer B, Baumeister W (2000) Characterization of the 20S proteasome from the *actinomycete Frankia*. Mol Microbiol 35:368–377

Pühler G, Pitzer F, Zwickl P, Baumeister W (1994) Proteasomes: multisubunit proteinases common to Thermoplasma and eukaryotes. System Appl Microbiol 16:734–741

Realini C, Rogers SW, Rechsteiner M (1994) KEKE motifs. Proposed roles in protein–protein association and presentation of peptides by MHC class I receptors. FEBS Lett 348:109–113

Rechsteiner M (1998) The 26S proteasome. In: Peters J-M, Harris JR, Finley D (eds) Ubiquitin and the biology of the cell. Plenum Press, New York

Rockel B, Walz J, Hegerl R, Peters J, Typke D, Baumeister W (1999) Structure of VAT, a CDC48/p97 ATPase homologue from the *archaeon Thermoplasma acidophilum* as studied by electron tomography. FEBS Lett 451:27–32

Rohrwild M, Pfeifer G, Santarius U, Muller SA, Huang HC, Engel A, Baumeister W, Goldberg AL (1997) The ATP-dependent HslVU protease from *Escherichia coli* is a four-ring structure resembling the proteasome. Nature Struct Biol 4:133–139

Ruepp A, Graml W, Santos-Martinez M-L, Koretke KK, Volker C, Mewes HW, Frishman D, Stocker S, Lupas AN, Baumeister W (2000) The genome sequence of the thermoacidophilic scavenger *Thermoplasma acidophilum*. Nature 407:508–513

Saier MH (1999) Genome archeology leading to the characterization and classification of transport proteins. Curr Opin Microbiol 2:555–561

Scheffzek K, Ahmadian MR, Kabsch W, Wiesmuller L, Lautwein A, Schmitz F, Wittinghofer A (1997) The Ras-RasGAP complex: structural basis for GTPase activation and its loss in oncogenic Ras mutants. Science 277:333–338

Schmidt M, Lupas AN, Finley D (1999) Structure and mechanism of ATP-dependent proteases. Curr Opin Chem Biol 3:584–591

Schmidtke G, Schmidt M, Kloetzel PM (1997) Maturation of mammalian 20S proteasome: purification and characterization of 13S and 16S proteasome precursor complexes. J Mol Biol 268:95–106

Seemuller E, Lupas A, Stock D, Löwe J, Huber R, Baumeister W (1995) Proteasome from *Thermoplasma acidophilum*: a threonine protease. Science 268:579–582

Seemuller E, Lupas A, Baumeister W (1996) Autocatalytic processing of the 20S proteasome. Nature 382:468–471

Swaffield JC, Purugganan MD (1997) The evolution of the conserved ATPase domain (CAD): reconstructing the history of an ancient protein module. J Mol Evol 45:549–563

Tamura T, Nagy I, Lupas A, Lottspeich F, Cejka Z, Schoofs G, Tanaka K, De Mot R, Baumeister W (1995) The first characterization of a eubacterial proteasome: the 20S complex of *Rhodococcus*. Curr Biol 5:766–774

Tanaka K (1994) Role of proteasomes modified by interferon-gamma in antigen processing. J Leukoc Biol 56:571–575

To WY, Wang CC (1997) Identification and characterization of an activated 20S proteasome in *Trypanosoma brucei*. FEBS Lett. 1997 404:253–262

Ustrell V, Realini C, Pratt G, Rechsteiner M (1995) Human lymphoblast and erythrocyte multicatalytic proteases: differential peptidase activities and responses to the 11S regulator. FEBS Lett 376:155–158

van Poelje PD, Snell EE (1988) Amine cations promote concurrent conversion of prohistidine decarboxylase from *Lactobacillus* 30a to active enzyme and a modified proenzyme. Proc Natl Acad Sci USA 85:8449–8453, 86:1223

Voges D, Zwickl P, Baumeister W (1999) The 26S proteasome: a molecular machine designed for controlled proteolysis. Annu Rev Biochem 68:1015–1068

Waldmann T, Lupas A, Kellermann J, Peters J, Baumeister W (1995) Primary structure of the thermosome from *Thermoplasma acidophilum*. Biol Chem Hoppe Seyler 376:119–126

Walz J, Erdmann A, Kania M, Typke D, Koster AJ, Baumeister W (1998) 26S proteasome structure revealed by three-dimensional electron microscopy. J Struct Biol 121:19–29

Wei N, Deng XW (1999) Making sense of the COP9 signalosome. A regulatory protein complex conserved from *Arabidopsis* to human. Trends Genet 15:98–103

Wei N, Tsuge T, Serino G, Dohmae N, Takio K, Matsui M, Deng XW (1998) The COP9 complex is conserved between plants and mammals and is related to the 26S proteasome regulatory complex. Curr Biol 8:919–922

Whitby FG, Masters EI, Kramer L, Knowlton JR, Yao Y, Wang CC, Hill CP (2000) Structural basis for the activation of 20S proteasomes by 11S regulators. Nature 408:115–120

Wolf S, Nagy I, Lupas A, Pfeifer G, Cejka Z, Muller SA, Engel A, De Mot R, Baumeister W (1998) Characterization of ARC, a divergent member of the AAA ATPase family from *Rhodococcus erythropolis*. J Mol Biol 277:13–25

Wollenberg K, Swaffield JC (2001) Evolution of proteasomal ATPases. Mol Biol Evol 18:962–974

Xiong H, Pegg AE (1999) Mechanistic studies of the processing of human S-adenosylmethionine decarboxylase proenzyme. Isolation of an ester intermediate. J Biol Chem 274:35059–35066

Xu MQ, Comb DG, Paulus H, Noren CJ, Shao Y, Perler FB (1994) Protein splicing: an analysis of the branched intermediate and its resolution by succinimide formation. EMBO J 13:5517–5522

Yao Y, Huang L, Krutchinsky A, Wong ML, Standing KG, Burlingame AL, Wang CC (1999) Structural and functional characterizations of the proteasome-activating protein PA26 from *Trypanosoma brucei*. J Biol Chem 274:33921–33930

York IA, Goldberg AL, Mo XY, Rock KL (1999) Proteolysis and class I major histocompatibility complex antigen presentation. Immunol Rev 172:49–66

Yu RC, Hanson PI, Jahn R, Brunger AT (1998) Structure of the ATP-dependent oligomerization domain of N-ethylmaleimide sensitive factor complexed with ATP. Nature Struct Biol 5:803–811, 924

Zhang Z, Krutchinsky A, Endicott S, Realini C, Rechsteiner M, Standing KG (1999) Proteasome activator 11S REG or PA28: recombinant REG α/REG β hetero-oligomers are heptamers. Biochemistry 38:5651–5658

Zühl F, Seemuller E, Golbik R, Baumeister W (1997) Dissecting the assembly pathway of the 20S proteasome. FEBS Lett 418:189–194

Zwickl P, Ng D, Woo KM, Klenk HP, Goldberg AL (1999) An archaebacterial ATPase, homologous to ATPases in the eukaryotic 26S proteasome, activates protein breakdown by 20S proteasomes. J Biol Chem 274:26008–26014

The 20S Proteasome

P. ZWICKL

1 Introduction 23
2 Occurrence and Subunit Composition of 20S Proteasomes 23
3 Structural Features of the 20S Proteasome 25
4 Catalytic Mechanism. 28
5 Processing and Assembly 29
6 Proteolytic Activity and Degradation Products 31
7 The Archaeal and Bacterial AAA ATPases 34
8 Summary 35
References 36

1 Introduction

Over recent years the 20S proteasome has been the subject of intensive research and as a result we have obtained a detailed view of its molecular structure and cellular function. Here I will review our knowledge of the prokaryotic and eukaryotic 20S proteasome and its activators in prokaryotes. The eukaryotic 19S regulator which assembles with the 20S core to form the 26S proteasome is reviewed in the following contribution (see the chapter by Glickman and Maytal, this volume).

2 Occurrence and Subunit Composition of 20S Proteasomes

As the number of sequenced genomes grows, a progressively clearer picture of the species distribution of proteasomes is emerging. In general, it appears that proteasomes are ubiquitous and essential in eukaryotes (HEINEMEYER 2000);

Department of Molecular Structural Biology, Max Planck Institute for Biochemistry, Am Klopferspitz 18a, 82152 Martinsried, Germany

ubiquitous but not essential in archaea (RUEPP et al. 1998); and rare and nonessential in bacteria, where several other energy-dependent proteases are found (DE MOT et al. 1999).

While in most archaeal genomes only one α- and one β-subunit gene have been found, a few species contain a second α or a second β gene (see Table 1) (ZWICKL et al. 2000; MAUPIN-FURLOW et al. 2000). To date, proteasomes have been purified from a few species and only two of those have either a duplicated proteasomal α or β gene (see Table 1). Proteasomes purified from the thermoacidophilic *Pyrococcus furiosus* apparently contained only one β-subunit (BAUER et al. 1997), although two β genes have been identified in the genomes of three distinct *Pyrococccus* species (see Table 1). In the genome of the halophilic archaeon *Haloferax volcanii* three proteasomal genes have been found, one coding for a β-subunit, and two coding for α-type subunits. 20S proteasomes isolated from *Haloferax* cells contain two distinct particle populations, one built of α_1 and β and one containing all three subunits (α_1, α_2, and β) (WILSON et al. 1999). Surprisingly, the complete genome of the related halophilic organism *Halobacterium* species NRC-1 revealed only one α- and one β-subunit (NG et al. 2000). Notably, the sequenced genomes of the crenarchaeotes *Aeropyrum pernix*, *Pyrobaculum aerophilum* and *Sulfolobus solfataricus* contain a single α-type and two β-type genes (KAWARABAYASI et al. 1999; MALLICK et al. 2000; SHE et al. 2001). Whereas both β-type subunits from the *Pyrococcus* species share the N-terminal active-site threonine, this residue is replaced in one of the two β-type subunits from *Aeropyrum*, *Pyrobaculum* and *Sulfolobus* by an alanine (ZWICKL et al. 2000; MAUPIN-FURLOW et al. 2000). Thus, if both β-type genes are expressed one of the two β-type subunits is inactive. Archaeal proteasomes

Table 1. Proteasomes in Archaea[a]

Organism	Cloned genes	Purified proteasome	Reference
Euryarchaeota			
Archaeoglobus fulgidus	α+β	−	KLENK et al. 1997
Halobacterium species NRC-1	α+β	−	NG et al. 2000
Haloferax volcanii	2α+β	+	WILSON et al. 1999
Methanobacterium thermoautotrophicum	α+β	−	SMITH et al. 1997
Methanococcus jannaschii	α+β	−	BULT et al. 1996
Methanosarcina thermophila	α+β	+	MAUPIN-FURLOW and FERRY 1995
Pyrococcus abysii	α+2β[b]	−	http://www.genoscope.cns.fr/Pab
Pyrococcus furiosus	α+2β[b]	+	BAUER et al. 1997; ROBB et al. 2001
Pyrococcus horikoshii	α+2β[b]	−	KAWARABAYASI et al. 1998
Thermoplasma acidophilum	α+β	+	DAHLMANN et al. 1989
Thermoplasma volcanium	α+β	−	KAWASHIMA et al. 2000
Crenarchaeota			
Aeropyrum pernix	α+2β[c]	−	KAWARABAYASI et al. 1999
Pyrobaculum aerophilum	α+2β[c]	−	Fitz-Gibbon, personal communication
Sulfolobus solfataricus	α+2β[c]	−	SHE et al. 2001

[a] Archaea split into two branches the Euryarchaeota and the Crenarchaeota.
[b] Both β-type subunits have the N-terminal active-site residue threonine 1.
[c] Only one of the two β-type subunits has the N-terminal active-site residue threonine 1.

containing both active and inactive β-type subunits would have a reduced number of active sites, an evolutionary precedent for eukaryotic proteasomes, which have only six active sites as compared with 14 in the archaetypal *Thermoplasma acidophilum* proteasome.

In bacteria, genuine proteasomes of the α+β-type have hitherto been found only in species belonging to the order *Actinomycetales*. With the exception of *Rhodococcus erythropolis*, where the 20S proteasomes are built of two α- and two β-type subunits (TAMURA et al. 1995), proteasomes from all other species, are composed of a single α- and a single β-type subunit (DE MOT et al. 1999; ZWICKL et al. 2000). Many other bacteria contain a gene, called hslV in *E. coli*, that encodes a protein which is closely related to the proteasomal β-type subunits (LUPAS et al. 1994) and forms a ring-shaped dodecamer (BOCHTLER et al. 1997). Occurrence of proteasomes and the HslV protease seems to be mutually exclusive; either one or the other, or neither of them is present in the same species (DE MOT et al. 1999; ZWICKL et al. 2000).

In eukaryotes, such as *Saccharomyces cerevisiae*, *Caenorhabditis elegans*, *Drosophila melanogaster* and *Oryza sativa*, seven different α and seven different β genes have been found (HUGHES 1997; BOUZAT et al. 2000; SASSA et al. 2000). In higher eukaryotes with an adaptive immune system, γ-interferon stimulates expression of three additional β-type subunits, which can replace the closely related, constitutively expressed active β-type subunits (ROCK and GOLDBERG 1999; RECHSTEINER et al. 2000). This results in a modulation of proteolytic specificity. Interestingly, a superfamily of 23 proteasomal genes has been identified in *Arabidopsis thaliana*, which can be grouped into 14 distinct subfamilies encoding isoforms (PARMENTIER et al. 1997; FU et al. 1998). The functional significance of these isoforms is unclear at present. Alignment of all the available sequences of proteasome subunits shows that the two families, α and β, originate from a gene duplication event, which must have taken place early in evolution (see chapter by Volker and Lupas, this volume; ZWICKL et al. 1992; HUGHES and YEAGER 1997; BOUZAT et al. 2000; ZWICKL et al. 2000).

3 Structural Features of the 20S Proteasome

In spite of the differences in subunit complexity, the quaternary structure is highly conserved in 20S proteasomes from all three domains of life, archaea, bacteria and eukaryotes. The 28 subunits, 14 of the α-type and 14 of the β-type, are grouped into four seven-membered rings, which collectively form a barrel-shaped complex with a length of 15nm and a diameter of 11nm (Fig. 1A). The two adjacent β-subunit rings, enclose the central cavity (CC) with a diameter of approximately 5nm, which harbours the active sites. It is connected via two narrow constrictions with two slightly smaller outer cavities, the 'antechambers' (AC), which are formed jointly by one α and one β ring. An axial pore in the α-rings gives access to the AC (Fig. 1B).

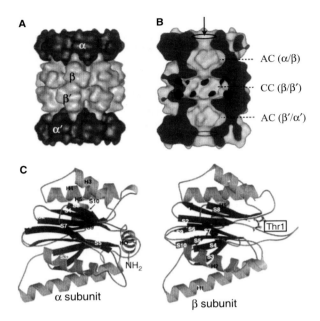

Fig. 1A–C. Structure of the 20S proteasome. **A** Low-resolution model (1.2nm) of the 20S proteasome derived from the crystal structure of the *Thermoplasma* proteasome (LÖWE et al. 1995). The α-subunits form the heptameric outer rings, the β-subunits the inner rings. **B** The same structure cut open along the seven-fold axis to display the two antechambers (*AC*) and the central chamber (*CC*) with the 14 active sites (*black*). The channel openings at the two ends of the cylinder are 1.3nm in diameter. **C** Fold of α- and β-subunits of the *Thermoplasma* proteasome (LÖWE et al. 1995). A pair of five-stranded β-sheets is flanked on both sides by α-helices. Helices (*H*) and strands (*S*) are numbered H0–H5 and S1–S10. The β-subunits lack helix H0, which occupies the cleft on one side of the β-sheet sandwich in the α subunits. The active-site threonine (*Thr1*) of β subunits is shown as a ball-and-stick representation

In prokaryotic proteasomes, built of 14 α and 14 β subunits, the rings are homomeric and the complex is described by an $\alpha_7\beta_7\beta_7\alpha_7$ stoichiometry. Correspondingly, the stoichiometry of eukaryotic 20S proteasomes is $\alpha_{1-7}\beta_{1-7}\beta_{1-7}\alpha_{1-7}$. Each of the 14 different subunits is present in two copies within one complex and occupies a precisely defined position. Thus the multiple axes of symmetry of the *Thermoplasma* proteasome are reduced to a C2 symmetry in the eukaryotic proteasome; each of the 14 different eukaryotic subunits is found twice within the complex and occupies well defined positions (SCHAUER et al. 1993; LÖWE et al. 1995; KOPP et al. 1997; GROLL et al. 1997). To indicate subunit positions within the rings, a systematic nomenclature for proteasome subunits has been proposed: Subunits are numbered α1 to α7 and β1 to β7, and those related by C2 symmetry are distinguished by the prime symbol (α1' to α7' and β1' to β7'). In mammals, the complexity of proteasomes is further enhanced by the fact that after γ-interferon stimulation three of the constitutive subunits (β1, β2, and β5) can be replaced by closely related subunits to form immunoproteasomes; according to the new nomenclature the inducible subunits are termed β1i, β2i, and β5i (GROLL et al. 1997).

As only three subunits (β1 or β1i, β2 or β2i, β5 or β5i) of the β-type subunits present in a single proteasome particle are proteolytically active, eukaryotic proteasomes contain a total of six active sites, whereas prokaryotic proteasomes, built of identical copies of β subunits, have 14 active sites.

Another feature that appeared to distinguish eukaryotic from prokaryotic proteasomes relates to the path of substrate entry into the inner cavities. Whereas the crystal structure of the *Thermoplasma* proteasome revealed the existence of an approximately 1.3-nm-wide axial channel in the α rings (LÖWE et al. 1995), and electron microscopy of 20S proteasomes incubated with gold-labelled substrates demonstrated that it is in fact the entry site (WENZEL and BAUMEISTER 1995), there was no obvious passageway for substrate entry or exit in the structure of the yeast 20S proteasome (GROLL et al. 1997). Meanwhile, it has become clear that the yeast particle had been crystallized in a closed form, in which the N-terminal tails of the α-subunits interdigitate to form a plug, whereas these tails are disordered in the archaeal proteasome and, therefore, the gate appears to be open. Deletion of the nine N-terminal residues of one subunit (α3) of the yeast proteasome results in disorder of a number of other subunits and creates an opening which is very similar to that seen in the *Thermoplasma* proteasome (GROLL et al. 2000a). Whereas eukaryotic wild-type proteasomes are almost 'latent', these mutant proteasomes have a greatly enhanced peptidase activity, probably because access is facilitated. Strikingly, mutation of a single residue of α3, Asp9 to Ala, in a highly conserved region found in all α-type subunits is sufficient to yield proteasomes with enhanced peptidase activity (GROLL et al. 2000a). Previously, mutational studies have indicated a role for one of the six ATPases present in the 19S regulator, Rpt2, in the control of peptide hydrolysis (RUBIN et al. 1998). Construction of a triple mutant, i.e. simultaneous deletion of the N-terminal tail of α3 and α7 and inactivation of the ATP-binding site of the 19S ATPase Rpt2, demonstrated that the function of Rpt2 is to open the channel in the α ring of the proteasome (KÖHLER et al. 2001). In vivo, associations with regulatory complexes, e.g. the 19S regulatory complex, the PA28 activator, or the PA26 activator, are known to activate 20S proteasomes (see the chapters by Glickman and Maytal, and Hill et al., this volume). All of these regulatory complexes interact with the terminal α rings and are likely to function by mechanisms that open the gate for substrate uptake (PICKART and VANDEMARK 2000; RECHSTEINER et al. 2000; WHITBY et al. 2000). Comparison of the size of degradation products from wild-type and open-channel proteasomes showed that the channel also determines the size of the products. The wild-type proteasome produced peptides with a mean length of 7.6 residues, whereas the mean size of products from the open-channel proteasome was 9.6 residues (KÖHLER et al. 2001). Thus the channel in the 20S proteasome is not only regulating access of substrates but also exit of products and their size.

As anticipated from their sequence similarity, the (non-catalytic) α- and the (catalytic) β-type subunits have the same fold (LÖWE et al. 1995; GROLL et al. 1997): a four-layer α+β structure with two antiparallel five-stranded β sheets, flanked on one side by two, on the other side by three α helices (Fig. 1C). In the β-type subunits, the β-sheet sandwich is closed at one end by four hairpin loops and open at the opposite end to form the active-site cleft; the cleft is oriented towards the inner surface of the central cavity. In the α-type subunits an additional helix formed by an N-terminal extension crosses the top of the β-sheet sandwich and fills this cleft. Initially, the proteasome fold was believed to be unique; however it turned out to be prototypical of a new superfamily of proteins referred to as Ntn (N-terminal nucleophile) hydrolases (BRANNIGAN et al. 1995). Beyond the common fold, members of this family share the mechanisms of the nucleophilic attack and self-processing (DODSON and WLODAWER 1998; OINONEN and ROUVINEN 2000).

4 Catalytic Mechanism

Site-directed mutagenesis and the crystal structure analysis of a proteasome-inhibitor complex identified the amino-terminal threonine (Thr1) of *Thermoplasma* β-subunits as both the catalytic nucleophile and the primary proton acceptor (SEEMÜLLER et al. 1995; LÖWE et al. 1995): The hydroxyl group of Thr1 attacks the carbonyl carbon of the scissile peptide bond, while its amino group serves as the primary proton acceptor, which enhances the nucleophilicity by stripping the proton from the side-chain hydroxyl; for steric reasons a water molecule is likely to mediate the proton transfer. Such a 'single-residue' active site is a characteristic feature of Ntn hydrolases, as is the autocatalytic removal of a prosequence, which is necessary to expose the nucleophilic group. For this autocatalytic cleavage Thr1Oγ of the proteasome β-subunit precursor adds to the carbonyl carbon of Gly-1, and a water molecule is supposed to fulfil the base function of the Thr1 α-amino group in the mature enzyme (DITZEL et al. 1998). Both, Gly-1 and Thr1 are invariant residues of all 'active' proteasome subunits. In other Ntn hydrolases the nucleophile is provided either by a threonine, cysteine or serine residue, and the residues at the (−1) position are variable. In proteasomes, replacement of the active-site threonine by serine does not alter the rates of hydrolysis of small fluorogenic peptide substrates, but the mutant is significantly slower in degrading larger peptides and proteins (SEEMÜLLER et al. 1995; MAUPIN-FURLOW et al. 1998; KISSELEV et al. 2000). Self-processing of β-subunit precursors is inefficient with serine at the +1 position (SEEMÜLLER et al. 1996; CHEN and HOCHSTRASSER 1996; HEINEMEYER et al. 1997). Conversely, substitution of the threonine by a cysteine yields proteasomes that are unable to cleave any substrate, but the self-cleavage reaction remains unaffected by this mutation (SEEMÜLLER et al. 1996). Besides the N-terminal threonine, several other residues of proteasome β-subunits (Glu17, Lys33, Asp166) are required for proteolytic activity, although their precise contribution to the catalytic mechanism awaits further clarification (LÖWE et al. 1995; SEEMÜLLER et al. 1996; CHEN and HOCHSTRASSER 1996; SCHMIDTKE et al. 1996; HEINEMEYER et al. 1997; MAUPIN-FURLOW et al. 1998; MAYR et al. 1998).

The catalytic role of the N-terminal threonine of β-subunits was soon extended to eukaryotic proteasomes by the observation that binding of lactacystin, a

Streptomyces metabolite, to the mammalian β-type subunit β5/X irreversibly inhibits its proteolytic activity (FENTEANY et al. 1995). Meanwhile, a large number of mutational studies with other archaeal (MAUPIN-FURLOW et al. 1998), bacterial (MAYR et al. 1998), yeast (CHEN and HOCHSTRASSER 1996; ARENDT and HOCHSTRASSER 1997; HEINEMEYER et al. 1997; GROLL et al. 1999) and mammalian β-type subunits (SCHMIDTKE et al. 1996; SALZMANN et al. 1999) have confirmed a common proteolytic mechanism for all kinds of proteasomes. It remains enigmatic, however, why in eukaryotic proteasomes four out of seven β subunits lack the N-terminal active site threonine, and therefore are presumably inactive.

5 Processing and Assembly

Proteolytically active β-type subunits are synthesized in an inactive precursor form containing N-terminal extensions of variable lengths, the propeptides which must be removed post-translationally to allow the formation of active sites. This process is tied in with the assembly of the 20S proteasome in such a manner that activation is delayed until assembly is complete and the active sites are sequestered from the cellular environment. Cleavage of the propeptide proceeds autocatalytically, relying on the active-site threonine, and the invariant glycine at position −1 appears to be the prime determinant of the cleavage site (SCHMIDTKE et al. 1996; SEEMÜLLER et al. 1996; CHEN and HOCHSTRASSER 1996). It is as yet unclear whether the reaction is intramolecular (*cis*) or intermolecular (*trans*) or whether both mechanisms apply. Experiments with *Thermoplasma* proteasomes, in which inactive mutant β-subunits were found to be correctly processed in the presence of wild-type β-subunits indicated an intermolecular mechanism (SEEMÜLLER et al. 1996). An inactive mutant form of the mammalian subunit β1i/LMP2, in contrast, is not correctly processed, in spite of being incorporated into the complex along with active neighbours (SCHMIDTKE et al. 1996). This finding would argue against an intermolecular mechanism of processing, which is indeed difficult to reconcile with our current understanding that processing occurs at a late stage of proteasome assembly.

Interestingly, mature eukaryotic proteasomes contain partially processed 'precursor' subunits; the inactive subunits β6 and β7, both synthesized in a precursor form, have their N termini removed. However, cleavage of these subunits (and of inactive mutant subunits) occurs some distance upstream of the consensus cleavage site of the active subunits (Gly−1/Thr1) leaving short propeptide remnants. The existence of such processing intermediates has led to the proposal of a two-step model for precursor processing: The propeptide is trimmed in size in a first intermolecular step, and removed completely in a second intramolecular step (SCHMIDTKE et al. 1996). Several reports, in particular a recent study combining mutational, biochemical and crystallographic data demonstrated that processing of naturally inactive or mutationally inactivated subunits relies indeed on the nearest neighbour in the complex carrying an active site (SCHMIDTKE et al. 1996; HEINE-

MEYER et al. 1997; DITZEL et al. 1998; GROLL et al. 2000b). Whether such an intermolecular cleavage merely serves to remove a bulky protrusion from inactive subunits, or whether such a trimming step is in a more intricate manner linked to the subsequent intramolecular cleavages is as yet not clear.

The pathways of 20S proteasome assembly, and also the roles of the β propeptides appear to differ somewhat between archaea, bacteria and eukaryotes. The α-subunits of *Thermoplasma* and of other archaea spontaneously form seven-membered rings in the absence of β-subunits (ZWICKL et al. 1994; MAUPIN-FURLOW et al. 1998; WILSON et al. 2000). β-subunits, in turn, remain monomeric, unprocessed and inactive, and do not even fold properly in the absence of α-subunits. Thus, it appears that the α rings serve as a template upon which the β-subunits assemble. Coexpression of both archaeal genes yields fully assembled and functional proteasomes, irrespective of the presence or absence of the relatively short (6–10 residues) β-subunit propeptides. Proteasomes can even be reassembled after dissociation, also demonstrating that the propeptides are not required for assembly (GRZIWA et al. 1994; WILSON et al. 1999). In contrast, neither α nor β-subunits of the bacterial *Rhodococcus* proteasome assemble by themselves, but complexes are formed as soon as both types of subunits are allowed to interact. This suggests that assembly starts with the formation of an α/β heterodimer, although half proteasomes, built of one ring of α-subunits and one ring of β-subunit precursors are the first intermediates that have been captured so far. Dimerization of half proteasomes, which are still inactive, triggers processing of the β-precursor subunits and fully assembled, proteolycially active proteasomes are formed. Although the β propeptides are not essential for the formation of *Rhodococcus* proteasomes, they have a strong effect on assembly efficiency. The 65-residues-long propeptide of *Rhodococcus* β1, which can be provided *in trans*, has two functions: it supports the initial folding of the β-subunit, and it promotes the final maturation of proteasomes after the docking of two half proteasomes (ZÜHL et al. 1997). The chaperone-like manner by which the propeptide accelerates subunit folding, resembles that of the well studied propeptides of some extracellular bacterial proteases, such as α-lytic protease and subtilisin (see, for example, BAKER et al. 1993).

Assembly of eukaryotic proteasomes is obviously a more complicated process. The correct positioning of each of the 14 different, but closely related subunits must be orchestrated. As some of the eukaryotic α-type subunits are capable of assembling into homomeric ring structures similar to *Thermoplasma* α-subunits it appears possible that α rings have a scaffolding function for the incorporation of β-subunits (GERARDS et al. 1997; YAO et al. 1999). Detailed studies of precursor complexes of mouse 20S proteasomes with subunit specific antibodies indeed identified intermediates that contain all α-type subunits and a subset of 'early' β-type subunits (β2, β3 and β4). Incorporation of the four remaining β-subunits yields half-proteasomes and triggers their dimerization; subsequent autocatalytic processing and conformational changes are supposed to complete proteasome maturation (NANDI et al. 1997). Previously described 13S–16S precursor complexes are probably complete or incomplete half-proteasome intermediates (YANG et al. 1995; SCHMIDTKE et al. 1997).

The eukaryotic β propeptides are not only highly variable in length and sequence, but also their importance for the assembly process varies. Those of yeast β5/Doa3 (CHEN and HOCHSTRASSER 1996; HEINEMEYER et al. 1997) and mammalian β5i/LMP7 (CERUNDOLO et al. 1995), both more than 70 residues long, are essential for subunit incorporation suggesting an intramolecular chaperone function resembling that of the *Rhodococcus* β1 propeptide. Of less importance are the propeptides of the human β1i/LMP2 and yeast β2 subunits, which were shown to improve the efficiency of subunit incorporation (GROETTRUP et al. 1997; SCHMIDT et al. 1999; JÄGER et al. 1999). Interestingly, propeptides do not seem to have a function in correct positioning of β-type subunits, as the replacement of the propeptide of β5i/LMP7 with the propeptide of β2i/LMP2 does not change the position of the β5i/LMP7 subunit within the particle (SCHMIDTKE et al. 1996). Recently, removal of the propeptides of yeast β1, β2 and β5 has shown that another critical function of β propeptides might be to protect the catalytic threonine against Nα-acetylation (ARENDT and HOCHSTRASSER 1999; JÄGER et al. 1999; KIMURA et al. 2000).

Whereas the efficient assembly of prokaryotic proteasomes proceeds autonomously, assembly of eukaryotic proteasomes appears to require extrinsic maturation factors. The chaperone protein Hsc73 was found to associate with mammalian proteasome precursor complexes, but not with mature proteasomes (SCHMIDTKE et al. 1997). Similarly, Ump1, a small protein contained in half-proteasome precursor complexes of yeast proteasomes, is supposed to have a chaperone-like function. Ump1 is essential for maturation of wild-type proteasomes, but not for mutants lacking the β5 propeptide suggesting a special interaction between Ump1 and the propeptide (RAMOS et al. 1998). Mammalian homologues of Ump1 have recently been described (BURRI et al. 2000; WITT et al. 2000).

6 Proteolytic Activity and Degradation Products

When tested with short fluorogenic peptides, proteasomes display 'chymotrypsin-like' (cleavage after hydrophobic residues), 'trypsin-like' (cleavage after basic residues) and 'peptidylglutamyl-peptide hydrolysing' (PGPH) activity (cleavage after acidic residues), as defined on the basis of the P1 residue of the substrate (the residue located N-terminal to the scissile bond) (CARDOZO 1993; ORLOWSKI and WILK 2000). Two additional activities, a 'branched-chain amino acid preferring' activity (BrAAP) and 'small neutral amino acid preferring' activity (SNAAP) have been described for mammalian proteasomes (ORLOWSKI et al. 1993). Each of these types of activity has been associated with distinct β-type subunits through a number of mutagenesis, inhibition, and X-ray diffraction studies (HILT et al. 1993; ENENKEL et al. 1994; HEINEMEYER et al. 1993, 1997; CHEN and HOCHSTRASSER 1996; ARENDT and HOCHSTRASSER 1997; DICK et al. 1998; MCCORMACK et al. 1998; GROLL et al. 1999): The chymotryptic, tryptic and PGPH activities have been assigned to the subunits β5, β2, and β1, respectively. BrAAP activity resides in both, β5 and β1, and SNAAP activity correlates with β2.

The barrel-shaped architecture of the 20S proteasome allows substrate proteins to be degraded in a processive manner, i.e. without the release of degradation intermediates (AKOPIAN et al. 1997). It is noteworthy that the cleavage sites found when longer peptides or proteins are used as substrates do not reflect the aforementioned specificities (WENZEL et al. 1994; EHRING et al. 1996). The *Thermoplasma acidophilum* proteasome for example, when assayed with short fluorogenic peptides displays solely a 'chymotryptic' activity, but chymotryptic cleavage sites are found only rarely in degradation products from polypeptides (WENZEL et al. 1994). This suggests that residues beyond P1 (P2, P3, P4, ...) may determine the site of cleavage (for review see ORLOWSKI and WILK 2000). Moreover, the association with regulatory complexes has been reported to modulate the nature of the degradation products of the 20S proteasome (KISSELEV et al. 1999b; EMMERICH et al. 2000). In any case, the complex rules are by no means fully understood and proteins are cleaved in an apparently non-specific manner (DOLENC et al. 1998).

Nevertheless, the peptide products generated by the 20S proteasome fall into a relatively narrow size range of 6–10 amino acid residues. This observation led to the proposal that proteasomes may possess an intrinsic molecular ruler. At the time it was considered that the distance between neighbouring active sites could provide the mechanistic basis for such a ruler (WENZEL et al. 1994). Indeed, the crystal structure of the *Thermoplasma* proteasome revealed a distance of 2.8nm between neighbouring active sites, which corresponds to a hepta- or octapeptide in an extended conformation, and seemed to provide strong evidence in support of the molecular ruler hypothesis (LÖWE et al. 1995). On the other hand, recent more comprehensive analyses of product lengths, whilst in agreement with an average length of eight residues (± 1 residue), showed larger size variations, which are difficult to reconcile with a purely geometry-based ruler which should yield products more focused in length (KISSELEV et al. 1998, 1999b; NUSSBAUM et al. 1998). Moreover, a reduction in the number of active sites to four or two in mutant yeast proteasomes had little effect on the size of the peptides that were generated (DICK et al. 1998). It is therefore unlikely that the distance between active sites is a major determinant of the product size. Studies with synthetic peptides varying in length but displaying the same (repetitive) pattern of cleavage sites indicated that, below a certain threshold in length (<12–14 residues), degradation is decelerated, possibly because the products have a higher probability of exiting the proteolytic nanocompartment (DOLENC et al. 1998). Although they might re-enter and be degraded further, this appears to be a slow and inefficient process and therefore products smaller than 12–14 residues accumulate. It should be noted here that a diffusion-controlled mechanism was considered as another option when the molecular ruler hypothesis was originally put forward (WENZEL et al. 1994).

Recently, a cyclical 'bite-chew' mechanism for protein breakdown was proposed for the eukaryotic proteasome (KISSELEV et al. 1999a). The model suggested that certain active sites regulate each other allosterically. Substrates of the chymotrypsin-like site stimulate the PGPH activity 5- to 35-fold and additionally activate the second chymotrypsin-like site. Moreover, substrates of the PGPH sites were reported to inhibit the chymotrypsin-like activity. These allosteric effects were proposed to be indicative of an ordered, cyclical mechanism for protein degradation. The model was challenged, when it was found that ritonavir, an inhibitor of human immunodeficiency virus-1 protease, inhibits the chymotrypsin-like activity and stimulates the trypsin-like activity of the proteasome (SCHMIDTKE et al. 1999). Moreover, it was shown that the mutual regulation of the chymotrypsin-like and PGPH activities by their substrates, as suggested by the bite-chew model, was not affected by selective inhibitors of the respective active sites (SCHMIDTKE et al. 2000). The data of Schmidtke et al. suggest that proteasome substrates can bind not only to an active but also to a non-catalytic site, which led to the proposal of a kinetic two-site modifier model (SCHMIDTKE et al. 2000). More recently it was shown that a selective inhibition of the PGPH activity by α',β'-epoxyketone inhibitors did not inhibit protein degradation and the inhibition of the chymotryptic activity by PGPH substrates did not require the binding to the PGPH site or hydrolysis of the PGPH substrate. This findings support the existence of an inter-subunit regulation through binding to non-catalytic site(s) (MYUNG et al. 2001).

The immune system has taken advantage of the fact that proteasomes generate peptides with an average length of 8–12 residues. However, in the cellular environment the peptide products of the proteasome are not stable and the vast majority undergo further cleavage by peptidases for completion of the catabolic conversion of proteins into free amino acids (ROCK and GOLDBERG 1999). Nevertheless, some peptides escape further degradation and are presented by major histocompatibility complex (MHC) class I molecules on the cell surface eliciting an immune response. It has been shown that leucine aminopeptidase, which is up-regulated by γ-interferon, assists in antigen production by N-terminal trimming of peptides generated by the proteasome (BENINGA et al. 1998). On the other hand thimet-oligopeptidase, which is down-regulated by γ-interferon, mediates the destruction of antigenic peptides. Thus, γ-interferon appears to increase the supply of peptides by stimulating their generation and decreasing their destruction (YORK et al. 1999). In this respect it is notable that 26S immunoproteasomes release two to four times more of the N-extended versions of immunogenic peptides than do 26S proteasomes (CASCIO et al. 2001). The N-extended versions of the immunogenic peptides might have a higher chance to escape the destruction of the epitope by cytosolic aminopeptidases (CASCIO et al. 2001). Accordingly, two cytosolic proteases, puromycin-sensitive aminopeptidase and bleomycin hydrolase have been shown to remove N-terminal amino acids from the vesicular stomatitis virus nucleoprotein; in conjunction with proteasomal cleavage at the C terminus, the immunogenic epitope is generated (STOLTZE et al. 2000). Obviously, the proteasome system alone is not sufficient to generate the necessary supply of MHC class I antigens (STOLTZE et al. 2000).

Apart from the immune system, there is little use for the degradation products released by the proteasome and several peptidases exist which complete their conversion into free amino acids. One of them is tricorn peptidase, originally found in *Thermoplasma* cells where it exists in the form of a giant icosahedral complex of approximately 15MDa (TAMURA et al. 1996; WALZ et al. 1997). It has been shown by in vitro studies that tricorn in cooperation with an array of aminopeptidases converts oligopeptides of 8–12 residues efficiently into free amino acids (TAMURA et al. 1998). Tricorn peptidase has been found in a number of archaeal and bacterial species, but it is not ubiquitous; functional homologues are found in other species (TAMURA et al. 2001). In eukaryotes inhibition of the 20S proteasome up-regulates tripeptidylpeptidase II, another huge peptidase complex which is also regarded as a functional homologue of tricorn peptidase (see the chapter by Niedermann, this volume; GEIER et al. 1999; WANG et al. 2000).

7 The Archaeal and Bacterial AAA ATPases

The sequencing of the genome of the methanogenic archaeon *Methanococcus jannaschii* revealed the existence of a gene, S4, with a high sequence similarity to the ATPases of the eukaryotic 19S regulator (BULT et al. 1996). The deduced 50kDa protein has an N-terminal coiled-coil, a hallmark of proteasomal AAA ATPases, and a C-terminal AAA domain. The protein was expressed in *Escherichia coli* and purified as a 650kDa complex with nucleotidase activity. When mixed with proteasomes from *Thermoplasma acidophilum*, degradation of substrate proteins was stimulated up to 25-fold; hence the complex was named PAN, for proteasome activating nucleotidase (ZWICKL et al. 1999). As expected, the *Methanococcus* PAN stimulates also the endogenous proteasome from *Methanococcus*, but it remains notoriously difficult to isolate a stable protease–ATPase complex formed by both molecules (WILSON et al. 2000). *Methanococcus* PAN was recently shown to recognize ssrA-tagged green fluorescent protein and mediate its energy-dependent unfolding and subsequent translocation into the 20S proteasome for degradation (BENAROUDJ and GOLDBERG 2000). The ssrA tag is an 11-residue, hydrophobic sequence which is added to the C terminus of incompletely translated proteins (KEILER et al. 1996), thus targeting them for degradation by ATP-dependent proteases (KARZAI et al. 2000). Interestingly, the ssrA tag is found in many bacteria but not in archaea (KEILER et al. 2000). This suggests that the hydrophobic nature of the peptide, which was shown to be essential for recognition by the ClpA and ClpX ATPases in *E. coli* (KARZAI et al. 2000), is sufficient for the recognition by the *Methanococcus* PAN complex.

Homologues of *Methanococcus* PAN have been found in most but not in all archaeal genomes (see Table 2). No PAN homologue exists for example in the closely related euryarchaeal species *Thermoplasma acidophilum* (RUEPP et al. 2000) and *Thermoplasma volcanium* (KAWASHIMA et al. 2000), and in the more distantly related crenarchaeal species *Pyrobaculum aerophilum* (S. Fitz-Gibbon, personal communication). Therefore, the role of PAN in activating the 20S proteasome must be assumed by different molecules in these organisms, most probably more divergent members of the AAA ATPase family. The complete sequence of the *Thermoplasma* genome revealed three candidate proteins: VAT, a two domain AAA ATPase, which is closely related to yeast Cdc48 and human p97 and two one-

Table 2. Distribution of genes for PAN and VAT in archaeal genomes

Organism	Number of cloned Homologues	
	PAN	VAT
Euryarchaeota[a]		
Thermoplasma acidophilum	–	1
Thermoplasma volcanium	–	1
Methanococcus jannaschii	1	1
Methanobacterium thermoautothrophicum	1	1
Archaeoglobus fulgidus	1	2
Pyrococcus furiosus	1	2
Pyrococcus horikoshii	1	2
Pyrococcus abysii	1	2
Halobacterium species NRC-1	2	3
Crenarchaeota[a]		
Aeropyrum pernix	1	2
Sulfolobus solfataricus	1	2
Pyrobaculum aerophilum	–	2

[a] Archaea split into two branches the Euryarchaeota and the Crenarchaeota.

domain AAA ATPases, VAT2 and Lon2 (RUEPP et al. 2000). While the latter proteins have not yet been characterized, a chaperone-like activity was demonstrated for *Thermoplasma* VAT (GOLBIK et al. 1999). Still it remains to be shown that VAT can stimulate the ATP-dependent degradation of proteins by the *Thermoplasma* 20S proteasomes. In this context it is noteworthy that the eukaryotic homologues of VAT (Cdc48, p97), had previously been implicated in the degradation of substrate proteins via the ubiquitin-proteasome pathway (JOHNSON et al. 1995; GHISLAIN et al. 1996; DAI et al. 1998; MAYR et al. 1999; KOEGL et al. 1999; HOPPE et al. 2000). Different from the N-terminal coiled-coil found in PAN and the 19S ATPases, both VAT and p97/Cdc48 have a N-terminal substrate binding domain which is built of a double-ψ β-barrel and a novel six-stranded β-clam fold (COLES et al. 1999; ZHANG et al. 2000).

A recurring feature of all bacteria possessing genuine 20S proteasomes is the existence of a gene, ARC, encoding a more distant member of the AAA family of ATPases, which is found upstream of the proteasome operons (NAGY et al. 1998; DE MOT et al. 1999). The recombinant ARC ATPase from *Rhodococcus erythropolis* is a complex of two six-membered rings with ATPase activity; like other proteasomal ATPases it has an N-terminal coiled-coil domain (WOLF et al. 1998). Although a functional interaction between the ARC complex and the *Rhodococcus* 20S proteasome has not yet been demonstrated, it is likely that it has a role in ATP-dependent protein degradation.

8 Summary

In contrast to our detailed knowledge of prokaryotic proteasomes, we have only a limited understanding of the prokaryotic regulators and their functional interaction with the proteasome. Most probably, we will soon learn more about the molecular structure and the mechanism of action of the prokaryotic regulators. Nevertheless, it still remains to be unravelled which signals or/and modifications transform an endogenous prokaryotic protein into a substrate of the proteasomal degradation machinery.

References

Akopian TN, Kisselev AF, Goldberg AL (1997) Processive degradation of proteins and other catalytic properties of the proteasome from *Thermoplasma acidophilum*. J Biol Chem 272:1791–1798
Arendt CS, Hochstrasser M (1997) Identification of the yeast 20S proteasome catalytic centers and subunit interactions required for active-site formation. Proc Natl Acad Sci USA 94:7156–7161
Arendt CS, Hochstrasser M (1999) Eukaryotic 20S proteasome catalytic subunit propeptides prevent active site inactivation by N-terminal acetylation and promote particle assembly. EMBO J 18:3575–3585
Baker D, Shiau K, Agard DA (1993) The role of proregions in protein folding. Curr Opin Cell Biol 5: 966–970
Bauer MW, Bauer SH, Kelly RM (1997) Purification and characterization of a proteasome from the hyperthermophilic archaeon *Pyrococcus furiosus*. Appl Eviron Microbiol 63:1160–1164
Benaroudj N, Goldberg AL (2000) PAN, the proteasome-activating nucleotidase from archaebacteria, is a protein-unfolding molecular chaperone. Nature Cell Biol 2:833–839
Beninga J, Rock KL, Goldberg AL (1998) Interferon-γ can stimulate post-proteasomal trimming of the N-terminus of an antigenic peptide by inducing leucine aminopeptidase. J Biol Chem 273:18734–18742
Bochtler M, Ditzel L, Groll M, Huber R (1997) Crystal structure of heat shock locus V (HslV) from *Escherichia coli*. Proc Natl Acad Sci USA 94:6070–6074
Bouzat JL, McNeil LK, Robertson HM, Solter LF, Nixon JE, Beever JE, Gaskins HR, Olsen G, Subramaniam S, Sogin ML, Lewin HA (2000) Phylogenomic analysis of the alpha proteasome gene family from early-diverging eukaryotes. J Mol Evol 51:532–543
Brannigan JA, Dodson G, Duggleby HJ, Moody PC, Smith JL, Tomchick DR, Murzin AG (1995) A protein catalytic framework with an N-terminal nucleophile is capable of self-activation. Nature 378:416–419
Bult CJ, White O, Olsen GJ, Zhou L, Fleischmann RD, Sutton GG, Blake JA, FitzGerald LM, Clayton RA, Gocayne JD, Kerlavage AR, Dougherty BA, Tomb JF, Adams MD, Reich CI, Overbeek R, Kirkness EF, Weinstock KG, Merrick JM, Glodek A, Scott JL, Geoghagen NSM, Venter JC (1996) Complete genome sequence of the methanogenic archaeon, *Methanococcus jannaschii*. Science 273:1058–1073
Burri L, Hockendorff J, Boehm U, Klamp T, Dohmen RJ, Levy F (2000) Identification and characterization of a mammalian protein interacting with 20S proteasome precursors. Proc Natl Acad Sci USA 97:10348–10353
Cardozo C (1993) Catalytic components of the bovine pituitary multicatalytic proteinase complex (proteasome). Enzyme Protein 47:296–305
Cascio P, Hilton C, Kisselev AF, Rock KL, Goldberg AL (2001) 26S proteasomes and immunoproteasomes produce mainly N-extended versions of an antigenic peptide. EMBO J 20:2357–2366
Cerundolo V, Kelly A, Elliott T, Trowsdale J, Townsend A (1995) Genes encoded in the major histocompatibility complex affecting the generation of peptides for TAP transport. Eur J Immunol 25:554–562
Chen P, Hochstrasser M (1996) Autocatalytic subunit processing couples active site formation in the 20S proteasome to completion of assembly. Cell 86:961–972
Coles M, Diercks T, Liermann J, Gröger A, Rockel B, Baumeister W, Koretke KK, Lupas A, Peters J, Kessler H (1999) The solution structure of VAT-N reveals a 'missing link' in the evolution of complex enzymes from a simple element. Curr Biol 9:1158–1168

Dai RM, Chen EY, Longo DL, Gorbea CM, Li CCH (1998) Involvement of valosin-containing protein, an ATPase co-purified with IκBα and 26S proteasome, in ubiquitin-proteasome-mediated degradation of IκBα. J Biol Chem 273:3562–3573

De Mot R, Nagy I, Walz J, Baumeister W (1999) Proteasomes and other self-compartmentalizing proteases in prokaryotes. Trends Microbiol 7:88–92

Dick TP, Nussbaum AK, Deeg M, Heinemeyer W, Groll M, Schirle M, Keilholz W, Stevanovic S, Wolf DH, Huber R, Rammensee HG, Schild H (1998) Contribution of proteasomal β-subunits to the cleavage of peptide-substrates analyzed with yeast mutants. J Biol Chem 273:25637–25646

Ditzel L, Huber R, Mann K, Heinemeyer W, Wolf DH, Groll M (1998) Conformational constraints for protein self-cleavage in the proteasome. J Mol Biol 279:1187–1191

Dodson G, Wlodawer A (1998) Catalytic triads and their relatives. Trends Biochem Sci 23:347–352

Dolenc I, Seemüller E, Baumeister W (1998) Decelerated degradation of short peptides by the 20S proteasome. FEBS Lett 434:357–361

Ehring B, Meyer TH, Eckerskorn C, Lottspeich F, Tampe R (1996) Effects of major-histocompatibility-complex-encoded subunits on the peptidase and proteolytic activities of human 20S proteasomes. Cleavage of proteins and antigenic peptides. Eur J Biochem 235:404–415

Emmerich NPN, Nussbaum AK, Stevanovic S, Priemer M, Toes REM, Rammensee HG, Schild H (2000) The human 26S and 20S proteasomes generate overlapping but different sets of peptide fragments from a model protein substrate. J Biol Chem 275:21140–21148

Enenkel C, Lehmann H, Kipper J, Guckel R, Hilt W, Wolf DH (1994) PRE3, highly homologous to the human major histocompatibility complex-linked LMP2 (RING12) gene, codes for a yeast proteasome subunit necessary for the peptidylglutamyl-peptide hydrolyzing activity. FEBS Lett 341:193–196

Fenteany G, Standaert RF, Lane WS, Choi S, Corey EJ, Schreiber SL (1995) Inhibition of proteasome activities and subunit-specific amino-terminal threonine modification by lactacystin. Science 268:726–731

Fu HY, Doelling JH, Arendt CS, Hochstrasser M, Vierstra RD (1998) Molecular organization of the 20S proteasome gene family from *Arabidopsis thaliana*. Genetics 149:677–692

Geier E, Pfeifer G, Wilm M, Lucchiari-Hartz M, Baumeister W, Eichmann K, Niedermann G (1999) A giant protease with potential to substitute for some functions of the proteasome. Science 283:978–981

Gerards WL, Enzlin J, Haner M, Hendriks IL, Aebi U, Bloemendal H, Boelens W (1997) The human α-type proteasomal subunit HsC8 forms a double ringlike structure, but does not assemble into proteasome-like particles with the β-type subunits HsDelta or HsBPROS26. J Biol Chem 272:10080–10086

Ghislain M, Dohmen RJ, Levy F, Varshavsky A (1996) Cdc48p interacts with Ufd3p, a WD repeat protein required for ubiquitin-mediated proteolysis in *Saccharomyces cerevisiae*. EMBO J 15:4884–4899

Golbik R, Lupas AN, Koretke KK, Baumeister W, Peters J (1999) The Janus Face of the archaeal Cdc48/p97 homologue VAT: protein folding *versus* unfolding. Biol Chem 380:1049–1062

Groettrup M, Standera S, Stohwasser R, Kloetzel PM (1997) The subunits MECL-1 and LMP2 are mutually required for incorporation into the 20S proteasome. Proc Natl Acad Sci USA 94:8970–8975

Groll M, Bajorek M, Kohler A, Moroder L, Rubin DM, Huber R, Glickman MH, Finley D (2000a) A gated channel into the proteasome core particle. Nature Struct Biol 7:1062–1067

Groll M, Ditzel L, Löwe J, Stock D, Bochtler M, Bartunik HD, Huber R (1997) Structure of 20S proteasome from yeast at 2.4Å resolution. Nature 386:463–471

Groll M, Heinemeyer W, Jager S, Ullrich T, Bochtler M, Wolf DH, Huber R (1999) The catalytic sites of 20S proteasomes and their role in subunit maturation: a mutational and crystallographic study. Proc Natl Acad Sci USA 96:10976–10983

Groll M, Kim KB, Kairies N, Huber R, Crews CM (2000b) Crystal structure of epoxomicin: 20S proteasome reveals a molecular basis for selectivity of α',β'-epoxyketone proteasome inhibitors. J Am Chem Soc 122:1237–1238

Grziwa A, Maack S, Pühler G, Wiegand G, Baumeister W, Jaenicke R (1994) Dissociation and reconstitution of the *Thermoplasma* proteasome. Eur J Biochem 223:1061–1067

Heinemeyer W (2000) Active Sites and Assembly of the 20S Proteasome. In: Hilt W, Wolf DH (eds) Proteasomes: The World of Regulatory Proteolysis. Eurekah.com/Landes Bioscience, Georgetown, pp 48–70

Heinemeyer W, Fischer M, Krimmer T, Stachon U, Wolf DH (1997) The active-sites of the eukaryotic 20S proteasome and their involvement in subunit precursor processing. J Biol Chem 272:25200–25209

Heinemeyer W, Gruhler A, Mohrle V, Mahe Y, Wolf DH (1993) PRE2, highly homologous to the human major histocompatibility complex-linked RING10 gene, codes for a yeast proteasome subunit necessary for chrymotryptic activity and degradation of ubiquitinated proteins. J Biol Chem 268:5115–5120

Hilt W, Enenkel C, Gruhler A, Singer T, Wolf DH (1993) The PRE4 gene codes for a subunit of the yeast proteasome necessary for peptidylglutamyl-peptide-hydrolyzing activity. Mutations link the proteasome to stress- and ubiquitin-dependent proteolysis. J Biol Chem 268:3479–3486

Hoppe T, Matuschewski K, Rape M, Schlenker S, Ulrich HD, Jentsch S (2000) Activation of a membrane-bound transcription factor by regulated ubiquitin/proteasome-dependent processing. Cell 102:577–586

Hughes AL (1997) Evolution of the proteasome components. Immunogenetics 46:82–92

Hughes AL, Yeager M (1997) Molecular evolution of the vertebrate immune system. Bioessays 19: 777–786

Jäger S, Groll M, Huber R, Wolf DH, Heinemeyer W (1999) Proteasome β-type subunits: Unequal roles of propeptides in core particle maturation and a hierarchy of active site function. J Mol Biol 291: 997–1013

Johnson ES, Ma PC, Ota IM, Varshavsky A (1995) A proteolytic pathway that recognizes ubiquitin as a degradation signal. J Biol Chem 270:17442–17456

Karzai AW, Roche ED, Sauer RT (2000) The SsrA-SmpB system for protein tagging, directed degradation and ribosome rescue. Nature Struct Biol 7:449–455

Kawarabayasi Y, Sawada M, Horikawa H, Haikawa Y, Hino Y, Yamamoto S, Sekine M, Baba S, Kosugi H, Hosoyama A, Nagai Y, Sakai M, Ogura K, Otsuka R, Nakazawa H, Takamiya M, Ohfuku Y, Funahashi T, Tanaka T, Kudoh Y, Yamazaki J, Kushida N, Oguchi A, Aoki K, Kikuchi H (1998) Complete sequence and gene organization of the genome of a hyper-thermophilic archaebacterium, *Pyrococcus horikoshii* OT3. DNA Res 5:55–76

Kawarabayasi Y, Hino Y, Horikawa H, Yamazaki S, Haikawa Y, Jin-no K, Takahashi M, Sekine M, Baba S, Ankai A, Kosugi H, Hosoyama A, Fukui S, Nagai Y, Nishijima K, Nakazawa H, Takamiya M, Masuda S, Funahashi T, Tanaka T, Kudoh Y, Yamazaki J, Kushida N, Oguchi A, Aoki K, Kubota K, Nakamura Y, Nomura N, Sako Y, Kikuchi H (1999) Complete genome sequence of an aerobic hyper-thermophilic crenarchaeon, *Aeropyrum pernix* K1. DNA Res 6:83–101

Kawashima T, Amano N, Koike H, Makino S, Higuchi S, Kawashima-Ohya Y, Watanabe K, Yamazaki M, Kanehori K, Kawamoto T, Nunoshiba T, Yamamoto Y, Aramaki H, Makino K, Suzuki M (2000) Archaeal adaptation to higher temperatures revealed by genomic sequence of *Thermoplasma volcanium*. Proc Natl Acad Sci USA 97:14257–14262

Keiler KC, Shapiro L, Williams KP (2000) tmRNAs that encode proteolysis-inducing tags are found in all known bacterial genomes: a two-piece tmRNA functions in *Caulobacter*. Proc Natl Acad Sci USA 97:7778–7783

Keiler KC, Waller PR, Sauer RT (1996) Role of a peptide tagging system in degradation of proteins synthesized from damaged messenger RNA. Science 271:990–993

Kimura Y, Takaoka M, Tanaka S, Sassa H, Tanaka K, Polevoda B, Sherman F, Hirano H (2000) N^α-acetylation and proteolytic activity of the yeast 20S proteasome. J Biol Chem 275:4635–4639

Kisselev AF, Akopian TN, Castillo V, Goldberg AL (1999a) Proteasome active sites allosterically regulate each other, suggesting a cyclical bite-chew mechanism for protein breakdown. Mol Cell 4: 395–402

Kisselev AF, Akopian TN, Goldberg AL (1998) Range of sizes of peptide products generated during degradation of different proteins by archaeal proteasomes. J Biol Chem 273:1982–1989

Kisselev AF, Akopian TN, Woo KM, Goldberg AL (1999b) The sizes of peptides generated from protein by mammalian 26 and 20S proteasomes – implications for understanding the degradative mechanism and antigen presentation. J Biol Chem 274:3363–3371

Kisselev AF, Songyang Z, Goldberg AL (2000) Why does threonine, and not serine, function as the active site nucleophile in proteasomes? J Biol Chem 275:14831–14837

Klenk H-P, Clayton RA, Tomb J-F, White O, Nelson KE, Ketchum KA, Dodson RJ, Gwinn M, Hickey EK, Peterson JD, Richardson DL, Kerlavage AR (1997) The complete genome sequence of the hyperthermophilic, sulphate-reducing archaeon *Archaeoglobus fulgidus*. Nature 390:364–370

Koegl M, Hoppe T, Schlenker S, Ulrich HD, Mayer TU, Jentsch S (1999) A novel ubiquitination factor, E4, is involved in multiubiquitin chain assembly. Cell 96:635–644

Köhler A, Cascio P, Leggett DS, Woo KM, Goldberg AL, Finley D (2001) The axial channel of the proteasome core particle is gated by the Rpt2 ATPase and controls both substrate entry and product release. Mol Cell 7:1143–1152

Kopp F, Hendil KB, Dahlmann B, Kristensen P, Sobek A, Uerkvitz W (1997) Subunit arrangement in the human 20S proteasome. Proc Natl Acad Sci USA 94:2939–2944

Löwe J, Stock D, Jap B, Zwickl P, Baumeister W, Huber R (1995) Crystal structure of the 20S proteasome from the archaeon *T. acidophilum* at 3.4Å resolution. Science 268:533–539

Lupas A, Zwickl P, Baumeister W (1994) Proteasome sequences in eubacteria. Trends Biochem Sci 19:533–534

Mallick P, Goodwill KE, Fitz-Gibbon S, Miller JH, Eisenberg D (2000) Selecting protein targets for structural genomics of *Pyrobaculum aerophilum*: validating automated fold assignment methods by using binary hypothesis testing. Proc Natl Acad Sci USA 97:2450–2455

Maupin-Furlow JA, Ferry JG (1995) A proteasome from the methanogenic archaeon *Methanosarcina thermophila*. J Biol Chem 270:28617–28622

Maupin-Furlow JA, Aldrich HC, Ferry JG (1998) Biochemical characterization of the 20S proteasome from the methanoarchaeon *Methanosarcina thermophila*. J Bacteriol 180:1480–1487

Maupin-Furlow JA, Wilson HL, Kaczowka SJ, Ou MS (2000) Proteasomes in the Archaea: from structure to function. Front Biosci 1:D837–D866A

Mayr J, Seemüller E, Müller SA, Engel A, Baumeister W (1998) Late events in the assembly of 20S proteasomes. J Struct Biol 124:179–188

Mayr PSM, Allan VJ, Woodman PG (1999) Phosphorylation of p97(VCP) and p47 in vitro by p34(cdc2) kinase. Eur J Cell Biol 78:224–232

McCormack TA, Cruikshank AA, Grenier L, Melandri FD, Nunes SL, Plamondon L, Stein RL, Dick LR (1998) Kinetic studies of the branched chain amino acid preferring peptidase activity of the 20S proteasome: Development of a continuous assay and inhibition by tripeptide aldehydes and *clasto*-lactacystin β-lactone. Biochemistry 37:7792–7800

Myung J, Kim KB, Linsten K, Dantuma NP, Crews CM (2001) Lack of proteasome active site allostery as revealed by subunit-specific inhibitors. Mol Cell 7:411–420

Nagy I, Tamura T, Vanderleyden J, Baumeister W, De Mot R (1998) The 20S proteasome of *Streptomyces coelicolor*. J Bacteriol 180:5448–5453

Nandi D, Woodward E, Ginsburg DB, Monaco JJ (1997) Intermediates in the formation of mouse 20S proteasomes – implications for the assembly of precursor β-subunits. EMBO J 16:5363–5375

Ng WV, Kennedy SP, Mahairas GG, Berquist B, Pan M, Shukla HD, Lasky SR, Baliga NS, Thorsson V, Sbrogna J, Swartzell S, Weir D, Hall J, Dahl TA, Welti R, Goo YA, Leithauser B, Keller K, Cruz R, Danson MJ, Hough DW, Maddocks DG, Jablonski PE, Krebs MP, Angevine CM, Dale H, Isenbarger TA, Peck RF, Pohlschroder M, Spudich JL, Jung K-H, Alam M, Freitas T, Hou S, Daniels CJ, Dennis PP, Omer AD, Ebhardt H, Lowe TM, Liang P, Riley M, Hood L, DasSarma S (2000) Genome sequence of *Halobacterium* species NRC-1. Proc Natl Acad Sci USA 97:12176–12181

Nussbaum AK, Dick TP, Keilholz W, Schirle M, Stevanovic S, Dietz K, Heinemeyer W, Groll M, Wolf DH, Huber R, Rammensee HG, Schild H (1998) Cleavage motifs of the yeast 20S proteasome β subunits deduced from digests of enolase 1. Proc Natl Acad Sci USA 95:12504–12509

Oinonen C, Rouvinen J (2000) Structural comparison of Ntn-hydrolases. Protein Sci 9:2329–2337

Orlowski M, Cardozo C, Michaud C (1993) Evidence for the presence of five distinct proteolytic components in the pituitary multicatalytic proteinase complex. Properties of two components cleaving bonds on the carboxyl side of branched chain and small neutral amino acids. Biochemistry 32:1563–1572

Orlowski M, Wilk S (2000) Catalytic activities of the 20S proteasome, a multicatalytic proteinase complex. Arch Biochem Biophys 383:1–16

Parmentier Y, Bouchez D, Fleck J, Genschik P (1997) The 20S proteasome gene family in *Arabidopsis thaliana*. FEBS Lett 416:281–285

Pickart CM, VanDemark AP (2000) Opening doors into the proteasome. Nature Struct Biol 7:999–1001

Ramos PC, Höckendorff J, Johnson ES, Varshavsky A, Dohmen RJ (1998) Ump1p is required for proper maturation of the 20S proteasome and becomes its substrate upon completion of the assembly. Cell 92:489–499

Rechsteiner M, Realini C, Ustrell V (2000) The proteasome activator 11S REG (PA28) and Class I antigen presentation. Biochem J 345:1–15

Robb FT, Maeder DL, Brown JR, DiRuggiero J, Stump MD, Yeh RK, Weiss RB, Dunn DM (2001) Genomic sequence of hyperthermophile, *Pyrococcus furiosus*: implications for physiology and enzymology. Methods Enzymol 330:134–157

Rock KL, Goldberg AL (1999) Degradation of cell proteins and the generation of MHC class I-presented peptides. Annu Rev Immunol 17:739–779

Rubin DM, Glickman MH, Larsen CN, Dhruvakumar S, Finley D (1998) Active site mutants in the six regulatory particle ATPases reveal multiple roles for ATP in the proteasome. EMBO J 17:4909–4919

Ruepp A, Eckerskorn C, Bogyo M, Baumeister W (1998) Proteasome function is dispensable under normal but not under heat-shock conditions in *Thermoplasma acidophilum*. FEBS Lett 425:87–90

Ruepp A, Graml W, Santos-Martinez ML, Koretle KK, Volker C, Mewes HW, Frishman D, Stocker S, Lupas AN, Baumeister W (2000) The genome sequence of the thermoacidophilic scavenger *Thermoplasma acidophilum*. Nature 407:508–513

Salzmann U, Kral S, Braun B, Standera S, Schmidt M, Kloetzel PM, Sijts A (1999) Mutational analysis of subunit i beta 2 (MECL-1) demonstrates conservation of cleavage specificity between yeast and mammalian proteasomes. FEBS Lett 454:11–15

Sassa H, Oguchi S, Inoue T, Hirano H (2000) Primary structural features of the 20S proteasome subunits of rice (*Oryza sativa*). Gene 250:61–66

Schauer TM, Nesper M, Kehl M, Lottspeich F, Müller-Taubenberger A, Gerisch G, Baumeister W (1993) Proteasomes from *Dictyostelium discoideum*: characterization of structure and function. J Struct Biol 111:135–147

Schmidt M, Zantopf D, Kraft R, Kostka S, Preissner R, Kloetzel PM (1999) Sequence information within proteasomal prosequences mediates efficient integration of β-subunits into the 20S proteasome complex. J Mol Biol 288:117–128

Schmidtke G, Emch S, Groettrup M, Holzhütter HG (2000) Evidence for the existence of a non-catalytic modifier site of peptide hydrolysis by the 20S proteasome. J Biol Chem 275:22056–22063

Schmidtke G, Holzhutter HG, Bogyo M, Kairies N, Groll M, de Giuli R, Emch S, Groettrup M (1999) How an inhibitor of the HIV-I protease modulates proteasome activity. J Biol Chem 274:35734–35740

Schmidtke G, Kraft R, Kostka S, Henklein P, Frommel C, Löwe J, Huber R, Kloetzel PM, Schmidt M (1996) Analysis of mammalian 20S proteasome biogenesis: The maturation of β-subunits is an ordered two-step mechanism involving autocatalysis. EMBO J 15:6887–6898

Schmidtke G, Schmidt M, Kloetzel PM (1997) Maturation of mammalian 20S proteasome: Purification and characterization of 13S and 16S proteasome precursor complexes. J Mol Biol 268:95–106

Seemüller E, Lupas A, Baumeister W (1996) Autocatalytic processing of the 20S proteasome. Nature 382:468–470

Seemüller E, Lupas A, Stock D, Löwe J, Huber R, Baumeister W (1995) Proteasome from *Thermoplasma acidophilum*: A threonine protease. Science 268:579–582

She Q, Singh RK, Confalonieri F, Zivanovic Y, Allard G, Awayez M, Chan-Weiher CC-Y, Clausen IG, Curtis BA, De Moors A, Erauso G, Fletcher C, Gordon PMK, Heikamp-de Jong I, Jeffries AC, Kozera CJ, Medina N, Peng X, Thi-Ngoc HP, Redder P, Schenk ME, Theriault C, Tolstrup N, Charlebois RL, Doolittle WF, Duguet M, Gaasterland T, Garrett RA, Ragan MA, Sensen CW, Van der Oost J (2001) The complete genome of the crenarchaeon *Sulfolobus solfataricus* P2. Proc Natl Acad Sci USA 98:7835–7840

Smith DS, Doucette-Stamm LA, Deloughery C, Lee H, Dubois J, Aldrege T, Bashirzadeh R, Blakely D, Cook R, Gilbert K, Harrison D, Hoang L (1997) Complete genome sequence of *Methanobacterium thermoautotrophicum* ΔH: functional analysis and comparative genomics. J Bacteriol 179:7135–7155

Stoltze L, Schirle M, Schwarz G, Schroter C, Thompson MW, Hersh LB, Kalbacher H, Stevanovic S, Rammensee HG, Schild H (2000) Two new proteases in the MHC class I processing pathway. Nature Immunol 1:413–418

Tamura N, Lottspeich F, Baumeister W, Tamura T (1998) The role of tricorn protease and its aminopetidase-interacting factors in cellular protein degradation. Cell 95:1–20

Tamura N, Pfeifer G, Baumeister W, Tamura T (2001) Tricorn protease in Bacteria: characterization of the enzyme from *Streptomyces coelicolor*. Biol Chem 382:449–458

Tamura T, Nagy I, Lupas A, Lottspeich F, Cejka Z, Schoofs G, Tanaka K, De Mot R, Baumeister W (1995) The first characterization of a eubacterial proteasome: the 20S complex of *Rhodococcus*. Curr Biol 5:766–774

Tamura T, Tamura N, Cejka Z, Hegerl R, Lottspeich F, Baumeister W (1996) Tricorn protease – the core of a modular proteolytic system. Science 274:1385–1389

Walz J, Tamura T, Tamura N, Grimm R, Baumeister W, Koster AJ (1997) Tricorn protease exists as an icosahedral supermolecule in vivo. Mol Cell 1:59–65

Wang EW, Kessler BM, Borodovsky A, Cravatt BF, Bogyo M, Ploegh HL, Glas R (2000) Integration of the ubiquitin-proteasome pathway with a cytosolic oligopeptidase activity. Proc Natl Acad Sci USA 97:9990–9995

Wenzel T, Baumeister W (1995) Conformational constraints in protein degradation by the 20S proteasome. Nature Struct Biol 2:199–204

Wenzel T, Eckerskorn C, Lottspeich F, Baumeister W (1994) Existence of a molecular ruler in proteasomes suggested by analysis of degradation products. FEBS Lett 349:205–209

Whitby FG, Masters EI, Kramer L, Knowlton JR, Yao Y, Wang CC, Hill CP (2000) Structural basis for the activation of 20S proteasomes by 11S regulators. Nature 408:115–120

Wilson HL, Aldrich HC, MaupinFurlow J (1999) Halophilic 20S proteasomes of the archaeon *Haloferax volcanii*: Purification, characterization, and gene sequence analysis. J Bacteriol 181:5814–5824

Wilson HL, Ou MS, Aldrich HC, Maupin-Furlow J (2000) Biochemical and physical properties of the *Methanococcus jannaschii* 20S proteasome and PAN, a homolog of the ATPase (Rpt) subunits of the eucaryal 26S proteasome. J Bacteriol 182:1680–1692

Witt E, Zantopf D, Schmidt M, Kraft R, Kloetzel PM, Krüger E (2000) Characterisation of the newly identified human Ump1 homologue POMP and analysis of LMP7(β5i) incorporation into 20S proteasomes. J Mol Biol 301:1–9

Wolf S, Nagy I, Lupas A, Pfeifer G, Cejka Z, Müller SA, Engel A, De Mot R, Baumeister W (1998) Characterization of ARC, a divergent member of the AAA ATPase family from *Rhodococcus erythropolis*. J Mol Biol 277:13–25

Yang Y, Früh K, Ahn K, Peterson PA (1995) In vivo assembly of the proteasomal complexes, implications for antigen processing. J Biol Chem 270:27687–27694

Yao Y, Toth CR, Huang L, Wong ML, Dias P, Burlingame AL, Coffino P, Wang CC (1999) α5 subunit in *Trypanosoma brucei* proteasome can self-assemble to form a cylinder of four stacked heptamer rings. Biochem J 344:349–358

York IA, Goldberg AL, Mo XY, Rock KL (1999) Proteolysis and class I major histocompatibility complex antigen presentation. Immunol Rev 172:49–66

Zhang XD, Shaw A, Bates PA, Newman RH, Gowen B, Orlova E, Gorman MA, Kondo H, Dokurno P, Lally J, Leonard G, Meyer H, van Heel M, Freemont PS (2000) Structure of the AAA ATPase p97. Mol Cell 6:1473–1484

Zühl F, Seemüller E, Golbik R, Baumeister W (1997) Dissecting the assembly pathway of the 20S proteasome. FEBS Lett 418:189–194

Zwickl P, Goldberg AL, Baumeister W (2000) Proteasomes in Prokaryotes. In: Hilt W, Wolf DH (eds) Proteasomes: The World of Regulatory Proteolysis. Eurekah.com/Landes Bioscience, Georgetown, pp 8–20

Zwickl P, Grziwa A, Pühler G, Dahlmann B, Lottspeich F, Baumeister W (1992) Primary structure of the *Thermoplasma* proteasome and its implications for the structure, function, and evolution of the multicatalytic proteinase. Biochemistry 31:964–972

Zwickl P, Kleinz J, Baumeister W (1994) Critical elements in proteasome assembly. Nature Struct Biol 1:765–770

Zwickl P, Ng D, Woo KM, Klenk H-P, Goldberg AL (1999) An archaebacterial ATPase, homologous to the ATPases in the eukaryotic 26S Proteasome, activates protein breakdown by the 20S proteasome. J Biol Chem 274:26008–26014

Regulating the 26S Proteasome

M.H. Glickman[1] and V. Maytal[2]

1 Introduction	43
2 Regulating Proteasomal Functions	46
2.1 Substrate Selection	46
2.2 Product Generation	48
3 The 20S Proteolytic Core of the Proteasome (CP)	49
4 The 19S Regulatory Particle of the Proteasome (RP)	52
4.1 The Base of the Regulatory Particle	54
4.2 The Lid, and Similarities to Other Regulatory Complexes	57
4.3 Alternative Regulatory Complexes of the Proteasome	60
5 Changes in Proteasome Composition and Associated Proteins/Factors	61
5.1 Diversity in Proteasome Composition	61
5.2 Auxiliary Factors and Transiently Associated Proteins	63
6 Summary	65
References	66

1 Introduction

Numerous regulatory pathways are regulated by timely removal of critical proteins. These include proteins involved in the cell cycle, transcriptional regulation, DNA repair, development and differentiation, long-term memory, circadian rhythms, stress response, transcriptional silencing, cell-surface signaling, antigen presentation, and in combating cancer or viral infection (Hershko and Ciechanover 1998; Spataro et al. 1998; Ciechanover et al. 2000). Other proteins that must be degraded properly are damaged, abnormal, or foreign (viral) proteins. Remarkably, in eukaryotes, almost all such proteins are degraded in an ATP-dependent manner by a single, highly conserved, 2.5MDa multisubunit enzyme, the proteasome (Voges et al. 1999; Ferrell et al. 2000; Glickman 2000). The majority of substrates for degradation are first covalently attached to a chain of multiple ubiquitin molecules – a highly conserved 76 amino acid protein – recognized by the

---
[1] Department of Biology, The Technion, Israel Institute of Technology, 32000 Haifa Israel
[2] Institute for Catalysis (ICST), The Technion, Israel Institute of Technology, 32000 Haifa Israel

proteasome, and hydrolyzed into short peptides (HERSHKO and CIECHANOVER 1998; CIECHANOVER et al. 2000; KORNITZER and CIECHANOVER 2000; WILKINSON 2000).

The proteasome holoenzyme (also known as the 26S proteasome) is an approximately 2.5MDa complex made up of two copies each of at least 31 different subunits which are highly conserved among all eukaryotes (Table 1). The overall structure can be divided into two major subcomplexes: the 20S core particle (CP)

Table 1. Principal subunits of the proteasome and Cop9 signalosome (CSN)

	Proteasome			Motifs/ comments	CSN	
	Human subunits	Yeast subunits	Yeast–human identity (%)		Human subunits	Identity to human lid subunits (%)
19S	S2/p97	Rpn1	39	LRR		
	S1/p112	Rpn2	39	LRR		
	S3/p58	Rpn3	33	PCI	CSN3	26
	p55	Rpn5	40	PCI	CSN4	23
	S9/p44.5	Rpn6	41	PCI	CSN2	27
	S10a/p44	Rpn7	36	PCI	CSN1	23
	S12/p40	Rpn8	47	MPN	CSN6	23
	S11/p40.5	Rpn9	29	PCI	CSN7	23
	S5a/Mbp1	Rpn10	48	vWA, UIM		
	S13/Poh1	Rpn11	66	MPN	CSN5	38
	S14/p31	Rpn12	32	PCI	CSN8	< 20
	S7/Mss1	Rpt1	76	AAA		
	S4/p56	Rpt2	71	AAA		
	S6/p48/Tbp7	Rpt3	66	AAA		
	S10b/p42	Rpt4	67	AAA		
	S6'/p50/Tbp1	Rpt5	68	AAA		
	S8/p45/Trip1	Rpt6	74	AAA		
20S	α1 hs/iota	α1 sc/C7	51			
	α2 hs/C3	α2 sc/Y7	57			
	α3 hs/C9	α3 sc/Y13	53			
	α4 hs/C6	α4 sc/Pre6	59			
	α5 hs/zeta	α5 sc/Pup2	62			
	α6 hs/C2	α6 sc/Pre5	52			
	α7 hs/C8	α7 sc/C1	51			
	β1 hs/Y/delta	β1 sc/Pre3	55	Threonine protease, post-acidic		
	β2 hs/Z	β2 sc/Pup1	54	Thr protease, post-basic		
	β3 hs/C10	β3 sc/Pup3	54			
	β4 hs/C7	β4 sc/Pre1	44			
	β5 hs/X	β5 sc/Pre2	66	Thr protease, post-hydrophobic		
	β6 hs/C5	β6 sc/C5	45			
	β7 hs/β	β7 sc/Pre4	43			

Amino acid sequences of all subunits were found in GenBank at NCBI using basic blast search (http://www.ncbi.nlm.nih.gov/). Transiently or loosely associated subunits are not listed. Homologies between yeast and human proteasome subunits, as well as between human CSN and human lid subunits were determined using MegAlign by the Jotun Hein Method (gap penalty, 11; gap length, 3). For additional information on these proteins and their structural motifs, see text. For additional/previous names of *S. cerevisiae* RP-subunits or (hs)CSN-subunits refer to FINLEY et al. 1998; GLICKMAN et al. 1998; DENG et al. 2000.

that contains the protease subunits, and the 19S regulatory particle (RP) that regulates the function of the CP (Fig. 1). The CP is a barrel-shaped structure made up of four rings of seven subunits each. The two inner β-rings contain the proteolytic active sites facing inwards into a sequestered proteolytic chamber (LOEWE et al. 1995; GROLL et al. 1997). One or two regulatory particles attach to the surface of the outer α-rings of the CP to form the 26S proteasome holoenzyme (Fig. 1). The RP itself can be further dissected into two multisubunit substructures, a lid and a base (GLICKMAN et al. 1998b). Six homologous ATPases (Rpt1–6) are present in the base together with three non-ATPase subunits (Rpn1, 2, and 10). The ATPases in the base most probably unfold substrates and translocate them into the CP (BRAUN et al. 1999; HORWICH et al. 1999). The lid of the RP is a 400-kDa complex made up of the remaining eight non-ATPase subunits which can be released from the proteasome or rebind under certain conditions (Fig. 1). The role of the lid is still unclear, though it is necessary for proper degradation of polyubiquitinated proteins (GLICKMAN et al. 1998a). All subunits of the lid subcomplex contain one of two structural motifs: six subunits contain a PCI domain (*p*roteasome, *C*OP9, *eIF*3), while the other two (Rpn8 and Rpn11) contain an MPN (*M*pr1, *P*ad1 *N*-terminal)

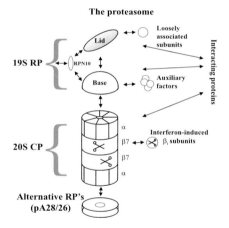

Fig. 1. The proteasome is a dynamic structure built in a modular manner. One or two regulatory particles (*RP*) attach to the outer surface of the core particle (*CP*). The CP is made up of four heptameric rings; the two outer α-rings are identical, as are the two inner β-rings. The β-subunits contain the protease active sites facing inwards into the sequestered proteolytic chamber. Upon interferon-γ induction, three β-subunits can be replaced by βi (LMP) homologs that alter the proteolytic specificities of the proteasome. The 19S RP is comprised of two eight-subunit subcomplexes, the lid and the base. The base contains all six proteasomal ATPases, attaches to the α-ring of the CP and activates peptide hydrolysis. The lid is necessary for proper multiubiquitin-protein degradation, but can disassociate from the proteasome, resulting in a truncated base-CP complex. Rpn10 can interact with both lid and base and stabilizes the interaction between the two. Rpn10 is also found outside of the proteasome. Numerous associated proteins and auxiliary factors – such as chaperones and components of the ubiquitination machinery – can interact with the RP, and serve to fine-tune proteolytic properties, shuttle substrates to the proteasome, localize the proteasome or target it to concentrations of substrates. Alternative regulatory complexes (such as PA28 and PA26) can also attach to the surface of either α-ring, activating peptidase activity

domain (Table 1) (ARAVIND and PONTING 1998; GLICKMAN et al. 1998a; HOFMANN and BUCHER 1998). The lid shares significant genetic and structural features with other cellular complexes indicating that there might be structure/function relationships common to a number of different regulatory pathways.

Archaea and some bacteria contain stripped-down versions of the proteasome, which play a regulatory role in stress response (DE MOT et al. 1999; MAUPIN-FURLOW et al. 2000; ZWICKL et al. 2000). Ubiquitin or the ubiquitin-conjugating system are not found in these species, therefore targeting substrates to these proteasomes probably entails recognizing unfolded and damaged proteins directly. It is likely that the proteasome is a modular system that was refined over the course of evolution, and adapted to its indispensable regulatory roles in eukaryotes. For instance, the eukaryotic proteasome has been adapted to degrading proteins in a ubiquitin-dependent fashion by the addition of regulatory factors that assemble in different layers onto the proteolytic core of the proteasome, as well as by increasing the diversity of the principal subunits themselves. In eukaryotes, in addition to hydrolyzing ubiquitinated proteins into amino acids, the proteasome can also proteolyse selected non-ubiquitinated proteins, process proteins into truncated versions, and possibly refold misfolded proteins.

2 Regulating Proteasomal Functions

2.1 Substrate Selection

The most important role of the proteasome in eukaryotes is to proteolyze proteins that are covalently attached to a chain of multiple ubiquitin proteins. These include regulatory proteins such as cyclins, cyclin-dependent kinase (CDK) inhibitors, transcription factors, cell surface receptors, oncoproteins, tumor suppressors, enzymes, and structural proteins (CIECHANOVER et al. 2000; KORNITZER and CIECHANOVER 2000). These proteins encode within them specific signals for degradation that can be switched on or off (for instance by phosphorylation) thus causing proteins to be removed from the cell or not. Each of the different classes of signals is recognized by a specific constellation of ubiquitinating enzymes that covalently link the carboxyl group at the C terminus of ubiquitin (G76) via an amide bond with an amino group on the substrate. This amino group can be either a specific lysine side chain (HERSHKO and CIECHANOVER 1998; CIECHANOVER et al. 2000) or the actual N terminus of the protein (BREITSCHOPF et al. 1998; AVIEL et al. 2000). During this ubiquitinating process, multiple ubiquitin molecules are added to the substrate, usually in the form of a polymeric chain between the carboxyl group at G76 of one ubiquitin and the amino side chain at K48 of another ubiquitin. It seems increasingly unlikely that covalent attachment of such a chain alters the tertiary structure of the substrate into a conformation that has a higher affinity for the proteasome, and it is now commonly believed that the regulatory particle of the proteasome contains a subunit or surface that can specifically recognize these K48–G76 isopeptide-linked ubiquitin chains either directly or via auxiliary poly-ubiquitin-binding proteins (PICKART 1997; THROWER et al. 2000; WILKINSON 2001; see also section 5.2).

The quaternary structure of ubiquitin polymers and the exact special relationship between each ubiquitin molecule in K48–G76 chains is critical for their ability to bind to the proteasome. Recognition of ubiquitin by the proteasome seems to be specific for K48–G76-linked polyubiquitin chains of at least four ubiquitin molecules (THROWER et al. 2000). For instance, although mono-ubiquitination may serve as a cellular targeting or localization signal, it does not seem to target proteins to the proteasome (STROUS and GOVERS 1999; THROWER et al. 2000). Likewise, modification of proteins by poly-ubiquitin chains linked via lysine 63 instead of lysine 48 plays a role in signaling, DNA repair, and perhaps additional tasks, but probably does not target proteins for degradation by the proteasome (SPENCE et al. 1995, 2000; HOFMANN and PICKART 1999). It is important to note that recognition of a protein by the proteasome is in itself insufficient for a protein to be efficiently proteolyzed. For instance, there are examples of both a non-ubiquitinated protein as well as a penta-ubiquitinated protein that are efficiently recognized by purified proteasomes in vitro but not efficiently proteolyzed (BRAUN et al. 1999; THROWER et al. 2000), and in vivo at least one case of a K48–G76 linked polyubiquitinated protein has been shown not to be rapidly degraded by the proteasome (KAISER et al. 2000).

Simple versions of the proteasome are present in archaea and certain bacteria, however ubiquitin and the ubiquitin conjugating system are not, indicating that the proteasome has the ability to proteolyze at least some non-ubiquitinated proteins (MAUPIN-FURLOW et al. 2000). As purified core particles from both prokaryotes and eukaryotes can hydrolyze a number of non-ubiquitinated unfolded proteins in vitro (KISSELEV et al. 1998, 1999), and the eukaryotic 26S proteasome can bind non-ubiquitinated unfolded proteins (BRAUN et al. 1999), it is possible that at least some misfolded/damaged/non-native proteins could be targeted directly to the proteasome. Fittingly, even though the proteasome is nonessential in the archaea and bacteria in which it is found – probably because prokaryotes contain multiple ATP-dependent complex proteases with redundant functions, such as the Lon protease, ClpA/P, HslV/U and the proteasome (DE MOT et al. 1999; MAUPIN-FURLOW et al. 2000; ZWICKL et al. 2000) – it is probably involved in the heat shock response of these organisms by proteolyzing damaged or misfolded proteins (KNIPFER and SHRADER 1997; RUEPP et al. 1998; MAUPIN-FURLOW et al. 2000).

Finally, there is at least one example of a metabolic protein which is targeted to the proteasome in a specific, yet ubiquitin-independent manner – the enzyme ornithine decarboxylase (ODC) is recognized and hydrolyzed by the proteasome upon binding to the protein antizyme, without it being ubiquitinated (MURAKAMI et al. 2000). Antizyme itself is probably released from the proteasome and not degraded concomitantly with ODC (MURAKAMI et al. 2000); however, it is possible that the levels of antizyme could be independently regulated via proteasome degradation. As the machinery for ubiquitin-independent targeting to the proteasome is present in eukaryotes, one can assume that the case of ODC will not be a true exception and that antizyme, or other small molecules, could target a subset of substrates to the proteasome.

2.2 Product Generation

In most cases, the proteasomes cleaves protein substrates into small peptides varying between 3 and 23 amino acids in length (NUSSBAUM et al. 1998; KISSELEV et al. 1999). The median length of peptides generated by the proteasome is 7–9 amino acids; however, in total peptides of this length make up only about 15% of the peptides generated by the proteasome. This process is processive such that a protein is hydrolyzed within the proteasome to the final products before the next substrate enters, thus the pattern of peptides generated from a specific protein is stable over time (KISSELEV et al. 1998; NUSSBAUM et al. 1998). The peptide-products of the proteasome are short lived, and do not accumulate in the cell (TAMURA et al. 1998; SILVA et al. 1999). Most probably, most of these peptides are rapidly hydrolyzed by a sequence of downstream proteases and aminopeptidases. Candidates for these downstream proteases are THIMET, tricorn, multicorn, TPPII, and leucine aminopeptidase (BENINGA et al. 1998; OSMULSKI and GACZYNCKA 1998; TAMURA et al. 1998; SILVA et al. 1999; YAO and COHEN 1999; WANG et al. 2000). Some of the peptides that are generated by the proteasome can be transported through the endoplasmic reticulum (ER) to be presented to the immune system by major histocompatibility complex (MHC) class I (MICHALEK et al. 1993; STOLTZE et al. 2000).

An interesting feature of the proteasome is that not all substrates are hydrolyzed to completion. In some cases, the proteasome processes the substrate into a truncated form. Processing by the proteasome can serve as a potent regulatory tool for transforming a protein from one form into another, thus altering its cellular activities. The best-studied example is the processing of the p105 precursor of p50, a component of the nuclear factor (NF)-κB (PALOMBELLA et al. 1994). After p105 is ubiquitinated within its C terminus, the C-terminal domain is proteolyzed by the proteasome and the 50-kDa N-terminal region is released as a stable and activated protein. In the case of p105, the site of processing is determined in part by a glycine-rich region in the middle of the protein (ORIAN et al. 1999). Interestingly, the yeast proteasome processes p105 at a different location than does the mammalian proteasome, suggesting that the glycine-rich region is probably not a universal signal for processing (SEARS et al. 1998). Similarly, the p100 precursor of the p52 component of NF-κB2 is processed by the proteasome after a glycine-rich region (HEUSCH et al. 1999). A unique case of processing a membrane-bound protein by the proteasome is the activation of SPT23, a transcription factor that controls the levels of unsaturated fatty acids. After ubiquitination and proteolysis by the proteasome, the active p90 N-terminal region of Spt23 is liberated from the inactive membrane-bound p120 form (HOPPE et al. 2000). The assumption is that a loop in the full-length precursor is endoproteolytically clipped by the proteasome, releasing the N-terminal region and degrading the membrane bound C terminus. Despite some structural similarity between NF-κB and Spt23, the latter does not contain a glycine-rich region but does have an asparagine-enriched region that might be central to processing of the loop (HOPPE et al. 2000). In yet another example, the 155-kDa protein cubitus interuptus is a transcriptional activator in the Hedgehog signaling pathway. There is genetic evidence that upon ubiquitination, its C terminus is proteolyzed by the proteasome, while the 75-kDa N-terminal domain is released as a transcriptional repressor in the Hedgehog pathway (MANIATIS 1999; WANG et al. 1999). A fifth example of processing by the proteasome is the production of a specific 90-kDa truncated fragment from the larger fusion protein N-degron-β-galactosides (VAN NOCKER et al. 1996). Thus, the proteasome not only 'destroys' proteins, but can also activate proteins or drastically alter their cellular behavior.

It is possible that the proteasome might also function in processes that do not involve proteolysis at all. At least in vitro, the proteasome can bind an unfolded protein, accelerate its refolding, and release it in its native form (BRAUN et al. 1999). This chaperone-like activity has been mapped to the ATPase-containing base of the RP. There is evidence that in vivo the proteasome is involved in disassembly and rearrangement of the nuclear excision repair complex, without performing proteolysis (RUSSELL et al. 1999). Similarly, RP subunits, but not CP subunits, co localize in vivo together with heat shock proteins and chaperones at sites of misfolded androgen receptor aggregates (STENOIEN et al. 1999). These may be in vivo examples of the in vitro observed chaperone-like activity of the RP. As the ATPase-containing regulatory complexes ClpA and ClpX of the prokaryotic ATP-dependent protease ClpP have dual functions, serving either as aids in proteolysis or as chaperones (GOTTESMAN et al. 1997a,b; WEBER-BAN et al. 1999), it is possible that a similar distribution of functions occurs in the ATPase-containing RP of the proteasome. Finally, ubiquitin is also a product of the proteasome: ubiquitin, or ubiquitin attached to a residual peptide chain, is released from the proteasome and recycled back into the ubiquitin pathway (HOUGH et al. 1986; SWAMINATHAN et al. 1999). A deubiquitinating enzyme must be associated with the RP in order to remove or edit these polyubiquitin chains.

3 The 20S Proteolytic Core of the Proteasome (CP)

A detailed overview of the 20S CP appears in the chapter by Zwickl and Seemueller, this volume. Here we will focus on factors regulating the proteolytic activity of the CP within the context of the proteasome holoenzyme. The proteolytic activity of the proteasome is found within its core particle (CP), also known as the 20S proteasome or the multicatalytic protease (MCP) (BOCHTLER et al. 1999). Although purified free CP can hydrolyze small peptides and some unfolded proteins, it cannot degrade multiubiquitinated proteins. The structure of CP purified from yeast was determined by X-ray crystallography (GROLL et al. 1997). The CP is a hollow cylindrical structure composed of four heptagonal rings, stacked in C_2 symmetry (Fig. 1). Each of the two outer rings is comprised of seven genetically related and structurally similar α subunits, and each of the two inner rings is comprised of similarly conserved β subunits, thus the CP barrel as a whole also exhibits pseudo seven-fold symmetry (in addition to the C_2 symmetry). In eukaryotes, three of the seven β subunits have functional threonine protease active sites, meaning that each proteasome has six (three different) proteolytic active sites. In archaea, all β subunits are identical leading to 14 active sites in total. The protease active sites face an inner cavity within the β-rings that can be accessed through a narrow channel leading from the surface of the α-rings (LOEWE et al. 1995; GROLL et al. 1997).

The N termini of the α subunits obstruct access to this channel, suggesting that the proteasome channel is gated (GROLL et al. 1997, 2000). The N terminus of α3 is somewhat distinct from other α subunits in that it points directly across the α-ring surface towards the center of pseudo seven-fold symmetry with close contacts to every other α subunit. In order for substrates to enter the CP, and probably for products to exit as well, the blocking N-terminal residues of these α subunits must be moved and rearranged. Interestingly, the formation of a well-defined closed configuration of the gate involves a dramatic departure from the pseudo seven-fold symmetry of the CP. Even though the seven α subunits are genetically related (about 30% identical to each other in yeast) and structurally almost superimposable (GROLL et al. 1997), their N-terminal tails are different from each other both in sequence and relative length (GROLL et al. 2000). At the same time, these N-terminal tails are absolutely conserved among orthologs form numerous species, suggesting that the N termini play a critical structural role that has been maintained in core particles in all eukaryotes, and it is precisely the difference between them that is integral to their function.

The peptidase activity of the eukaryotic CP can be activated by a variety of treatments. For instance, activation occurs when the CP complexes with the RP to form the 26S proteasome holoenzyme (ADAMS et al. 1998; GLICKMAN et al. 1998b). Other endogenous activators of the CP include the interferon-induced PA28/11S REG complex (RECHSTEINER et al. 2000; STOHWASSER et al. 2000). Mild chemical treatments, such as exposure to low levels of sodium dodecylsulfate (SDS) are also effective (GLICKMAN et al. 1998b). A nine-residue deletion mutation of the N-terminal tail of the α3 subunit (α3ΔN) results in constitutively activated peptide hydrolysis by the CP (GROLL et al. 2000). While CP purified from wild-type yeast can be stimulated by additions of small amounts of SDS, CP purified from the *α3ΔN* strain are constitutively activated and cannot be activated further. This result indicates that the wild-type free CP is found in a repressed state, and activation reflects de-repression of the CP by the opening of a channel into the CP, allowing substrate access into the proteolytic chamber and revealing the intrinsic maximal peptidase activity. Thus the N termini of the α subunits in wild-type CP appear to regulate proteasome function by blocking the channel thereby inhibiting substrate access to the proteolytic active sites.

The N terminus of α3 forms interactions with N termini of other α subunits. Of special note is a salt bridge formed between Asp7 of α3 with Tyr4 and Arg6 of α4 (GROLL et al. 1997, 2000). Six of the seven α subunits contain adjacent tyrosine and asparagine residues in homologous locations within their N-terminal tails (all seven contain this conserved tyrosine, and three also have an additional arginine immediately following the asparagine residue); the conservation of this 'YD(R)' sequence suggests a key structural role for this element. In addition to the bridge between the N-terminal tails of α3 and α4, other pairs of α subunits could similarly interact through their YD(R) motifs. The direct contacts formed between these residues in adjacent subunits may explain their correlated evolutionary conservation. Likewise, the N terminus of the single α subunit in the archaeon *Thermoplasma acidophilum* contains a Tyr8-Asp9-Arg10 (YDR) sequence, making it likely that the identical neighboring α subunits in the archaeal CP also interact with each other via similar salt bridges. Thus in reality the α subunit N termini are not randomly ordered but interact carefully with each other to maintain a closed structure at the center of the α-ring in free CP.

One role of the RP might be to control opening and closing of this channel by forming competing interactions with the blocking α subunit N-terminal residues (GROLL et al. 2000). Indeed, a substitution mutation in the ATP-binding site of a single RP ATPase severely affects peptidase activity of the proteasome, probably by hampering the ability of the RP to properly gate the channel into the CP (RUBIN et al. 1998). Furthermore, the α3ΔN 'open gate' mutation has only a small effect on peptidase activity of proteasome holoenzymes. The specific activity of purified α3ΔN CP is similar to that of 26S proteasome purified from the same mutant strain, as well as being similar to the specific activity of wild-type 26S proteasome holoenzyme. This indicates that that attachment of the RP activates peptidase activity of the CP to a similar extent as genetically opening the channel as seen in the α3ΔN 'open gate' mutation. One possible mechanism for this activation is that attachment of regulatory particles to the surface of the α-ring forms competing interactions with the tyrosine, asparagine, or arginine residues situated in the α subunits, thus realigning their N termini to allow passage of substrates or products.

One reason for a gated channel in the CP could be to serve as a transition from one form of inhibition to another during assembly of the mature CP. In the final stage of CP assembly, self-compartmentalization is achieved by the association of two $\alpha_7\beta_7$ half-CPs at the β–β interface. These half CPs are inactive due to propeptides in the critical β subunits that mask the active site. As these half-CPs are joined, inhibition by β subunit N termini is relieved by autolysis (CHEN and HOCHSTRASSER 1996) while inhibition by the blocking N termini of the α subunits is imposed. Binding of the RP relieves this inhibition by opening the channel, thus giving rise to the proteolytically active form of the complex (GROLL et al. 2000).

A second reason for a gated channel could be to regulate generation of products by the proteasome. It is possible that under normal conditions, product release is slowed down by a gated channel in order to increase processivity and decrease average peptide length. Most of these short peptides are quickly removed from the cytoplasm. Under certain conditions (such as during an immune response), it might be beneficial to produce longer peptides that can play a regulatory role. As mentioned above, the majority of peptides generated by the proteasome contain fewer than eight amino acids. A fraction of the peptides that are 8–10 amino acids in length can be transported through the ER and presented to the immune system by MHC class I (GLICKMAN 2000; RECHSTEINER et al. 2000). An increase in average peptide length of proteasome products could increase the efficiency of antigen presentation, and by extension, the efficiency of combating viral infection. Upon interferon-γ induction – following viral infection for example – a regulator of the proteasome, PA28 (also known as the 11S regulator), plays a role in antigen processing by activating the peptidase activity of the 20S CP (SEEGER et al. 1997; RECHSTEINER et al. 2000; STOHWASSER et al. 2000). PA28 probably activates the CP by binding to the α-ring surface and rearranging the blocking N-terminal residues thus opening the gate and increasing the rate by which peptides exit the CP (see the chapter by Hill et al., this volume).

Overall, the 20S CP can cleave peptide bonds after any amino acid. However, each of the three active site-containing β-subunits cleaves preferentially after different amino acids, such that $\beta1$ cleaves after acidic or small hydrophobic amino acids, $\beta2$ cuts after basic or small hydrophobic amino acids, whereas $\beta5$ hydrolyzes the peptide bond after hydrophobic residues whether bulky or not (DICK et al. 1998). However the rules that govern the cleavage rate of the same peptide bond can be significantly altered when put into the context of the primary structure of the polypeptide (HOLZHUTTER et al. 1999). For instance, the specificity towards a peptide bond between two amino acids can be affected by the amino acids in flanking regions on either side up to eight amino acids away. The location and identity of each of these anchoring residues are different for different classes of peptide bonds (HOLZHUTTER et al. 1999; KUTTLER et al. 2000). An interesting feature of proteolysis by the proteasome is that the 20S CP and the proteasome holoenzyme generate different patterns of cleavage products (KISSELEV et al. 1999; EMMERICH et al. 2000), indicating that even the distal 19S RP possibly exerts some kind of allosteric affect on the structure of the CP. Furthermore, the CP contains specific 'non-catalytic' sites to which additional factors can bind and alter cleavage sites and product composition (SCHMIDTKE et al. 2000). Understanding the precise rules regulating the makeup of peptides generated by the proteasome will have far reaching consequences on predicting immunogenic peptides 'hidden' within viral or tumorigenic proteins.

4 The 19S Regulatory Particle of the Proteasome (RP)

The regulatory particle (RP; also known as the 19S or PA700) serves multiple roles in regulating proteasomal activity: selecting substrates, preparing them for degradation, translocating them into the CP, as well as probably influencing the nature of products generated by the CP. The subunit composition of the RP from different species is remarkably similar (Table 1) (DeMartino et al. 1994; Dubiel et al. 1995; Finley et al. 1998; Glickman et al. 1998b; Hoelzl et al. 2000). The RP is comprised of at least 17 different subunits with a total mass of close to 1MDa (DeMartino et al. 1994; Dubiel et al. 1995; Glickman et al. 1998b; Hoelzl et al. 2000), and can assemble at either end of the 20S CP to form the 26S proteasome (Rechsteiner et al. 1993; DeMartino et al. 1994; Peters et al. 1994; Glickman et al. 1998b). Purified proteasomes are always found as a mixture of free core particles, singly capped (RP_1CP), and doubly capped (RP_2CP) forms (Glickman et al. 1998b) whereas in vivo it seems that the majority of proteasomes are present as doubly capped forms (Russell et al. 1999). Six of the RP subunits are ATPases of the AAA family (Beyer 1997; Patel and Latterich 1998) and are designated in yeast as Rpt1–6 (for regulatory particle triple-A protein). The other subunits of the RP are designated in yeast Rpn1–12 (regulatory particle non-ATPase). The six Rpt subunits together with three of the Rpn proteins (Rpn1, Rpn2, and Rpn10) form the base of the RP that attaches to the α-ring of the CP, while the remaining eight Rpn subunits can dissociate together as a lid subcomplex (see Fig. 1).

The overall conservation of the RP subunits in eukaryotes is extraordinary. The Rpt subunits are the most conserved subunits of the RP, each of which is 66%–76% identical between yeast and humans, pointing to their central and enzymatic role in proteasome function (Table 1). The non-ATPase subunits show a lower yet significant amount of sequence identity, typically in the range of 33%–47% (Table 1). A number of species-specific or loosely associated RP subunits have been identified in certain purified preparations (see Sect. 5), however the basic 17 Rpt and Rpn subunits are present in all eukaryotic preparations. Some of the RP subunits show homology to each other (Glickman et al. 1998b). The six ATPases (Rpt1–6) are roughly 40% identical to each other over the length of the protein, with the AAA domain at the center showing a greater degree of identity (Rubin et al. 1998). The six ATPases are present in every molecule of proteasome and are not equivalent as similar mutations in each ATPase gene result in unique phenotypes (Glickman et al. 1998b; Rubin et al. 1998). Among the non-ATPase subunits, three pairs show close to 20% identity to each other: Rpn1 with Rpn2, Rpn5 with Rpn7, and Rpn8 with Rpn11. The same relationship is maintained among their mammalian counterparts. The sequence similarities between different Rpt and Rpn subunits raise the possibility that gene duplication played a major role in the evolution of the regulatory particle. For instance, a simple regulatory complex made up of six copies of a single AAA ATPase, probably in the form of a ring, is found in archaea and certain bacteria (Wolf et al. 1998; Zwickl et al. 1999; Wilson et al. 2000). Thus, the six eukaryotic Rpt subunits may have diverged from a small number of ATPase subunits in an evolutionary precursor, similar to the apparent divergence of the eukaryotic core particle's 14 subunits from a single α and a single β precursor found in the CP of archaea (Bochtler et al. 1999; Voges et al. 1999).

The modular structure of the RP probably arose concomitantly with the advent of small protein labels such as ubiquitin, which are apparently absent from prokaryotes. For instance, whereas free CP can hydrolyze peptides, addition of either the eukaryotic ATPase-containing base of the RP or an archaeal regulatory ring of ATPases (PAN) allows for proteolysis of certain non-ubiquitinated proteins (GLICKMAN et al. 1998a,b; ZWICKL et al. 1999). Further attachment of the lid to the RP then allows for ubiquitinated protein degradation (GLICKMAN et al. 1998a,b). Addition of a ubiquitin binding site in the RP in eukaryotes would enhance the specificity of the proteasome for those substrates targeted for degradation (PICKART 1997; THROWER et al. 2000). The poly-ubiquitin chain could anchor substrates to the RP while they are being unfolded and translocated into the CP.

The RP from a number of sources has also been shown to contain a ubiquitin hydrolase activity that can serve to edit these ubiquitin chains, or remove ubiquitin from protein substrates (LAM et al. 1997a,b; LAYFIELD et al. 1999; HOELZL et al. 2000; LI et al. 2000). In *Drosophila*, the RP subunit responsible for this activity was identified as p37a, however it is not yet clear whether it is present in stochiometric amounts, and an obvious ortholog of this protein is not present in *Saccharomyces cerevisiae*. The exact mechanistic role of this deubiquitinating activity in proteasome function is still not clear. It is possible that the poly-ubiquitin chain must be removed from the substrate by the RP before the substrate can be translocated via the narrow channel into the proteolytic chamber of the CP. However, it is also possible that removal of the poly-ubiquitin anchor by deubiquitination is antagonistic to degradation by releasing substrates that have been improperly ubiquitinated.

Finally, not only is ATP hydrolysis also required for assembly of the proteasome from isolated RP and CP (ARMON et al. 1990; HOFFMAN and RECHSTEINER 1994), proteolysis of proteins by the proteasome is also strictly ATP-dependent (HERSHKO et al. 1984; DEMARTINO et al. 1994; GLICKMAN et al. 1998b). In analogy to molecular motors and chaperones (ALBERTS and MIAKE-LYE 1992; AMOS and CROSS 1997; SZPIKOWSKA et al. 1998), it is likely that ATP hydrolysis cycles the RP between high and low affinity states, alternately binding and releasing substrate. The conformational changes associated with this ATPase cycle could be used in three processes: (a) gating the channel defined by the N termini of the core particle's α-ring subunits; (b) unfolding the substrate; and (c) threading the unfolded substrate through the channel into the lumen of the core particle. It would be interesting to know whether these functions of the ATPases are coupled, that is, whether upon binding of substrate the ATPases unfold it and gate the channel simultaneously, or whether separate ATPases play distinct roles. As archaeal proteasomes, and other ATP-dependent proteases do not contain a lid-like subcomplex, it seems that the above-proposed functions do not require the lid.

4.1 The Base of the Regulatory Particle

An important property of the base is that it is nearly as efficient as the RP itself in stimulating the degradation of peptides and a non-ubiquitinated protein substrate by the CP, suggesting a role for proteasomal ATPases in preparing substrates for degradation (GLICKMAN et al. 1998b). The six ATPases found in the base belong to the AAA family of ATPases, whose members are found in many multi-subunit cellular machines such as translocaters, transporters, membrane fusion complexes, and proteases (BEYER 1997; PATEL and LATTERICH 1998). The function of these ATPases can be deduced by analogy to other complexed ATP-dependent proteases. Prokaryotes contain a number of different compartmentalized protease complexes such as Lon/La, ClpAP, ClpXP, and FtsH, which are regulated in an energy-dependent fashion, although their subunits are genetically unrelated to proteasome subunits from eukaryotes (GOTTESMAN et al. 1997a; LUPAS et al. 1997b; DE MOT et al. 1999). Bacteria do contain an ATP-dependent protease, HslVU, with a core protease ring (HslV) that shares significant homology with the β-subunits of the CP from eukaryotes. HslV can associate with a ring of ATPases (HslU) to form an ATP-dependent protease complex; however, HslU is not an Rpt-like ATPase but is rather related to the bacterial ClpX ATPases (ROHRWILD et al. 1997). Possibly, HslVU represents a dead-end in the evolution of the proteasome from a CP-like protease. Homologs of the *RPT* family of proteasomal ATPases are found in archaea (ZWICKL et al. 1999) and in some prokaryotes [such as *Rodococuss erythropolis* and *Mycobacterium leprae* (WOLF et al. 1998)].

The single *RPT* homolog in *Methanococcus jannaschii* (PAN) or *Rhodococcus erythropolis* (ARC) can form homo-hexameric rings (WOLF et al. 1998; ZWICKL et al. 1999; WILSON et al. 2000). Moreover, the hexameric ring of PAN from *Methanococcus*, perhaps the most rudimentary form of the RP, can regulate at least some functions of the CP and confer the capability to proteolyse certain non-ubiquitinated proteins by the archaeal CP (ZWICKL et al. 1999; WILSON et al. 2000). PAN rings do indeed attach to the outer surface of the α-ring of the CP (WILSON et al. 2000). By analogy, the six different Rpt subunits may also form a ring within the eukaryotic RP that contacts the surface of the CP in a similar fashion, explaining why attachment of the RP base from *S. cerevisiae* can increase casein hydrolysis activity by the CP (GLICKMAN et al. 1998a). A distinction between proteasomes from archaea and those from yeast, however, is that the putative regulatory ring of ATPases from archaea does not significantly enhance the peptidase activity of the CP (ZWICKL et al. 1999), while both the RP and the RP base from yeast do (GLICKMAN et al. 1998a,b). This might reflect a greater need for gating the channel leading into the proteasome from eukaryotic sources due to the asymmetric arrangement of the core particle's seven different α-ring subunits (LOEWE et al. 1995; GROLL et al. 1997, 2000). In support of a role for the RP ATPases in gating the channel of eukaryotic CPs, a single ATPase mutant showed a dramatic inhibition of peptidase activity (RUBIN et al. 1998), indicating that even the entry of small peptides – which do not need to be unfolded – can be controlled by the RP.

Unfolding of protein substrates is required as the estimated inner diameter of the CP is too narrow for proteins to enter in their native state. ATP binding and hydrolysis by the ATPases may trigger cycles of higher and lower affinities of the RP for substrate proteins, thus stabilizing them in an unfolded state. A possible role for the base in promoting substrate unfolding is suggested by its location, covering the entry ports into the CP, and by the presence of six ATPases within this complex. In vitro, both the RP and the base can bind unfolded proteins, and release them in an ATP-dependent fashion (BRAUN et al. 1999). The refolding activity of the base may be an independent function of the RP or it may reflect a normal function of the base as part of the unfolding process of proteolytic substrates. For instance, the directionality of the reaction – whether the base serves to fold or unfold – may be dependent on the structure of the substrate as it binds the RP (i.e., native or denatured). The directionality could also be influenced by the end products: proteolysis is irreversible, driving the reaction towards unfolding and translocation into the CP. Substrates are thus translocated from one domain within the proteasome (RP) to another (CP) probably in an ATP-dependent manner.

Rpn10, Rpn1, and Rpn2, make up the RP base together with the six Rpt ATPases. Both Rpn1 and Rpn2 contain multiple leucine-rich repeats (LRR), a domain suggested for protein–protein interaction (LUPAS et al. 1997a) and the six ATPases of the base are also likely to function through protein–protein interactions. Similarly to the regulatory domain of simple ATP-dependent proteases in prokaryotes, which interact directly with substrates (GOTTESMAN et al. 1997b), it is plausible that these eight components of the base engage in direct interactions with substrates. Rpn10 contains a ubiquitin binding site found within its C terminus in a hydrophobic patch rich with alternating leucines and alanines (FU et al. 1998; HOFMANN 2001). The C-termini of Rpn10 homologs, such as p54 from *Drosophila m.* and S5a from mammals, contain two such hydrophobic repeats both of which have been shown to bind ubiquitin (HARACSKA and UDVARDY 1997; YOUNG et al. 1998). However, as *RPN10* or its homologs are nonessential in *S. cerevisiae, S. pombe* or moss, it was suggested that it does not function as the sole ubiquitin recognition site in the proteasome (VAN NOCKER et al. 1996; GIROD et al. 1999; WILKINSON et al. 2000). Interestingly the two other non-ATPase subunits that reside in the base, Rpn1 and Rpn2 (as well as their homologs in other eukaryotes), also contain leucine-rich repeats at their C termini (LUPAS et al. 1997a; YOUNG et al. 1998). As proteins are targeted for degradation via a multiubiquitin chain, and not by a single ubiquitin molecule (BEAL et al. 1996, 1998; PIOTROWSKI et al. 1997), it is not unreasonable that more than one proteasomal subunit should be necessary for recognition and binding. Indeed, Rpn10 in combination with a malfunctional version of Rpn1 is essential (WILKINSON et al. 2000).

Support for this hypothesis comes from Rad23, a molecule that potentially links the proteasome to the DNA repair pathway (SCHAUBER et al. 1998; RUSSELL et al. 1999). Rad23 (as well as its human counterpart hH23) can bind tightly to yeast or mammalian proteasomes (SCHAUBER et al. 1998; HIYAMA et al. 1999). As Rad23 contains a domain that shares homology with ubiquitin, its association may be with the ubiquitin-binding site in the proteasome. Indeed, Rad23 can bind Rpn10/s5a through its ubiquitin-like domain, indicating that its binding could be similar to ubiquitin binding to the proteasome. Rad23 can also bind to the proteasome independently of Rpn10/s5a (LAMBERTSON et al. 1999) indicating that there might be an additional site for binding of ubiquitin or ubiquitin-like proteins in the RP. A yeast two-hybrid screen shows that Rad23 interacts with Rpn1

(CAGNEY 2000). Thus, it is possible that the non-ATPase subunits of the base have affinity for ubiquitin or ubiquitin-like molecules. It is important to mention, however, that there is no evidence that mono-ubiquitin targets proteins to the proteasome (STROUS and GOVERS 1999; THROWER et al. 2000). Therefore, there is no reason to believe at this stage that the binding of Rad23, with its single ubiquitin-like domain, points directly to the poly-ubiquitin recognition site in the RP. It is very plausible that recognition of poly-ubiquitin chains deploys different machinery.

Finally, in addition ubiquitin binding, Rpn10 has additional roles in proteasome function. Even though the ubiquitin binding site is stringently conserved in the C terminus of Rpn10 homologs from a number of species (HARACSKA and UDVARDY 1997; FU et al. 1998; YOUNG et al. 1998), deletion of the C terminus of Rpn10 has no discernible phenotype in *S. cerevisiae*. Deleting only the N terminus of Rpn10, on the other hand, causes a phenotype similar to deletion of the entire rpn10 ORF (FU et al. 1998). This indicates that Rpn10 functions in the proteasome independently of its ability to bind ubiquitin chains. One of these roles is to stabilize the interactions between the lid and the base subcomplexes of the RP; the lid detaches to a greater extent from proteasomes purified from the Δrpn10 strain (GLICKMAN et al. 1998a). In some instances, Rpn10 (with a short deletion in its N terminus) copurifies with the base (GLICKMAN et al. 1998a), in other cases Rpn10 cofractionates with the lid (SAEKI et al. 2000), however purified lid does not contain Rpn10 (GLICKMAN et al. 1998a; BRAUN et al. 1999; HOELZL et al. 2000). Rpn10 is also the only proteasome subunit found in significant amounts outside of the proteasome or its subcomplexes (VAN NOCKER et al. 1996; HARACSKA and UDVARDY 1997). Obviously, Rpn10 can interact with both the lid and the base of the RP and plays an important role in stabilizing the interaction between the two. Most likely, it also has an as yet unknown function outside of the proteasome as well.

4.2 The Lid, and Similarities to Other Regulatory Complexes

The lid consists of eight subunits arranged in a disk-like shape that can detach from the RP base as a discreet complex and reattach to it (GLICKMAN et al. 1998a; BRAUN et al. 1999; KAPELARI et al. 2000). Rpn10 aids in the correct assembly of the RP by linking the lid and the base via its N terminus, though Rpn10 alone is neither essential nor sufficient for this task (GLICKMAN et al. 1998a). The lid subunits are all non-ATPases and their function or enzymatic activities are not yet known. Proteolysis of ubiquitinated proteins by the proteasome is required for lid attachment, which may indicate interaction with the multiubiquitin chain either through a ubiquitin binding, or deubiquitinating subunit (GLICKMAN et al. 1998a). In comparison, certain non-ubiquitinated proteins are efficiently proteolyzed by eukaryotic proteasomes from which the lid has been detached (GLICKMAN et al. 1998a), as well as by naturally 'lidless' archaeal or prokaryotic proteasomes (DE MOT et al. 1999; MAUPIN-FURLOW et al. 2000). The simplest explanation is that a 'lid' was added to the rudimentary RP concomitantly with the appearance of ubiquitin as a targeting system in eukaryotes. The apparent functions of the base and the lid are consistent with their location within the RP: the distal positioning of the lid may ensure that the ubiquitin chain of the conjugate does not occlude access of the target protein to the channel. The substrate is then properly positioned to be unfolded by the base and translocated into the CP. Alternatively, the lid could play a structural role, such as defining a cavity within the RP, and it is a conformational change of the base caused by attachment of the lid that allows proper binding of poly-ubiquitinated substrates.

Each of the eight subunits of the lid subcomplex contains one of two structural motifs that are also found in several other protein complexes, such as the eukaryotic regulator of translation eIF3 (eukaryotic initiation factor 3), and the COP9 signalosome (CSN) (ARAVIND and PONTING 1998; GLICKMAN et al. 1998a; HOFMANN and BUCHER 1998). One motif is the PCI domain that is up to 200 residues long and is predicted to form an α-helical structure found at the C termini of Rpn3, Rpn5, Rpn6, Rpn7, Rpn9 and Rpn12 (Table 1). The second is the MPN domain, a 120-aa-long sequence that forms an α/β structure at the N termini of Rpn8 and Rpn11. None of the RP-base or CP proteins contain these motifs, further differentiating these complexes and their evolutionary origins.

The eIF3 of human is a 600-kDa subcomplex that contains around 10 subunits, binds to the ribosome, and serves as an organization locus for the binding of mRNA and other proteins that trigger the initiation of protein translation events. Yeast eIF3 is somewhat smaller and contains fewer subunits. Both the human and yeast eIF3 complexes include subunits with PCI and MPN motifs, as well as proteins with homology to a specific lid component, (hs)s12/(sc)Rpn8 (ASANO et al. 1997; HOFMANN and BUCHER 1998). The CSN is a highly conserved complex that plays a role in regulating the development of eukaryotes; for instance, it is essential for light signal transduction in plants (CHAMOVITZ et al. 1996) and for development of *Drosophila* embryos (FREILICH et al. 1999). The CSN has been purified from a number of metazoans as a 450-kDa complex, probably part of a larger, as yet unidentified, complex (SEEGER et al. 1998; WEI et al. 1998; SERINO et al. 1999). Like the lid of the proteasome, the CSN is also an eight-subunit complex, six of which share a PCI domain, while the other two share an MPN domain (DENG et al. 2000). A number of subunits have also been identified in *S. pombe* indicating that the CSN also plays a role in single cellular organisms (MUNDT et al. 1999). Both the lid and the CSN complexes contain the same number of subunits, have a similar molecular weight, roughly the same architecture, share the same structural motifs, and each subunit in each complex has a specific paralog in the other (Table 1) (GLICKMAN et al. 1998a; DENG et al. 2000; KAPELARI et al. 2000). Thus these remarkably similar complexes have different biological roles.

In addition to structural and genetic homology between the lid and the CSN, the two complexes also have similar intracellular distribution, cellular activity, share substrates and take part in similar regulatory pathways. Both particles were found to be concentrated at the nuclear periphery, but also present in the cytosol and the nucleus (ENENKEL et al. 1998; SEEGER et al. 1998). Furthermore, it is known that the interaction of the CSN with certain transcription factors alters their stability, function, or activity. For example, CSN5/Jab1 interacts with c-jun and junD and enhances specific c-jun or junD associated AP-1 activity by promoting c-jun phosphorylation and increasing c-jun stability (CLARET et al. 1996; NAUMANN et al. 1999). In general, many CSN substrates are also substrates of the proteasome. These include transcription factors such as members of the AP-1, interferon regulatory factor (IRF), NFκB, or Id (transcriptional *i*nhibitors of *d*ifferentiation) families, cell cycle factors such as $p27^{kip1}$, and receptors such as SRC^{-1} (steroid receptor coactivator 1) and the lutropin/choriogonadotropin receptor precursor (BOUNPHENG et al. 1999; TOMODA et al. 1999; BOUSIOTIS et al. 2000; CHAUCHEREAU et al. 2000; DENG et al. 2000; LI et al. 2000). These examples suggest that the CSN cooperates with the 26S proteasome in regulating protein degradation. Indeed, reports suggest interactions between subunits of these complexes (KARNIOL et al. 1998; KWOK et al. 1999).

One manner in which it is becoming increasingly clear that CSN signaling cross-talks with proteasome-dependent degradation, is the affect CSN or its subunits have on the stability and potency of specific members of transcription factor families. The interaction of CSN with certain transcription factors alters their ability to heterodimerize with their various partners. In a number of cases, this also has a direct affect on the stability of these transcription factors. For instance, the Jun proteins dimerize with the Fos proteins to form the AP-1 transcriptional activator. Each type of heterodimer activates a set of only partially overlapping AP-1 target genes. Csn5/Jab1 interacts with c-Jun and JunD, but not with JunB or v-Jun, and enhances the specific c-Jun- or JunD-associated AP-1 activity (CLARET et al. 1996). Unphosphorylated c-Jun is normally an unstable protein that is rapidly removed from the cell by the proteasome (TREIER et al. 1994). Interaction of the CSN with c-Jun increases its stability and elevates AP-1 activity, probably by promoting c-Jun phosphorylation (SEEGER et al. 1998; NAUMANN et al. 1999). Thus, through selective interactions with the Jun proteins, the CSN can increase the specificity of target-gene activation. A similar mechanism seems to regulate Id activity. The four Id proteins are a family of negative regulators of transcription that heterodimerize with transcription factors and prevent them from binding to their target DNA. Most Id proteins – as well as many of their partners such as MyoD and E12 – are substrates of the proteasome and are rapidly degraded under certain conditions (BOUNPHENG et al. 1999). CSN5/Jab1 interacts with Id3 possibly affecting its stability or that of its partners (BOUNPHENG et al. 1999). A recent example highlights a possible link between interferon regulatory factor (IRF)-signaling and the CSN; IRFs are a family of nine transcription factors that are recruited upon interferon signaling to activate target-genes involved in combating viral infection, including even the catalytic βi subunits of the CP in the immunoproteasome (Foss and PRYDZ 1999; MAMANE et al. 1999). The CSN binds to IRF-8/ICSBP via its CSN2 subunit and promotes its phosphorylation on a unique serine residue thus enhancing its ability to bind to IRF-1 (COHEN et al. 2000). Once again, IRF-1 (as well as some other members of the IRF family) is normally an unstable protein and a substrate for the proteasome; interestingly, the C-terminal domain that controls its stability is also the same region to which IRF-8 binds (NAKAGAWA and YOKASAWA 2000). Thus it is possible that the complex of IRF-8 with IRF-1 is protected from degradation by the proteasome as the same module in IRF-1 is essential both for its stability and for IRF-8 binding. From these examples, a pattern evolves in which the CSN presents, or prevents the presentation of, certain substrates to the proteasome. It is tempting to speculate that the shared homology and common architecture of the proteasome lid and the CSN are major features in their ability to share substrates.

4.3 Alternative Regulatory Complexes of the Proteasome

In addition to the 19S RP, another regulatory complex called the 11S REG or PA28 can activate peptidase activity of the 20S CP (DUBIEL et al. 1992; MA et al. 1992). PA28 increases V_{max} for hydrolysis of certain peptides by the 20S CP by up to 100-fold, but in contrast with the 19S RP, it does not promote protein degradation by the CP (DUBIEL et al. 1992; MA et al. 1992). PA28 is a hetero-heptamer (or possibly a hexamer) of two similar proteins, α and β, that can attach to either one or both outer α-rings of the 20S CP. This complex can be found in the cytoplasm of vertebrate cells, and cellular levels of both the α- and β-subunits of PA28 increase significantly by interferon-γ linking PA28 with the immune response (SEEGER et al. 1997; RECHSTEINER et al. 2000; STOHWASSER et al. 2000). Additional complexes related to PA28/11S REG are also found in eukaryotes: REGγ (also called the Ki antigen or PA28γ) is a homo-heptamer found in numerous metazoans (even those lacking an adaptive immune system), and a more distantly related heptamer-ring called PA26 is found only in *Trypanosoma brucei* (REALINI et al. 1997; YAO et al. 1999). *S. cerevisiae* appears to lack homologs of this class of proteasome activator; however, as PA28 and PA26 are highly divergent, it is possible that functional homologs with only low-level sequence similarity are present in yeast or additional eukaryotes.

Based on the crystal structure of a complex of PA26 from *T. brucei* with a 20S CP, PA26 attaches to the α-ring of the CP. Each of the 20S CP α-subunits is contacted equally by the PA26 heptamer, and the region most likely to serve as a site for PA26 binding is the H1 and H0 helices immediately C terminal to the N termini of the α-subunits that protrude into the center of the ring (see the chapter by Hill et al., this volume). Upon binding of PA26, the N termini of the different α-subunits are disordered and pulled into the central cavity of PA26 and away from the center of symmetry defining the α-ring. It is assumed that similarly to PA26, PA28 also activates the CP by binding to the α-ring surface and rearranging the blocking N-terminal residues thus opening the gate and increasing the rate by which peptides enter the CP (see the chapter by Hill et al., this volume). However, because peptides are normally the products of the CP, it is possible that PA28 also affects the rate at which peptides are generated by the CP and exit the proteolytic chamber (STOHWASSER et al. 2000).

A significant portion of the antigenic peptides to be presented by MHC class I is generated by the proteasome (RECHSTEINER et al. 2000). The proteasome removes foreign proteins – such as those of viral origin – from the cell by hydrolyzing them into short peptides. Only a fraction of the peptides generated by the proteasome are of the correct length to be transported through the ER and presented to the immune system by MHC class I. The regulatory role of PA28 in antigen processing might be to affect the nature of products generated by the proteasome by altering the peptidase activity of the 20S CP in order to increase the efficiency of antigen presentation and, by extension, combat of viral infection. Indeed, hybrid proteasomes with one 19S RP and one PA28 at opposite ends of the CP have been identified (HENDIL et al. 1998; TANAHASHI et al. 2000) suggesting that an intricate process of selecting substrates by the RP and regulating products by the PA28 is necessary for efficient antigenic peptide production.

5 Changes in Proteasome Composition and Associated Proteins/Factors

In addition to the different types of regulatory particles that attach to the outer surfaces of the core particle's α-rings, numerous proteins can bind to the proteasome with different affinities and ratios. Some of the proteins that are found loosely associated with proteasomes are likely to be substrates. Additional proteins that are found to interact with the proteasome could be factors that target these substrates to the proteasome or that stabilize proteasome–substrate interactions. Specifically, chaperones and components of the ubiquitin system are thought to aid in shuttling substrates to the proteasome. Cellular proteins could also bind to the proteasome in order to fine-tune proteolysis, link the proteasome to other signaling pathways, or to aid in proteasome assembly and localization. For instance, in yeast, a number of nonessential gene products can associate with the proteasome, including Rpn4/Son1/Ufd5 (FUJIMURO et al. 1998b), p27/Nas2 (WATANABE et al. 1998), Doa4 (PAPA and HOCHSTRASSER 1993), p28/Nas6 (HORI et al. 1998), and Rad23 (SCHAUBER et al. 1998). An exhaustive list of cellular factors that interact with the proteasome has been complied in a recent review (FERRELL et al. 2000). Here we will focus on a few interactions that seem to specifically regulate proteasomal functions.

5.1 Diversity in Proteasome Composition

The structure and subunit composition of proteasomes purified from different species or by different protocols are almost identical (DEMARTINO et al. 1994; DUBIEL et al. 1995; GLICKMAN et al. 1998b; HOELZL et al. 2000). In a few preparations, a unique subunit was identified that is not present in genomes of other species or has not been identified in other preparations (for example, Son1/Ufd5 in

S. cerevisiae, S5b in human, and p37a in *Drosophila*); these factors could be transiently associated, substochiometric, or species-specific (DEVERAUX et al. 1995; FUJIMURO et al. 1998b; HOELZL et al. 2000).

Development-dependent changes account for some of the diversity in proteasome composition. Higher levels of proteasome and RP components are found in stationary yeast compared to rapidly growing yeast (FUJIMURO et al. 1998a). The growing-to-stationary phase transition also induces the assembly of proteasome holoenzymes from its CP and RP subcomplexes, probably favoring doubly capped (RP_2CP) proteasomes over singly capped (RP_1CP) and free core particles (FUJIMURO et al. 1998a). A similar observation was made for elevated proteasome levels in *Manduca sexta* during cell death. Complicating matters, however, is evidence that the ratio of one ATPase to another in the proteasome may change during the course of programmed cell death. Some proteasomal ATPases (Rpt subunits) as well as s5a/Rpn10 show increased levels compared to other RP-subunits during cell death in *Manduca sexta*, indicating that proteasome function might be fine-tuned during the course of development by altering its subunit composition (DAWSON et al. 1995, 1997). Most probably, however, basal expression levels of the principal 31 proteasomal genes are quite similar based on the seemingly stochiometric levels of proteasomal subunits in purified samples. Likewise, almost all CP and RP genes contain an identical proteasome-associated control element (PACE) in their promoter situated at the same distance upstream from the start codon indicating that they might be under similar transcriptional control (MANNHAUPT et al. 1999). A notable exception is Rpn10 that does not contain this element and, as mentioned above, is the only proteasomal gene found in significant levels unbound to the proteasome.

Changes in the broader sense of proteasome composition are achieved by the attachment of auxiliary factors and transiently associated factors in response to specific signals. A number of proteins specifically associate with the proteasome during different developmental stages, or under different physiological conditions. For instance, in yeast, Nob1 cofractionates almost exclusively with the proteasome but only in growing and dividing cells, whereas the levels of Nob1 in stationary or resting cells is below the limit of detection (TONE et al. 2000). Because Nob1 is an essential gene, it would seem that Nob1 plays an essential role in regulating proteasomal function in growing cells (TONE et al. 2000). A blast search indicates that proteins with varying levels of homology to Nob1 are present in the genomes of numerous archaea and eukaryotes. Of the known mammalian RP-subunits, only S5b/p50.5 (DEVERAUX et al. 1995) appears to have no ortholog in the genome of budding yeast. Furthermore, no evidence for an S5b homolog has been found in proteasomes purified from other sources. In vitro experiments do show, however, that S5b interacts with components of the base of the mammalian RP (GORBEA et al. 2000). Its exact role and how integral it is to proteasome composition remains to be clarified. In budding yeast, the protein Son1/Ufd5/Rpn4 can copurify with the proteasome by gel filtration or immuno-precipitation, though it is not found in purified preparations (FINLEY et al. 1998; FUJIMURO et al. 1998b; GLICKMAN et al. 1998a). However, as Rpn4 has been found to be a DNA-binding protein that acts as a trans-activator of proteasome and ubiquitin pathways genes, it is not obvious whether Rpn4 is indeed an integral component of the proteasome (MANNHAUPT et al. 1999).

The best-studied change in proteasome subunit composition is the replacement of the proteolytically active CP β-subunits with alternative, so called LMP or βi, subunits upon interferon signaling. Each of these βi-subunits is genetically homologous to a specific constitutively expressed β-subunit, and can be incorporated into the corresponding position within the β-ring of newly assembled proteasomes. β1i/LMP2 replaces its constitutively expressed β1 homolog, β5i/LMP7 replaces β5, and β2i/LMP10 replaces β2 (BOCHTLER et al. 1999). The subunit composition of proteasomal core particles is thus altered upon induction by γ-interferon and they are therefore often referred to as 'immunoproteasomes'. Immunoproteasomes are thought to generate peptides that are more appropriate for antigen presentation. Due to amino acid differences in the S1 pocket (that defines substrate specificity) of the βi-subunits compared to their constitutive β counterparts, a drastic decrease in post-acidic, and to some extent post-basic, endopeptidase activity is measured for immunoproteasomes; immunoproteasomes cleave peptides mainly after hydrophobic positions (GACZYNSKA et al. 1993). Because peptides are anchored in the MHC-I groove preferentially by hydrophobic terminal residues, immunoproteasomes could be instrumental in the generation of antigens for MHC class I presentation (MICHALEK et al. 1993; STOLTZE et al. 2000). Immunoproteasomes are a well-characterized example of signaling-dependent changes in proteasome structure as a method to tailor proteasome function to biological needs. Possibly, other such subunit replacements or additions occur in response to other signals.

Another group of proteins that is found tightly bound to the proteasome or its subunits are the factors that assist in proteasome assembly. In mammalian cells, a protein called p27 has been found associated with two of the proteasomal ATPases (s6'/Rpt5 and s10b/Rpt4) and has been proposed to aid in the assembly of the RP (DEMARTINO et al. 1996; ADAMS et al. 1997). The exact role of this modulator is unclear, however, especially as in yeast Nas2, the homolog of the mammalian p27 protein, it is nonessential (WATANABE et al. 1998). Another yeast protein, Ump1, is needed for maturation of 20S CPs from α- and β-subunits and is degraded upon completion of assembly, in effect becoming the proteasome's first substrate (RAMOS et al. 1998).

5.2 Auxiliary Factors and Transiently Associated Proteins

The proteasome may interact with the enzymatic complexes that present substrates to it for degradation. For instance, it is still not clear exactly how ubiquitinated proteins are brought to the proteasome. There are now several reports that there are direct contacts between the ubiquitin system and the proteasome suggesting that ubiquitination and degradation might not be completely independent processes. Ubiquitin-conjugating enzymes (E2s) such as Ubc1, Ubc2, and Ubc4, coimmunoprecipitate with the proteasome (TONGAONKAR et al. 2000). The interaction between Ubc4 and the proteasome seems to be enhanced under conditions of stress, when the need to remove damaged proteins increases (TONGAONKAR et al. 2000). Interactions of the ubiquitin-ligases (E3s) Ubr1 and Ufd4 mapped to components of the base of the RP, specifically to Rpt6, Rpt1, and Rpn2 (XIE and VARSHAVSKY 2000). Two additional E3s have been shown to interact with the proteasome via a ubiquitin-like protein, hPLIC/Dsk2 (KLEIJNEN 2000). Because the substrate specificity of the ubiquitin system is concentrated at the level of the E3 enzymes, interactions with E3s may confer substrate specificity on the proteasome.

The proteasome also interacts with the other end of the ubiquitin system – the deubiquitinating (DUB) enzymes. A number of reports indicate that the proteasome, and more specifically the RP, have DUB activity: i.e., the ability to hydrolyze peptide or isopeptide bonds after the C terminus of ubiquitin (LAM et al. 1997a,b; LAYFIELD et al. 1999; HOELZL et al. 2000; LI et al. 2000). The subunit responsible for this activity in *Drosophila* has been identified as p37a (HOELZL et al. 2000). P37a, UCH37, or a homolog might be the subunit responsible for the polyubiquitin chain editing function associated with proteasomes purified from different sources (LAM et al. 1997b; HOELZL et al. 2000; LI et al. 2000). A clear ortholog of p37a has not been identified in *S. cerevisiae*, suggesting that additional DUBs might be associated with the proteasome. In *S. cerevisiae*, Doa4, a deubiquitinating enzyme of the UBP family, has been proposed to function in conjunction with the proteasome by trimming multiubiquitin chains from proteasome-bound substrates (PAPA and HOCHSTRASSER 1993; PAPA et al. 1999). Doa4 interacts weakly and in substochiometric amounts with the proteasome (PAPA et al. 1999). Doa4 might serve to release ubiquitin and regenerate the proteasome for the next catalytic cycle (PAPA et al. 1999). Since the proteasome exhibits a wide range of DUB activity (LAYFIELD et al. 1999), it is possible that more than one DUB enzyme is associated with the proteasome.

Another class of transiently associated proteins is the auxiliary factors that either shuttle substrates to the proteasome, or target the proteasome to concentrations of substrates. P28 has been shown to interact with the proteasome in mammals, and is possibly a subunit (HORI et al. 1998). p28 has five to six ankyrin repeats that can serve to promote protein–protein interactions. p28 is probably identical to the oncogene gankyrin that is overexpressed in certain carcinomas, binds to the retinoblastoma (RB) gene product, and promotes the degradation of RB via its interaction with the proteasome (DAWSON et al. 1997; HIGASHITSUJI et al. 2000). Similar to a number of other 'loosely associated' proteasomal proteins, Nas6, the yeast homolog of mammalian p28, is nonessential (HORI et al. 1998). As many substrates of the proteasome contain ankyrin repeats, for example IκB and certain cyclins, it is possible that p28 is either a substrate of the proteasome itself, or a substrate-shuttle that can aid in the recognition and anchoring of substrates – such as RB – to the regulatory particle of the proteasome (DAWSON et al. 1997; HIGASHITSUJI et al. 2000).

Interactions with cellular factors could be a means of targeting the proteasome to sites of potential substrates. For instance, the proteasome interacts with components of the cell cycle machinery (KAISER et al. 1999), and proteasomal subunits colocalize in vivo together with heat shock proteins and chaperones at sites of misfolded androgen receptor aggregates (STENOIEN et al. 1999). Rad23 can bind tightly to both purified and partially purified proteasomes (SCHAUBER et al. 1998), linking the proteasome to the DNA repair pathway and targeting it to the nuclear excision repair complex (RUSSELL et al. 1999). As Rad23 includes a ubiquitin-like domain as well as polyubiquitin conjugate binding abilities, its association may reflect ubiquitin recognition by the proteasome (SCHAUBER et al. 1998; HIYAMA et al. 1999; RUSSELL et al. 1999; WILKINSON 2001). Also hPLIC/Dsk2 contains a ubiquitin-like domain and links the proteasome with various ubiquitin ligase enzymes (KLEIJNEN 2000). An additional protein that has a ubiquitin-like domain and that can bind to the proteasome is the chaperone cofactor BAG-1 (LUEDERS et al. 2000). Both BAG-1 and Rad23 contain noncleavable ubiquitin-like domains at their N terminals that are required for their attachment to the RP. Binding of BAG-1 promotes the association of Hsc70/Hsp70 with the proteasome. The interaction with BAG-1 points to the linkage between proteasome and chaperone functions.

The link between heat shock proteins, chaperones and the proteasome could also be mechanistic. Despite the fact that the RP of the proteasome might have intrinsic chaperone-like activity, increasing evidence links chaperones with proper proteasome function. In yeast, overexpression of (wild-type but not mutant) Hsp70 and DnaJ can suppress growth defects associated with mutant CP-subunits (OHBA 1997). One model proposes that molecular chaperones shuttle substrates to the proteasome (LEE et al. 1996). For instance, in some cases mutant membrane proteins are dislodged from the membrane by Hsp90 and are presented to the proteasome for degradation. Indeed, in other eukaryotes, the proteasome can be found in specific cellular structures that include ubiquitin also and the cell stress chaperones Hsp70 and Hsp90 indicating that the proteasome is possibly targeted to concentrations of misfolded proteins which much be removed (WIGLEY et al. 1999). Even in vitro it seems that some substrates are proteolyzed only with the help of chaperones (BERCOVICH et al. 1997) suggesting that 'chaperone presentation' to the proteasome might be a feature of protein degradation.

6 Summary

Despite the fact that the composition of proteasomes purified from different species is almost identical, and the basic components of the proteasome are remarkably conserved among all eukaryotes, there are quite a few additional proteins that show up in certain purifications or in certain screens. There is increasing evidence that the proteasome is in fact a dynamic structure forming multiple interactions with transiently associated subunits and cellular factors that are necessary for functions such as cellular localization, presentation of substrates, substrate-specific interactions, or generation of varied products. Harnessing the eukaryotic proteasome to its defined regulatory roles has been achieved by a number of means: (a) increasing the complexity of the proteasome by gene duplication, and differentiation of members within each gene family (namely the CP and RPT subunits); (b) addition of regulatory particles, complexes, and factors that influence both what enters and what exits the proteasome; and (c) signal-dependent alterations in subunit composition (for example, the CP β to βi exchange). It is not be surprising that the proteasome plays diverse roles, and that its specific functions can be fine-tuned depending on biological context or need.

Acknowledgements. We would like to thank the Israel Cancer Research Fund (ICRF) and the Yigal Alon foundation for Grants to M.G. The work of the laboratory is supported by grants from the Israel Academy of science, the Ministry of Science, and the foundation for promotion of research at the Technion.

References

Adams GM, Crotchett B, Slaughter CA, DeMartino GN, Gogol EP (1998) Formation of proteasome-PA700 complexes directly correlates with activation of peptidase activity. Biochemistry 37:12927–12932
Adams GM, Falke S, Goldberg AL, Slaughter CA, DeMartino GN, Gogol EP (1997) Structural and functional effects of PA700 and modulator protein on proteasomes. J Mol Biol 273:646–657
Alberts B, Miake-Lye R (1992) Unscrambling the puzzle of biological machines: the importance of details. Cell 68:415–420
Amos LA, Cross RA (1997) Structure and dynamics of molecular motors. Curr Opin Struct Biol 7:239–246
Aravind L, Ponting CP (1998) Homologues of 26S proteasome subunits are regulators of transcription and translation. Prot Sci 7:1250–1254
Armon T, Ganoth D, Hershko A (1990) Assembly of the 26S complex that degrades proteins ligated to ubiquitin is accompanied by the formation of ATPase activity. J Biol Chem 265:20723–20726
Asano K, Kinzy TG, Merrick WC, Hershey JWB (1997) Conservation and diversity of eukaryotic translation initiation factor 3. J Biol Chem 272:1101–1109
Aviel S, Winberg G, Massucci M, Ciechanover A (2000) Degradation of the Epstein–Barr virus latent membrane protein 1 (LMP1) by the ubiquitin-proteasome pathway. targeting via ubiquitination of the N-terminal residue. J Biol Chem 275:23491–23499
Beal R, Deveraux Q, Xia G, Rechsteiner M, Pickart C (1996) Surface hydrophobic residues of multiubiquitin chains essential for proteolytic targeting. Proc Natl Acad Sci USA 93:861–866
Beal RE, Toscano-Cantaffa D, Young P, Rechsteiner M, Pickart CM (1998) The hydrophobic effect contributes to polyubiquitin chain recognition. Biochemistry 37:2925–2934
Beninga J, Rock KL, Goldberg AL (1998) Interferon-gamma can stimulate post-proteasomal trimming of the N terminus of an antigenic peptide by inducing leucine aminopeptidase. J Biol Chem 273:18734–18742
Bercovich B, Stancovski I, Mayer A, Blumenfeld N, Laszlo A, Schwatz AL, Ciechanover A (1997) Ubiquitin-dependent degradation of certain protein substrates in vitro requires the molecular chaperone Hsc70. J Biol Chem 272:9002–9010
Beyer A (1997) Sequence analysis of the AAA protein family. Prot Sci 6:2043–2058
Bochtler M, Ditzel M, Groll M, Hartmann C, Huber R (1999) The Proteasome. Annu Rev Biophys Bio Mol Struct 28:295–317
Bounpheng MA, Dimas JJ, Dodds SG, Christy BA (1999) Degradation of Id proteins by the ubiquitin-proteasome pathway. FASEB J 13:2257–2264
Bousiotis VA, Freedman GJ, Taylor PA, Berezovskaya A, Grass I, Blazar RB, Nadler LM (2000) p27^{kip1} functions as an anergy factor inhibiting interleukin2 transcription and clonal expansion of alloreactive human and mouse T lymphocytes. Nature Med 6:290–297

Braun BC, Glickman MH, Kraft R, Dahlmann B, Kloetzel PM, Finely D, Schmidt M (1999) The base of the proteasome regulatory particle exhibits chaperone-like activity. Nature Cell Biol 1:221–226

Breitschopf K, Bengal E, Ziv T, Admon A, Ciechanover A (1998) A novel site for ubiquitination: the N-terminal residue, and not internal lysines of MyoD, is essential for conjugation and degradation of the protein. EMBO J 17:5964–5973

Cagney G, Uetz P, Fields S (2001) Two-hybrid analysis of the Saccharomyces cerevisiae 26S proteasome. Physiol Genomics 7:27–34

Chamovitz DA, Wei N, Osterlund MT, von Arnim AG, Staub JM, Deng XW (1996) The COP9 complex, a novel multisubunit nuclear regulator involved in light control of a plant development switch. Cell 86:115–121

Chauchereau A, Georgiakaki M, Perrin-Wolff M, Milgrom E, Loosfelt H (2000) JAB1 interacts with both the progesterone receptor and SRC-1. J Biol Chem 275:8540–8548

Chen P, Hochstrasser M (1996) Autocatalytic subunit processing couples active site formation in the 20S proteasome to completion of assembly. Cell 86:961–972

Ciechanover A, Orian A, Schwartz AL (2000) Ubiquitin-mediated proteolysis: biological regulation via destruction. Bioessays 22:442–451

Claret FX, Hibi M, Dhut S, Toda T, Karin M (1996) A new group of conserved coactivators that increase the specificity of AP-1 transcription factors. Nature 383:453–457

Cohen H, Azriel A, Cohen T, Meraro D, Hashmueli S, Bech-Otschir D, Kraft R, Dubiel W, Levi BZ (2000) Interaction between ICSBP and CSN2 (Trip15) – a possible link between IRF signaling and the COP9/Signalosome. J Biol Chem (in press)

Dawson S, Hastings R, Takayanagi K, Reynolds S, Low P, Billet M, Mayer RJ (1997) The 26S proteasome: regulation and substrate recognition. Mol Biol Rep 24:39–44

Dawson SP, Arnold JE, Mayer NJ, Reynolds SE, Billett MA, Gordon C, Colleaux L, Kloetzel PM, Tanaka K, Mayer RJ (1995) Developmental changes of the 26S proteasome in abdominal intersegmental muscles of Manduca sexta during programmed cell death. J Biol Chem 270:1850–1858

De Mot R, Nagy I, Walz J, Baumeister W (1999) Proteasomes and other self-compartmentalizing proteases in prokaryotes. Trends Microbiol 7:88–92

DeMartino GN, Moomaw CR, Zagnitko OP, Proske RJ, Ma CP, Afendis SJ, Swaffield JC, Slaughter CA (1994) PA700, an ATP-dependent activator of the 20S proteasome, is an ATPase containing multiple members of a nucleotide binding protein family. J Biol Chem 269:20878–20884

DeMartino GN, Proske RJ, Moomaw CR, Strong AA, Song X, Hisamatsu H, Tanaka K, Slaughter CA (1996) Identification, purification, and characterization of a PA700-dependent activator of the proteasome. J Biol Chem 271:3112–3118

Deng XW, Dubiel W, Wei N, Hofmann K, Mundt K, Colicelli J, Kato JY, Naumann M, Segal D, Seeger M, Carr A, Glickman MH, Chamovitz DA (2000) Unified nomenclature for the COP9 signalosome: an essential regulator of development. Trends Genet Sci 16:202–203

Deveraux Q, Jensen C, Rechsteiner M (1995) Molecular cloning and expression of a 26S proteasome subunit enriched in dileucine repeats. J Biol Chem 270:23726–23729

Dick TP, Nussbaum AK, Deeg M, Heinemeyer W, Groll M, Schirle M, Keilholz W, Stefanovic S, Wolf DH, Huber R, Rammenesee HG, Schild H (1998) Contribution of proteasomal beta-subunits to the cleavage of peptide substrates analyzed with yeast mutants. J Biol Chem 273:25637–25646

Dubiel W, Ferrell K, Rechsteiner M (1995) Subunits of the regulatory complex of the 26S proteasome. Mol Biol Rep 21:27–34

Dubiel W, Pratt G, Ferrell K, Rechsteiner M (1992) Purification of an 11S regulator of the multicatalytic protease. J Biol Chem 267:22369–22377

Emmerich NP, Nussbaum AK, Stevanovic S, Priemer M, Toes RE, Rammensee HG, Schild H (2000) The human 26S and 20S proteasomes generate overlapping but different sets of peptide fragments from a model protein substrates. J Biol Chem 275:21140–21148

Enenkel C, Lehmann A, Kloetzel PM (1998) Subcellular distribution of proteasomes implicates a major location of protein degradation in the nuclear envelope-ER network in yeast. EMBO J 17:6144–6154

Ferrell K, Wilkinson CR, Dubiel W, Gordon C (2000) Regulatory subunit interactions of the 26S proteasome, a complex problem. Trends Biochem Sci 25:83–88

Finley D, Tanaka K, Mann C, Feldmann H, Hochstrasser M, Vierstra R, Johnston S, Hampton R, Haber J, Silver P, Frontali L, Thorsness P, Varshavsky A, Byers B, Wolf D, Jentsch S, Sommer T, Baumeister W, Goldberg A, Fried V, Rubin D, Glickman M, Toh-e A (1998) Unified nomenclature for subunits of the S. cerevisiae proteasome regulatory particle. Trends Biochem Sci 271:244–245

Foss GS, Prydz H (1999) Interferon regulatory factor 1 mediates the interferon-gamma induction of the human immunoproteasome subunit MECL-1. J Biol Chem 274:35196–35202

Freilich S, Oron E, Orgad S, Segal D, Chamovitz DA (1999) The COP9 complex is essential for development of *Drosophila melanogaster*. Curr Biol 9:1187–1190

Fu H, Sadis S, Rubin DM, Glickman MH, van Nocker S, Finley D, Vierstra RD (1998) Multiubiquitin chain binding and protein degradation are mediated by distinct domains within the 26S proteasome subunit Mcb1. J Biol Chem 273:1970–1989

Fujimuro M, Takada H, Saeki Y, Toh-e A, Tanaka K, Yokosawa H (1998a) Growth-dependent change of the 26S proteasome in budding yeast. Biochem Biophys Res Commun 251:818–823

Fujimuro M, Tanaka K, Yokosawa H, Toh-e A (1998b) Son1p is a component of the 26S proteasome of the yeast *Sacharomyces cerevisiae*. FEBS Lett 423:149–154

Gaczynska M, Rock KL, Goldberg AL (1993) Gamma-interferon and expression of MHC genes regulate peptide hydrolysis by proteasomes. Nature 365:264–267

Girod PA, Fu H, Zryd JP, Vierstra RD (1999) Multiubiquitin chain binding subunit MCB1 (RPN10) of the 26S proteasome is essential for developmental progression in *Physcomitrella patens*. The Plant Cell 11:1457–1471

Glickman MH (2000) Getting in and out of the proteasome. Semin Cell Dev Biol 11:149–158

Glickman MH, Rubin DM, Coux O, Wefes I, Pfeifer G, Cjeka Z, Baumeister W, Fried VA, Finley D (1998a) A subcomplex of the proteasome regulatory particle required for ubiquitin-conjugate degradation and related to the COP9/Signalosome and eIF3. Cell 94:615–623

Glickman MH, Rubin DM, Fried VA, Finley D (1998b) The regulatory particle of the *S. cerevisiae* proteasome. Mol Cell Biol 18:3149–3162

Gorbea C, Taillander D, Rechsteiner R (2000) Mapping subunit contacts in the regulatory complex of the 26S proteasome. J Biol Chem 275:875–882

Gottesman S, Maurizi MR, Wickner S (1997a) Regulatory subunits of energy-dependent proteases. Cell 91(4):435–438

Gottesman S, Wickner S, Maurizi MR (1997b) Protein quality control: triage by chaperones and proteases. Genes Dev 11:815–823

Groll M, Bajorek M, Koehler A, Moroder L, Rubin D, Huber R, Glickman M, Finley D (2000) A gated channel into the core particle of the proteasome. Nature Struct Biol 7:1062–1067

Groll M, Ditzel L, Loewe J, Stock D, Bochtler M, Bartunik HD, Huber R (1997) Structure of 20S proteasome from yeast at a 2.4 Angstrom resolution. Nature 386:463–471

Haracska L, Udvardy A (1997) Mapping the ubiquitin-binding domains in the p54 regulatory complex subunit of the Drosophila 26S protease. FEBS Lett 412:331–336

Hendil KB, Khan S, Tanaka K (1998) Simultaneous binding of PA28 and PA700 activators to 20S proteasomes. Biochem J 332:749–754

Hershko A, Ciechanover A (1998) The ubiquitin system. Annu Rev Biochem 67:425–479

Hershko A, Leshinsky E, Ganoth D, Heller H (1984) ATP-dependent degradation of ubiquitin-protein conjugates. Proc Natl Acad Sci USA 81:1619–1623

Heusch M, Lin L, Geleziunas R, Greene WC (1999) The generation of nfkb2 p52: mechanism and efficiency. Oncogene 18:6201–6208

Higashitsuji H, Itoh K, Nagao T, Dawson S, Nonoguchi K, Kido T, Mayer RJ, Arii S, Fujita J (2000) Reduced stability of retinoblastoma protein by gankyrin, an oncogenic ankyrin-repeat protein overexpressed in hepatomas. Nature Med 6:96–99

Hiyama H, Yokoi M, Masutani C, Sugasawa K, Maekawa T, Tanaka K, Hoeijmaker JH, Hanaoka F (1999) Interaction of hHR23 with s5a. J Biol Chem 274:28019–28025

Hoelzl H, Kapelari B, Kellermann J, Seemuller E, Sumegi M, Udvardy A, Medalia O, Sperling J, Muller S, Engel A, Baumeister W (2000) The regulatory complex of *Drosophila melanogaster* 26S proteasomes. Subunit composition and localization of a deubiquitylating enzyme. J Cell Biol 150: 119–130

Hoffman L, Rechsteiner M (1994) Activation of the multicatalytic protease. J Biol Chem 269:16890–16895

Hofmann K, Bucher P (1998) The PCI domain: a common theme in three multi-protein complexes. Trends Biochem Sci 23:204–205

Hofmann RM, Pickart CM (1999) Noncanonical MMS2-encoded ubiquitin-conjugating enzyme functions in assembly of novel polyubiquitin chains for DNA repair. Cell 96:645–653

Hofmann K, Falquet L (2001) A ubiquitin-interacting motif conserved in components of the proteasomal and lysosomal degradation systems. Trends Biochem Sci 26(6):347–350

Holzhutter HG, Froemmel C, Kloetzel PM (1999) A theoretical approach towards the identification of cleavage-determining amino acid motifs of the 20S proteasome. J Mol Biol 286:1251–1265

Hoppe T, Matuschewski K, Rape M, Schlenker S, Ulrich HD, Jentsch S (2000) Activation of a membrane-bound transcription factor by regulated ubiquitin/proteasome-dependent processing. Cell 102:577–586

Hori T, Kato S, Saeki M, DeMartino GN, Slaughter CA, Takeuchi J, Toh-e A, Tanaka K (1998) cDNA cloning and functional analysis of p28 (Nas6p) and p40.5 (Nas7p), two novel regulatory subunits of the 26S proteasome. Gene 216:113–122

Horwich AL, Weber-Ban EU, Finley D (1999) Chaperone rings in protein folding and degradation. Proc Natl Acad Sci USA 96:11033–11040

Hough R, Pratt G, Rechsteiner M (1986)Ubiquitin-lysozyme conjugates. J Biol Chem 261:2400–2408

Kaiser P, Flick K, Wittenberg C, Reed S (2000) Regulation of transcription by ubiquitination without proteolysis: Cdc34/SCF(Met30)-mediated inactivation of the transcription factor Met4. Cell 102: 303–314

Kaiser P, Moncollin V, Clarke DJ, Watson MH, Bertolaet BL, Reed SI, Bailly E (1999) Cyclin-dependent kinase and Cks/Suc1 interact with the proteasome in yeast to control proteolysis of M-phase targets. Genes Dev 13:1190–202

Kapelari B, Bech-Otschir D, Hegerl R, Schade R, Dumdey R, Dubiel W (2000) Electron microscopy and subunit-subunit interaction studies reveal a first architecture of COP9 signalosome. J Mol Biol 300:1169–1178

Karniol B, Yahalom T, Kwok S, Tsuge T, Matsui M, Deng XW, Chamovitz DA (1998) The *Arabidopsis* homologue of an eIF3 complex subunit associates with the COP9 complex. FEBS Lett 439:173–179

Kisselev AF, Akopian TN, Goldberg AL (1998) Range of sizes of peptide products generated during degradation of different proteins by Archaeal proteasomes. J Biol Chem 273:1982–1989

Kisselev AF, Akopian TN, Woo KM, Goldberg AL (1999) The size of peptides generated from protein by mammalian 26S and 20S proteasomes. J Biol Chem 274:3363–3371

Kleijnen MF, Shiih AH, Zhou P, Kumar S, Soccio RE, Kedersha NL, Gill G, Howley PM (2000) The hPLIC proteins may provide a link between the Ubiquitination machinery and the proteasome. Mol Cell 6:409–419

Knipfer N, Shrader TE (1997) Inactivation of the 20S poteasome in *Mycobacterium smegmatis*. Mol Microbiol 25:375–383

Kornitzer D, Ciechanover A (2000) Modes of regulation of ubiquitin-mediated protein degradation. J Cell Physiol 182:1–11

Kuttler C, Nussbaum AK, Dick TP, Rammensee HG, Schild H, Hadeler KP (2000) An algorithm for the prediction of proteasomal cleavages. J Mol Biol 298:417–29

Kwok SF, Staub JM, Deng XW (1999) Characterisation of two subunits of *Arabidopsis* 19S proteasome regulatory complex and its possible interaction with the COP9 complex. J Mol Biol 285:85–95

Lam YA, DeMartino GN, Pickart CM, Cohen RE (1997a) Specificity of the ubiquitin isopeptidase in the PA700 regulatory complex of the 26S proteasome. J Biol Chem 272:28438–28446

Lam YA, Xu W, DeMartino GN, Cohen RE (1997b) Editing of ubiquitin conjugates by an isopeptidase in the 26S proteasome. Nature 385:737–740

Lambertson D, Chen L, Madura K (1999) Pleiotropic defects caused by loss of the proteasomal-interacting factors Rad23 and Rpn10 of *S. cerevisiae*. Genetics 153:69–79

Layfield R, Franklin K, Landon M, Walker G, Wang P, Ramage R, Brown A, Love S, Urquhart K, Muir T, Baker R, Mayer RJ (1999) Chemically synthesized ubiquitin extension proteins detect distinct catalytic capacities of deubiquitinating enzymes. Anal Biochem 274:40–49

Lee DH, Sherman MY, Goldberg AL (1996) Involvement of the molecular chaperone Ydj1 in the ubiquitin-dependent degradation of short-lived abnormal proteins in *Saccharomyces cerevisiae*. Mol Cell Biol 16:4773–4781

Li S, Liu X, Ascoli M (2000) p38^{JAB1} binds to the Intracellular Precursor of the Lutropin/Choriogonadotropin Receptor and promotes Its Degradation. J Biol Chem 275:13386–13393

Li T, Naqvi NI, Yang H, Teo TS (2000) Identification of a 26S proteasome-associated UCH in fission yeast. Biochem Biophys Res Commun 272:270–275

Loewe J, Stock D, Jap B, Zwickl P, Baumeister W, Huber R (1995) Crystal structure of the 20S proteasome from the archeon *T. acidophilum* at 3.4 Angstrom resolution. Science 268:533–539

Lueders J, Demand J, Hoehfeld J (2000) The ubiquitin-related BAG-1 provides a link between the molecular chaperones Hsc70/Hsp70 and the proteasome. J Biol Chem 275:4613–4617

Lupas A, Baumeister W, Hofmann K (1997a) A repetitive sequence in subunits of the 26S proteasome and 20S cyclosome (APC). Trends Biochem Sci 22:195–196

Lupas A, Flanagan JM, Tamura T, Baumeister W (1997b) Self compartmentalizing proteases. Trends Biochem Sci 22:399–404

Ma CP, Slaughter CA, DeMartino GN (1992) Identification, purification, and characterization of a protein activator (PA28) of the 20S proteasome. J Biol Chem 267:10515–10523

Mamane Y, Heylbroeck C, Genin P, Algarte M, Servant MJ, LePage C, DeLuca C, Kwon H, Lin R, Hiscott J (1999) IRFs: the next generation. Gene 237:1–14

Maniatis T (1999) A ubiquitin ligase complex essential for the NF-κB, Wnt/Wingless, and Hedgehog signaling pathways. Genes and Dev 13:505–514

Mannhaupt G, Schnall R, Karpov V, Vetter I, Feldmann H (1999) Rpn4 acts as a transcription factor by binding to PACE, a nonamer box found upstream of 26S proteasomal and other genes in yeast. FEBS Lett 450:27–34

Maupin-Furlow JA, Wilson HL, Kaczowka SJ, Ou MS (2000) Proteasomes in the archaea: from structure to function. Front Biosci 5:D837–D865

Michalek MT, Grant EP, Gramm C, Goldberg AL, Rock KL (1993) A role for the ubiquitin-dependent proteolytic pathway in MHC class 1 restricted antigen presentation. Nature 363:552–554

Mundt KE, Porte J, Murray JM, Brikos C, Christensen PU, Caspari T, Hagan IM, Millar JB, Simanis V, Hofmann K, Carr AM (1999) The COP9/signalosome complex is conserved in fission yeast and has a role in S phase. Curr Biol 9:1427–1430

Murakami Y, Matsufuji S, Hayashi SI, Tanahashi N, Tanaka K (2000) Degradation of ODC by the 26S proteasome. Biochem Biophys Res Commun 267:1–6

Nakagawa K, Yokasawa H (2000) Degradation of transcription factor IRF-1 by the ubiquitn-proteasome pathway. The C-terminal region governs protein stability. Eur J Biochem 267:1680–1686

Naumann M, Bech-Otschir D, Huang X, Ferrell K, Dubiel W (1999) COP9 signalosome-directed c-Jun activation/stabilization is independent of JNK. J Biol Chem 274:35297–35300

Nussbaum AK, Dick TP, Keilholz W, Schirle M, Stevanovic S, Dietz K, Heinemeyer W, Groll M, Wolf DH, Huber R, Rammensee HG, Schild H (1998) Cleavage motifs of the yeast 20S proteasome beta-subunits deduced from digests of enolase 1. Proc Natl Acad Sci USA 95:12504–12509

Ohba M (1997) Modulation of intracellular protein degradation by SSB1-SIS1 chaperon system in yeast S. cerevisiae. FEBS Lett 409:307–311

Orian A, Schwartz AL, Israel A, Whiteside S, Kahana C, Ciechanover A (1999) Structural motifs involved in ubiquitin-mediated processing of the NF-κB precursor p105: roles of the glycine-rich region and a downstream ubiquitination domain. Mol Cell Biol 19:3664–3673

Osmulski PA, Gaczyncka M (1998) A new large proteolytic complex distinct from the proteasome is present in the cytosol of fission yeast. Curr Biol 8:1023–1026

Palombella VJ, Rando OJ, Goldberg AL, Maniatis T (1994) The ubiquitin proteasome pathway is required for processing the NF-κB1 precursor protein and activation of NF-κB. Cell 78:773–785

Papa FR, Amerik AY, Hochstrasser M (1999) Interaction of the Doa4 deubiquitinating enzyme with the yeast 26S proteasome. Mol Biol Cell 10:741–756

Papa FR, Hochstrasser M (1993) The yeast *DOA4* gene encodes a deubiquitinating enzyme related to a product of the tre-2 oncogene. Nature 366:313–319

Patel S, Latterich M (1998) The AAA team: related ATPases with diverse functions. Trends Cell Biol 8:65–71

Peters JM, Franke WW, Kleinschmidt JA (1994) Distinct 19S and 20S subcomplexes of the proteasome and their distribution in the nucleus and the cytoplasm. J Biol Chem 269:7709–7718

Pickart C (1997) Targeting of substrates to the 26S proteasome. FASEB J 11:1055–1066

Piotrowski J, Beal R, Hoffman L, Wilkinson KD, Cohen RE, Pickart C (1997) Inhibition of the 26S proteasome by polyubiquitin chains synthesized to have defined lengths. J Biol Chem 272:23712–23721

Ramos PC, Hockendorff J, Johnson ES, Varshavsky A, Dohmen RJ (1998) Ump1p is required for proper maturation of the 20S proteasome and becomes its substrate upon completion of the assembly. Cell 92:489–499

Realini C, Jensen CC, Zhang Z, Johnsto SC, Knowlton JR, Hill CP, Rechsteiner M (1997) Characterization of recombinant REGα, REGβ, and REGγ proteasome activators. J Biol Chem 272:25483–25492

Rechsteiner M, Hoffman L, Dubiel W (1993) The multicatalytic and 26S proteases. J Biol Chem 268:6065–6068

Rechsteiner M, Realini C, Ustrell V (2000) The proteasome activator 11S REG (PA28) and class I antigen presentation. Biochem J 345:1–15

Rohrwild M, Pfeifer G, Santarius U, Mueller SA, Hauang HC, Engel A, Baumeister W, Goldberg AL (1997) The ATP-dependent HslV/U protease from *Escherichia coli* is a four-ring structure resembling the proteasome. Nature Struct Biol 4:133–139

Rubin DM, Glickman MH, Larsen CN, Dhruvakumar S, Finley D (1998) Active site mutants in the six regulatory particle ATPases reveal multiple roles for ATP in the proteasome. EMBO J 17:4909–4919

Ruepp A, Eckerskorn C, Bogyo M, Baumeister W (1998) Proteasome function is dispensable under normal but not under heat shock conditions in *Thermoplasma acidophilum*. FEBS Lett 425:87–90

Russell SJ, Reed SH, Huang W, Friedberg EC, Johnston SJ (1999)The 19S regulatory complex of the proteasome functions independently of proteolysis in nucleotide excision repair. Mol Cell 3:687–695

Russell SJ, Steger KA, Johnston SA (1999) Subcellular localization stochiometry and protein levels of 26S proteasome subunits in yeast. J Biol Chem 274:21943–21952

Saeki Y, Toh-e A, Yokosawa H (2000) Rapid isolation and characterization of the yeast proteasome regulatory complex. Biochem Biophys Res Commun 273:509–15

Schauber C, Chen L, Tongaonakar P, Vega I, Lambertson D, Potts W, Madura K (1998) Rad23 links DNA repair to the ubiquitin/proteasome pathway. Nature 391:715–718

Schmidtke G, Emch S, Groettrup M, Holzhutter HG (2000) Evidence for the existence of a non-catalytic modifier site of peptide hydrolysis by the 20S proteasome. J Biol Chem 275:22056–22063

Sears C, Olesen J, Rubin DM, Finley D, Maniatis T (1998) NF-κB p105 processing via the ubiquitin-proteasome pathway. J Biol Chem 273:1409–1419

Seeger M, Ferrell K, Frank R, Dubiel W (1997) HIV-1 Tat inhibits the 20S proteasome and its 11S regulator-mediated activation. J Biol Chem 272:8145–8148

Seeger M, Kraft R, Ferrel K, Bech-Otschir D, Dumdey R, Schade R, Gordon C, Naumann M, Dubiel W (1998) A novel protein complex involved in signal transduction possessing similarities to the 26S proteasome subunits. FASEB J 12:469–478

Serino G, Tsuge T, Kwok SF, Matsui M, Wei N, Deng XW (1999) Arabidopsis cop8 and fus4 mutations define the same locus that encodes subunit 4 of the COP9 signalosome. Plant Cell 11:1967–1980

Silva CL, Portaro FC, Bonato VL, deCamargo AC, Ferro ES (1999) Thimet oligopeptidase (EC 3.4.24.15), a novel protein on the route of MHC class I antigen presentation. Biochem Biophys Res Commun 255:591–595

Spataro V, Norbury C, Harris AL (1998) The ubiquitin-proteasome pathway in cancer. Br J Cancer 77:448–455

Spence J, Gali RR, Dittmar G, Sherman F, Karin M, Finley D (2000)Cell cycle-regulated modification of the ribosome by a variant multiubiquitin chain. Cell 102:67–76

Spence J, Sadis S, Haas AL, Finley D (1995) A ubiquitin mutant with specific defects in DNA repair and multiubiquitination. Mol Cell Biol 15:1265–1273

Stenoien DL, Cummings CJ, Adams HP, Manicini MG, Patel K, DeMartino GN, Marcelli M, Weigel NL, Mancini MA (1999) Polyglutamine-expanded androgen receptors form aggregates that sequester heat shock proteins, proteasome components and SRC-1, and are suppressed by the HDJ-2 chaperone. Hum Mol Genet 8:731–741

Stohwasser R, Salzmann U, Ruppert T, Kloetzel PM, Holzhuetter HG (2000) Kinetic evidences for facilitation of peptide channeling by the proteasomal activator PA28. Eur J Biochem 267:6221–6230

Stoltze L, Nussbaum AK, Sijts A, Emmerich NP, Kloetzel PM, Schild H (2000) The function of the proteasome system in MHC class I antigen processing. Immunol Today 21:317–319

Strous GJ, Govers R (1999) The ubiquitin-proteasome system and endocytosis. J Cell Sci 112:1417–1423

Swaminathan S, Amerik AY, Hochstrasser M (1999) The Doa4 deubiquitinating enzyme is required for ubiquitin homeostasis in yeast. Mol Biol Cell 10:2583–2594

Szpikowska BK, Swiderek KM, Sherman MA, Mas MT (1998) MgATP binding to the nucleotide-binding domains of the eukaryotic cytoplasmic chaperonin induces conformational changes in the putative substrate binding domains. Prot Sci 7:1524–1530

Tamura N, Lottspeich F, Baumeister W, Tamura T (1998) The role of tricorn protease and its amino-peptidase-interacting factors in cellular protein degradation. Cell 95:637–648

Tanahashi N, Murakami Y, Minami Y, Shimbara N, Hendil KB, Tanaka K (2000) Hybrid proteasomes: induction by interferon-gamma and contribution to ATP-dependent proteolysis. J Biol Chem 275:14336–14345

Thrower JS, Hoffman L, Rechsteiner M, Pickart C (2000) Recognition of the polyubiquitin proteolytic signal. EMBO J 19:94–102

Tomoda K, Kubota Y, Kato J (1999) Degradation of the cyclin dependent kinaseinhibitor p27/Kip1 is instigated by Jab1. Nature 398:160–165

Tone Y, Tanahashi N, Tanaka K, Fujimoro M, Yokosawa H, Toh-e A (2000) Nob1, a new essential protein associates with the 26S proteasome of growing *S. cerevisiae* cells. Gene 243:37–45

Tongaonkar P, Chen L, Lambertson D, Ko B, Madura K (2000) Evidence for an interaction between ubiquitin-conjugating enzymes and the 26S proteasome. Mol Cell Biol 2000:4691–4698

Treier M, Staszewski LM, Bohmann D (1994) Ubiquitin-dependent c-Jun degradation in vivo is mediated by the delta domain. Cell 78:787–798

van Nocker S, Sadis S, Rubin DM, Glickman MH, Fu H, Coux O, Wefes I, Finley D, Vierstra RD (1996) The multiubiquitin chain binding protein Mcb1 is a component of the 26S proteasome in *S. cerevisiae* and plays a nonessential, substrate-specific role in protein turnover. Mol Cell Biol 11:6020–6028

Voges D, Zwickl P, Baumeister W (1999) The 26S proteasome: a molecular machine designed for controlled proteolysis. Annu Rev Biochem 68:1015–1068

Wang EW, Kessler BM, Borodovsky A, Cravatt BF, Bogyo M, Ploegh HL, Glas R (2000) Integration of the ubiquitin-proteasome pathway with a cytosolic oligopeptidase activity. Proc Natl Acad Sci USA 97:9990–9995

Wang G, Wang B, Jiang J (1999) Protein kinase A antagonises Hedgehog signaling by regulating both the activator and repressor forms of Cubitus interruptus. Genes Dev 13:2828–1960

Watanabe TK, Saito A, Suzuki M, Fujiwara T, Takahashi E, Slaughter CA, DeMartino GN, Hendil KB, Chung CH, Tanahashi N, Tanaka K (1998) cDNA cloning and characterization of a human proteasomal modulator subunit p27. Genomics 50:241–250

Weber-Ban EU, Reid BG, Miranker AD, Horwich AL (1999) Global unfolding of a substrate protein by the Hsp100 chaperone ClpA. Nature 401:90–93

Wei N, Tsuge T, Serino G, Dohmae N, Takio K, Matsui M, Deng XW (1998)The COP9 complex is conserved between plants and mammals and is related to the 26S proteasome regulatory complex. Curr Biol 8:919–922

Wigley WC, Fabunmi RP, Lee MG, Marino CR, Muallem S, DeMartino GN, Thomas PJ (1999) Dynamic association of proteasomal machinery with the centrosome. J Cell Biol 145:481–490

Wilkinson CRM, Ferrell K, Penny M, Wallace M, Dubiel W, Gordon C (2000) Analysis of a gene encoding Rpn10 of the fission yeast proteasome reveals that the polyubiquitin-binding site of this subunit is essential when Rpn12/Mts3 activity is compromised. J Biol Chem 275:15182–15192

Wilkinson CR, Seeger M, Hartmann-Petersen R, Stone M, Wallace M, Semple C, Gordon C (2001) Proteins containing the UBA domain are able to bind to multi-ubiquitin chains. Nature Cell Biol 3:939–943

Wilkinson KD (2000)Ubiquitination and deubiquitination: Targeting of proteins for degradation by the proteasome. Semin Cell Dev Biol 11:141–148

Wilson HL, Ou MS, Aldrich HC, Maupin-Furlow J (2000) Biochemical and physical properties of the *M. jannaschii* 20S proteasome and PAN, a homolog of the ATPase (Rpt) subunits of the eucaryal 26S proteasome. J Bacteriol 186:1680–1692

Wolf S, Nagy I, Lupas A, Pfeifer G, Cejka Z, Mueller SA, Engel A, De Mot R, Baumeister W (1998) Characterization of ARC, a divergent member of the AAA ATPase family from *Rhodococcus erythropolis*. J Mol Biol 277:13–25

Xie Y, Varshavsky A (2000) Physical association of ubiquitin ligases and the 26S proteasome. Proc Natl Acad Sci USA 97:2497–2502

Yao T, Cohen RE (1999) Giant proteases: beyond the proteasome. Curr Biol 9:R551–R553

Yao Y, Huang L, Krutchinsky A, Wong ML, Standing KG, Burlingame AL, Wang CC (1999) Structural and functional characterizations of the proteasome-activating protein PA26 from *Trypanosoma brucei*. J Biol Chem 274:33921–33930

Young P, Deveraux Q, Beal RE, Pickart CM, Rechsteiner M (1998)Characterization of two polyubiquitin binding sites in the 26S protease subunit 5a. J Biol Chem 273:5461–5467

Zwickl P, Baumeister W, Steven A (2000) Dis-assembly lines: the proteasome and related ATPase-assisted proteases. Curr Opin Struct Biol 10:242–250

Zwickl P, Ng D, Woo KM, Klenk HP, Goldberg AL (1999) An archaebacterial ATPase, homologous to ATPases in the eukaryotic 26S proteasome, activates protein breakdown by 20S proteasomes. J Biol Chem 274:26008–26014

The 11S Regulators of 20S Proteasome Activity

C.P. HILL, E.I. MASTERS, and F.G. WHITBY

1 Introduction	73
2 Biochemical Properties	74
3 Oligomeric State	75
4 Three-Dimensional Structure and Functional Domains	78
5 Structural Basis for Proteasome Binding and Activation	79
6 Allosteric Communication Between 11S Regulators and 20S Proteasome Active Sites?	82
7 Speculation on the Biological Roles of 11S Regulators	83
8 Summary	85
References	86

1 Introduction

The 20S proteasome, an approximately 700-kDa complex of 28 protein subunits, performs most of the proteolysis that takes place in the cytosol and nucleus of eukaryotic cells. As discussed elsewhere in this volume, structural studies on the 20S proteasome revealed that the active sites are sequestered within a central catalytic chamber that substrates access by passing through a narrow opening (α annulus) (GROLL et al. 1997; LÖWE et al. 1995; WENZEL and BAUMEISTER 1995). This architecture explains how the 20S proteasome, which is an abundant and nonspecific protease, avoids unregulated degradation of inappropriate substrates. Because the α annulus is normally closed by N-terminal sequences of the 20S proteasome α subunits, this architecture also appears to restrict the release of degradation products from the proteasome interior.

Legitimate substrates are delivered to 20S proteasomes by activating complexes that bind to the proteasome α subunits and open the entrance gate. The best known of these activators is the 19S ATPase regulatory complex, which, as discussed elsewhere in this volume, binds 20S proteasome to form the 26S proteasome

Biochemistry Department, University of Utah Medical School, 50 N Medical Drive, Salt Lake City, UT 84132, USA

that is responsible for the degradation of polyubiquitinated proteins. In this chapter we will address a different family of activators, known as 11S regulators, that stimulate the degradation of peptides in biochemical assays by opening a channel between the buried active sites and the exterior of the proteasome. (For recent reviews see DeMartino and Slaughter 1999; Hoffman and Rechsteiner 1996; Kloetzel et al. 1999; Kuehn and Dahlmann 1997; Rechsteiner et al. 2000.) Unlike the 19S regulators, 11S regulators do not utilize ATP, do not promote the degradation of proteins, and do not recognize polyubiquitin. Because of these biochemical properties, it has been suggested that 11S regulators do not deliver substrate but instead function to facilitate the egress of product peptides from the interior of the 20S proteasome.

The presentation and emphasis of this chapter reflect our interests and prejudices as structural biologists. We begin by describing the biochemical properties of 11S regulators and the functional domains identified within their amino acid sequences. Discussion of oligomeric state will be followed by review of X-ray crystal structures and discussion of the possibility of allosteric communication between 11S regulators and the catalytic sites of the 20S proteasome. The final topic addressed will be the possible biological roles of 11S regulators, including mechanisms by which some family members may function in the production of peptides for display on major histocompatibility (MHC) class I molecules.

2 Biochemical Properties

11S regulators were first identified in bovine (Ma et al. 1992) and human (Dubiel et al. 1992) red blood cells as a result of their ability to stimulate the degradation of small fluorogenic peptides by 20S proteasome. The activity did not require ATP, and proteins were not degraded, even when unfolded or ubiquitinated (Dubiel et al. 1992; Ma et al. 1992; Shimbara et al. 1997; Tanahashi et al. 1997). It has recently been reported that ATP is required for association of 11S regulator with a complex of 20S proteasome and 19S regulator (Tanahashi et al. 2000), although this observation may result from the ATP-dependence of proteasome binding by the 19S regulator.

11S regulators are called either PA28 (*p*roteasome *a*ctivator of apparent subunit molecular weight *28*kDa) (Ma et al. 1992) or REG (11S *REG*ulator) (Dubiel et al. 1992; Realini et al. 1997). The homolog from the trypanosome *Trypanosoma brucei*, which has a slightly smaller subunit mass, is called PA26 (To and Wang 1997; Yao et al. 1999). The three 11S regulators identified in mammals have a broad tissue distribution (Jiang and Monaco 1997; Kuehn and Dahlmann 1996a; Ma et al. 1993; Soza et al. 1997) and are referred to as the α, β, and γ homologs, such that PA28α = REGα, PA28β = REGβ, and PA28γ = REGγ. The γ homolog was originally identified as the nuclear Ki antigen that is associated with systemic lupus erythematosus (Nikaido et al. 1990; Tojo et al. 1981) and later shown to have similar biochemical properties to the α and β homologs (REALINI et al. 1997; TANAHASHI et al. 1997). The α and β homologs share about 50% sequence identity with each other and about 30%–40% sequence identity with the γ protein (AHN et al. 1995). Simpler species, such as tick (PAESEN and NUTTALL 1996) and *Caenorhabditis elegans* (KANDIL et al. 1997), generally encode just one 11S regulator, which is usually most closely related to the γ homolog. It seems that all of the 11S regulators have similar biochemical properties: they bind to 20S proteasomes and activate peptidase activity but they do not utilize ATP or promote the degradation of proteins. Some studies found that the isolated β homolog failed to activate 20S proteasome (KUEHN and DAHLMANN 1996b; SONG et al. 1997). These misleading observations may have resulted from the relatively low concentration of regulator used in those studies and the inherently low heptamerization affinity for the β homolog (REALINI et al. 1997; ZHANG et al. 1998b). Remarkably, it appears that yeast do not contain an 11S regulator.

It has been reported that the 11S regulator purified from reticulocytes and erythrocytes is phosphorylated, and that dephosphorylation with alkaline phosphatase abolishes the ability to activate 20S proteasome peptidase activity (LI et al. 1996). This does not seem consistent, however, with the observation that bacterially expressed recombinant 11S regulators, which are not phosphorylated, have activities similar to those of the purified endogenous regulator (REALINI et al. 1997; SONG et al. 1997). It has also been reported that the rat 11S regulator α homolog binds to the B-Raf protein kinase, although B-raf does not appear to phosphorylate the regulator (KALMES et al. 1998). The significance of these observations is unclear.

Consistent with their copurification from tissues (MOTT et al. 1994), coimmunoprecipitation and Western blotting have demonstrated that the α and β 11S regulator homologs form hetero-oligomeric complexes (AHN et al. 1996; SONG et al. 1996). These hetero-oligomeric α/β complexes are more effective activators of the 20S proteasome than either of the isolated α or β homologs (KUEHN and DAHLMANN 1996b; REALINI et al. 1997; SONG et al. 1997). Mutagenesis studies have further demonstrated that both α and β subunits make essential contributions to the biochemical activity of the hetero-oligomeric complexes (ZHANG et al. 1998b). The importance of the α/β hetero-oligomer is underscored by the observation that both α and β proteins are undetectable in mice for which the β gene has been deleted (PRECKEL et al. 1999). In contrast to the hetero-oligomeric state of the α/β complex, the γ homolog and other 11S regulators are homo-oligomeric (REALINI et al. 1997; TANAHASHI et al. 1997; YAO et al. 1999).

3 Oligomeric State

The literature regarding the oligomeric state of 11S regulators is rather confusing. The true oligomeric state is almost certainly heptameric, although the early belief that 11S regulators are hexamers still persists. Thus, most recent publications treat this as a controversial or ambiguous issue by use of phrases such as 'hexameric or heptameric complexes'. This subject will be discussed in some detail since, in our opinion, the oligomeric state is of critical importance for understanding the mechanism of 11S regulator function and this issue should now be considered resolved in favor of the heptamer model.

The first estimates of 11S regulator oligomeric states were based upon gel filtration (MA et al. 1992) and appearance of negatively stained samples in electron micrographs (GRAY et al. 1994). These estimates, which were appropriately cautious, tended to favor the hexamer model without excluding the possibility of a heptamer. In 1996, three groups independently reported that the hetero-oligomeric α/β complex is an α3/β3 hexamer (AHN et al. 1996; KUEHN and DAHLMANN 1996b; SONG et al. 1996). Song et al. proposed the α3β3 hexamer model based on the observation of indistinguishable patterns obtained using antibodies to α or β homologs after partially cross-linked hetero-oligomers were visualized on SDS–PAGE (SONG et al. 1996). Ahn et al. based their conclusions on the observation that a population of hetero-oligomeric complexes contains an equal molar ratio of α and β homologs (AHN et al. 1996). In a similar fashion, Kuehn and Dahlmann based their conclusions on the observation that maximal activity for reconstituted hetero-oligomers is obtained when α and β subunits are mixed in equal amounts (KUEHN and DAHLMANN 1996b). An alternative interpretation is possible for all of these observations, and one of these studies acknowledged that the same observations would result from a population of heptameric complexes that contained a random distribution of α and β subunits (SONG et al. 1996). This stochastic model, which is consistent with the earlier observations and now seems to be correct (ZHANG et al. 1999), was judged at the time to be less likely than a hexamer of alternating subunits.

The initial bias against a stochastic model is understandable. Whereas the alternating α/β hexamer model predicts just one type of oligomer that contains only one type of subunit–subunit interaction, the stochastic model implies the presence of several different types of complexes and three different types of subunit interface (α–α, β–β, and α–β). There are relatively few well-defined examples of stochastically arranged complexes in biology, although one relevant example is provided by the 20S proteasome of *Rhodococcus*, which contains two distinct but related α-type and two distinct but related β-type subunits that are probably distributed randomly in the α and β rings (ZUHL et al. 1997).

The first clear evidence for the sevenfold arrangement of an 11S regulator came from equilibrium sedimentation analysis of the recombinant human α homolog (JOHNSTON et al. 1997). The crystal structure of the same protein subsequently showed the heptameric ring structure in atomic detail (KNOWLTON et al. 1997) (Fig. 1 shows the closely similar structure of PA26). Although these data were unequivocal, their relevance was questioned because these studies used the isolated recombinant α homolog, whereas the physiologically relevant form is a mixed α/β complex. Nevertheless, the similar biochemical properties of the isolated α homolog and the hetero-oligomeric α/β complex (REALINI et al. 1994a) strongly suggested that these complexes have the same basic architecture.

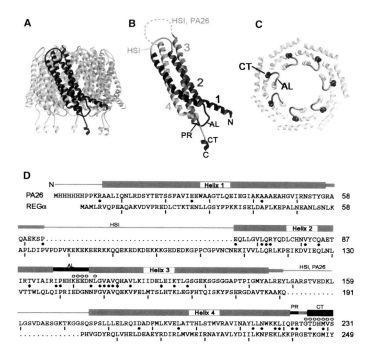

Fig. 1A–D. Structure of PA26 as seen in a complex with yeast 20S proteasome (WHITBY et al. 2000). The structure of the unliganded human α homolog is quite similar. **A** Side view of the PA26 heptamer. **B** Structure of the PA26 monomer. This corresponds to the orientation of the monomer highlighted in **A**. Helices are numbered 1–4. N and C termini are labeled N and C, respectively. The CT of PA26 forms a helix when bound to 20S proteasome, whereas the equivalent residues are disordered in the unliganded structure of the human α homolog. **C** Bottom view of the PA26 heptamer. AL and CT domains are highlighted. **D** Amino acid sequences of *T. brucei* PA26 and the human α homolog (REGα). Identical residues are indicated with a *filled circle* between the lines. Secondary structural elements and functional domains are indicated. Residues that contact the 20S proteasome are marked with an *open circle* above the PA26 sequence. (*CT*, C-terminal domain; *AL*, activation loop; *PR*, proline and arginine that contact the AL; *HSI*, homolog-specific insert of α, β, and γ homologs; *HSI, PA26*, region of PA26 that may be functionally equivalent to the HSI)

The stoichiometry of the α/β hetero-oligomer was resolved by use of mass spectrometry, which showed that α and β subunits do indeed form a stochastic heptamer (ZHANG et al. 1999). Recombinant hetero-oligomers were prepared by coexpressing the α and β homologs in *Escherichia coli*. Subsequent mass spectrometric analysis did not detect hexamers, but did reveal the formation of various α/β heptamers (α3β4, α4/β3, α5/β2, etc.) with the relative proportion of the different heptameric species dependent upon the relative coexpression levels of α and β homologs. More recently, equilibrium sedimentation, mass spectrometry, and electron microscopy have been used to show that the naturally homo-oligomeric 11S regulators, trypanosome PA26 (YAO et al. 1999) and the human γ homolog (M. Rechsteiner, personal communication), are heptamers. Very recently, crystal structure determination showed that PA26 is heptameric when bound to the 20S proteasome (WHITBY et al. 2000).

One reason why the hexamer model of 11S regulators has retained significant popularity may be due to analogy with the AAA family of ATPases. These enzymes are known to form hexameric ring structures that apparently disassemble protein complexes (VOGES et al. 1999). Family members include the six ATPases of the 19S proteasome regulator, which are therefore thought to form a six-membered ring that binds against the seven α subunits of the 20S proteasome. Furthermore, a 6:7 symmetry mismatch has been clearly demonstrated by electron microscopic analysis of the analogous ClpAP complexes found in bacteria (BEURON et al. 1998). The different symmetries of the 19S ATPase and 11S regulators may reflect the possibility that they perform different biological functions. It is well established that the 19S ATPase presents substrates to the proteasome, whereas, as discussed later, the 11S regulators may function to facilitate the egress of products.

4 Three-Dimensional Structure and Functional Domains

The crystal structure of the human 11S regulator α homolog (KNOWLTON et al. 1997) revealed that the monomer, which is comprised of four long alpha helices, assembles into a heptamer that surrounds a central pore. The pore appears suited to serve as a conduit for the passage of peptides into or out of the 20S proteasome, but could not support passage of folded protein. This structure (Fig. 1) provides a framework for discussing the three functional domains (C-terminal residues, activation loop, and homolog-specific insert) that have been identified from mutagenesis and biochemical studies.

The C-terminal residues (CT; Fig. 1) are especially important for binding to the 20S proteasome (MA et al. 1993; SONG et al. 1996, 1997; ZHANG et al. 1998; LI 2000) although it is remarkable that the sequence is highly variable in this region. These residues project away from the body of the protein and are disordered in the crystal structure, although, as discussed later, they become ordered upon binding to the 20S proteasome (WHITBY et al. 2000). Deletion of the C-terminal tyrosine of the α homolog or substitution of this residue for a charged side chain abolished binding to 20S proteasome. The identity of this residue is not critical, however, since the tryptophan variant binds and activates 20S proteasome to the same extent as wild type, and phenylalanine and serine substitutions are also partially active (SONG et al. 1997).

An internal stretch of nine amino acids, dubbed the activation loop (AL; Fig. 1), is especially important for activation of 20S proteasomes (ZHANG et al. 1998a). These residues, which are located between helices 2 and 3, are adjacent to the C-terminal tails in the structure. The activation loops and C-terminal tails cluster in a sevenfold symmetric array around the opening of the pore on the face of the heptamer that contacts the 20S proteasome. Two other residues, a proline and an arginine (PR; Fig. 1) that are located just before the C-terminal tail, are integral to the activation loop motif, as they form tertiary interactions that appear to stabilize the activation loop conformation. Indeed, this proline was shown to be critically important for activation in the same screen that identified the activation loop (ZHANG et al. 1998a). Activation and binding are not completely coupled, as mutation of a conserved activation loop asparagine to tyrosine inactivates all of the human homologs without apparently altering their ability to bind to 20S proteasome (ZHANG et al. 1998a). Not surprisingly, the activation loop sequence is almost completely invariant in 11S regulators from mammals and most other species, although the highly divergent 11S regulator from trypanosome (PA26) is a notable exception (WHITBY et al. 2000; YAO et al. 1999).

Whereas the mammalian 11S regulator sequences show significant conservation throughout most of their length, a highly variable region, known as the homolog-specific insert (HSI; Fig. 1) is located within a disordered loop between helices 1 and 2 at the opposite end of the structure from the activation loops and C-terminal tail (ZHANG et al. 1998a). Consistent with the structure, deletion of these residues has no measurable effect on binding or activation of 20S proteasomes (SONG et al. 1997; ZHANG et al. 1998c). The homolog-specific insert of the α subunit contains a 28-residue segment that is rich in alternating lysines and glutamates. It has been proposed that this 'KEKE' motif functions as a protein interaction domain that localizes 20S proteasome activity or delivers product peptides to a preferred location (REALINI et al. 1994b; RECHSTEINER et al. 2000). The KEKE motif binds calcium and calcium also inhibits 20S proteasomes that have been stimulated by the α homolog (REALINI and RECHSTEINER 1995), although the calcium concentration used in this study was relatively high and the significance of these observations is unclear (REALINI et al. 1997). An interesting variation occurs in PA26, which has a relatively short loop between helices 1 and 2 whereas the spatially adjacent loop between helices 3 and 4 is correspondingly long and unstructured (WHITBY et al. 2000). Thus, it seems possible that the loop between helices 3 and 4 of PA26 is functionally equivalent to the loop between helices 1 and 2 of the other homologs (HSI, PA26; Fig. 1).

5 Structural Basis for Proteasome Binding and Activation

In order to understand how 20S proteasomes are activated by 11S regulators, it is first necessary to understand how the unliganded proteasome is maintained in an inactive conformation. 20S proteasomes are assemblies of 28 protein subunits that are arranged in four stacked rings, with the outer rings each comprised of seven α subunits and the two inner rings each comprised of seven β subunits (GROLL et al. 1997). The seven different proteasome α subunits ($\alpha 1-\alpha 7$) share about 30% sequence identity with each other and occupy unique positions in a precise order around the ring. The seven different β subunits each also occupy unique positions.

Thus, the 20S proteasome has a pseudo sevenfold axis of symmetry along the length of the barrel-shaped structure and one exact twofold axis of symmetry in a perpendicular direction. The active sites are sequestered in a central proteolytic chamber, formed by the β subunits, that substrates enter by passing through a narrow opening in the α subunits known as the α annulus (GROLL et al. 1997; LÖWE et al. 1995; WENZEL and BAUMEISTER 1995). In the absence of an activating complex, the α annulus is blocked by N-terminal residues of α subunits (GROLL et al. 1997). A tight seal is achieved in this closed conformation by a dramatic departure from the sevenfold symmetry of α subunit N-terminal sequences. Three of the N-terminal sequences turn away from the proteasome surface while the other four pack across the substrate access gate in an irregular conformation that is stabilized by numerous hydrogen bond and van der Waals interactions (GROLL et al. 1997) (Fig. 2).

The crystal structure of a complex between the trypanosome 11S regulator, PA26, and the yeast 20S proteasome reveals conformational changes that explain how proteasomes are activated (WHITBY et al. 2000). Although this is a noncognate complex, there are good reasons to believe that the structure is relevant: (a) the crystal structure resembles negative-stained electron micrographs of cognate bovine 20S proteasome/PA28 complexes (GRAY et al. 1994); (b) analysis by velocity

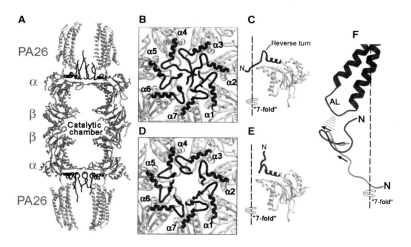

Fig. 2A–F. Structure of the PA26/20S proteasome complex. **A** PA26 and the 20S proteasome α and β subunits are indicated. The first ~35 residues of 20S proteasome α subunits are shown in a *darker* shade. **B** Top view of the 20S proteasome α subunits as seen in the unliganded 20S proteasome structure (GROLL et al. 1997). **C** Structure of the unliganded 20S proteasome α2 subunit. The α3, α4, and to a lesser extent α5 subunits, have similar conformations of N-terminal residues that block the entrance through the ring of α subunits. **D** Top view of the 20S proteasome α subunits as seen in the PA26/20S proteasome complex. **E** Structure of the 20S proteasome α2 subunit as seen in the PA26 complex. All other subunits adopt essentially identical conformations in the PA26 complex, whereas α1, α6, and α7 already adopt this conformation in the unliganded structure. **F** Mechanism of activation. Close approach of the PA26 activation loop induces a shift in the position of the α2 reverse turn (*upper arrow*). This 2.5Å displacement propagates to displace more N-terminal residues (*lower arrow*), and thereby disrupts the numerous contacts that stabilize the closed gate conformation

sedimentation demonstrated that PA26 binds to yeast 20S proteasome with a 2:1 stoichiometry in solution (L. Joss, personal communication); (c) PA26 stimulates an eightfold increase in the rate of hydrolysis of a small fluorogenic peptide by the yeast 20S proteasome (E.I. Masters, unpublished data); (d) there are many other cases of heterologous activation by 11S regulators. For example, PA26 activates 20S proteasome from rat (YAO et al. 1999) and cow (E.I. Masters, unpublished data); the human 11S α homolog activates 20S proteasome from cow (E.I. Masters, unpublished data) and yeast (M. Rechsteiner, personal communication); and cow red blood cell PA28 activates lobster 20S proteasome (MYKLES 1996).

The PA26/20S proteasome crystal structure showed that 11S regulators activate 20S proteasomes by inducing conformational changes in the N-terminal sequences of proteasome α subunits that result in opening of the channel through the α annulus. No other significant conformational changes are apparent in the 20S proteasome structure. Whereas N-terminal sequences of four of the α subunits pack to seal the entrance to the unliganded 20S proteasome, all seven of the α subunit N-terminal tails extend away from the proteasome and project into the PA26 pore in the complex. The open channel through PA26 appears suitable for the transit of model peptide substrates and small peptide products. It is conceivable that unfolded protein substrates may be threaded through the 11S regulator pore into the 20S proteasome, although the available biochemical evidence argues against this possibility (DUBIEL et al. 1992; MA et al. 1992).

The mechanism by which PA26 opens the 20S proteasome entrance/exit gate appears to have two components. These are the binding affinity provided by the PA26 C-terminal sequences and the imposition of more exact sevenfold symmetry by the PA26 activation loops.

The C-terminal residues are disordered in the structure of the unliganded human 11S α homolog (KNOWLTON et al. 1997) but become ordered in the PA26/proteasome complex where they insert into pockets formed between 20S proteasome α subunits (WHITBY et al. 2000). Most of the PA26 C-terminal tail side chains are in relatively open environments, and a major contact seems to be between the C-terminal main chain carboxylate of PA26 and main chain amides near the N-terminus of a helix in the yeast 20S proteasome. This explains why large variations in the amino-acid sequence are tolerated but the length of the C-terminal tail of 11S regulators is important (SONG et al. 1997). The heptameric assembly allows the inherently weak binding energy of one individual tail to be amplified up to seven times to give a substantial resultant binding affinity.

Binding affinity of the C-terminal tails is used to press the exactly sevenfold symmetric activation loops of PA26 against the 20S proteasome. This forces the α subunit N-terminal residues to adopt a more symmetric arrangement in which all seven α subunit N termini project away from the 20S proteasome surface into the large central pore of PA26. More specifically, residues of the activation loop apparently displace a reverse turn near the N terminus of proteasome α subunits that change conformation. This movement destabilizes the closed gate conformation by displacing the N-terminal tails and thereby disrupting numerous hydrogen bonds and van der Waals interactions that stabilize the closed gate conformation. The wide variation in amino acid sequence between PA26 and other 11S regulator activation loops indicates that although the ring of seven activation loops functions by imposing sevenfold symmetry on the N-terminal sequences of 20S proteasome α subunits, specific details of this interaction are probably not critical. Notably, however, the activation loop is almost invariant among the three 11S regulator homologs from human and other mammalian sources. This conservation may reflect structural roles that will only be apparent in a cognate regulator/proteasome complex.

6 Allosteric Communication Between 11S Regulators and 20S Proteasome Active Sites?

The available structural data strongly suggest that proteasome activation is primarily a function of regulating substrate access to the catalytic chamber. In summary: (a) all significant conformational changes observed upon binding PA26 are localized to the N-terminal residues of 20S proteasome α subunits (WHITBY et al. 2000); (b) conformational changes are not induced upon binding of small molecule inhibitors (GROLL et al. 1997; LOIDL et al. 1999; LÖWE et al. 1995); (c) a mutant yeast 20S proteasome, in which the gate structure is disrupted but otherwise has a structure identical to that of the wild-type, appears to be constitutively fully active in biochemical assays (GROLL et al. 2000).

In conflict with the structural data, a number of biochemical studies have been reported that suggest the presence of allosteric communication between binding of 11S regulator and the proteolytic active sites: (a) some studies reported that 11S regulator promotes an alternative 'dual cleavage' mechanism of catalysis by 20S proteasomes (DICK et al. 1996; SHIMBARA et al. 1997); (b) studies with short fluorogenic peptides suggested that the α and β 11S homologs activate all three 20S proteasome active sites, whereas the γ homolog appears to activate just one of the active sites (REALINI et al. 1997); (c) immunoprecipitation experiments suggested that binding of the 20S proteasome inhibitor lactacystin results in stabilization of the 20S proteasome/PA28 complex (PRECKEL et al. 1999).

It is not clear how the contradictory impressions gained from structural and biochemical studies will be resolved. There are limitations to the current level of structural analysis. For example, perhaps allosteric communication occurs for 20S proteasomes from higher organisms, whereas structural data are currently available only for 20S proteasomes from yeast and an archeon. Perhaps allosteric communication occurs only for cognate 11S regulator complexes, or perhaps the relevant conformational changes are so subtle that they are not visible at the currently available resolution. There may also be limitations with the biochemical analysis. For example, the apparent change in catalysis to a dual cleavage mechanism may in fact reflect an increased rate of access to the catalytic chamber. For example, a recent kinetic analysis concluded that recombinant mouse α/β homologs activate proteasomes by opening the gate (STOHWASSER et al. 2000). Similarly, the apparent activation of different catalytic sites may reflect the relative ease with which peptides pass through the 11S regulator pore to enter the interior of the proteasome.

The structural data clearly indicate that opening of the gate formed by N-terminal residues of the proteasome alpha subunits is a major component of activation by 11S regulators. Nevertheless, other factors may also contribute. There are reports of allosteric communication between different proteasome active sites (KISSELEV et al. 1999a) and between active sites and some other noncatalytic modifier site (SCHMIDTKE et al. 2000). Recently, Rechsteiner and colleagues have identified a point mutant in the 11S γ homolog that switches the sensitivity of different proteasome active sites towards an inhibitor to match that of the α/β proteins, apparently without affecting access of the inhibitor to the proteasome interior (M. Rechsteiner, personal communication).

7 Speculation on the Biological Roles of 11S Regulators

It is well established that the heteromeric α/β 11S regulator functions in the production of peptide ligands for MHC class I molecules. (For recent reviews see FRÜH and YANG 1999; GROETTRUP et al. 1996b; KLOETZEL et al. 1999; PAMER and CRESSWELL 1998; RECHSTEINER et al. 2000; ROCK and GOLDBERG 1999; TANAKA and KASAHARA 1998). An early indication of this role was the observation that the α and β homologs are substantially upregulated by the cytokine interferon-γ (INF-γ) (AHN et al. 1995, 1996; BROOKS et al. 2000; JIANG and MONACO 1997; KUEHN and DAHLMANN 1996b; REALINI et al. 1994a; TANAHASHI et al. 1997), which induces synthesis of several proteins involved in antigen presentation (YEWDELL and BENNINK 1992). More direct evidence was provided by the observations that overexpression of the α (GROETTRUP et al. 1996a), β, or α/β (SHIMBARA et al. 1997) homologs increased the presentation of MHC class I epitopes on cultured cells, although overexpression of α and β homologs does not guarantee improved antigen presentation (SCHWARZ et al. 2000). Upregulation of α and β homologs has been observed in HeLa cells in response to bacterial invasion (MAKSYMOWYCH et al. 1998). Up-regulation has also been observed upon maturation of specialized antigen-presenting dendritic cells (MACAGNO et al. 1999). Finally, mice with a disrupted β gene express undetectably low levels of both α and β homologs, and are impaired in the processing of antigenic epitopes (PRECKEL et al. 1999).

In contrast, the biological role of the homo-heptameric γ homolog is unclear. Whereas the α and β homologs are found in the cytosol and are induced by INF-γ, the γ homolog is predominantly nuclear (BROOKS et al. 2000; SOZA et al. 1997; WOJCIK et al. 1998) and is not induced significantly by INF-γ (AHN et al. 1995; BROOKS et al. 2000; TANAHASHI et al. 1997). Also, in contrast to the α and β homologs, mice deleted for the γ gene show a slight growth retardation but are not markedly impaired in MHC class I antigen presentation (MURATA et al. 1999).

Consistent with these observations, organisms that do not possess MHC class I molecules generally posses a single 11S regulator homolog that more closely resembles the γ rather than the α or β homologs of higher organisms. The various possible functions of 11S regulators include targeting of product peptides to the endoplasmic reticulum (REALINI et al. 1994b), localizing proteasomes within the cell (RECHSTEINER et al. 2000), promoting immunoproteasome assembly (PRECKEL et al. 1999), targeting of substrates for degradation or refolding via interaction with chaperones (MINAMI et al. 2000), and facilitating release of product peptides from the proteasome interior (WHITBY et al. 2000). Regardless of the precise biological role, the structural data strongly imply that a key function of 11S regulators is to open the gate separating the central proteolytic chamber of 20S proteasomes from the intracellular environment.

11S regulators and the 19S ATPase regulatory complex can bind simultaneously at opposite ends of the same 20S proteasome molecule (HENDIL et al. 1998; TANAHASHI et al. 2000). As the 19S ATPase actively delivers substrate to 20S proteasome, whereas 11S regulators appear to passively open the gate sealing the proteasome interior, one attractive possibility is that 11S regulators function to facilitate the release of product peptides from the proteasome. This model suggests a general role in accelerating throughput, consistent with the following observations: (a) 11S regulators appear to be concentrated at intracellular sites associated with high levels of protein turnover (FABUNMI et al. 2000; WIGLEY et al. 1999); (b) 11S regulators are upregulated under physiological conditions that increase protein turnover (ORDWAY et al. 2000); and (c) the induction of 11S regulator by INF-γ correlates with the increased rate of degradation of a proteasome substrate (TANAHASHI et al. 2000). As discussed below, the model that 11S regulators facilitate egress of product peptides suggests a mechanism by which the α/β 11S regulator may promote the production of antigenic peptides by 20S proteasomes.

MHC class I molecules bind a wide variety of peptide sequences, with the two main restrictions that the C-terminal side chain must be hydrophobic or basic and the peptide must be 9 or 10 residues in length (ENGELHARD 1994; RAMMENSEE et al. 1993; STERN and WILEY 1994). The first of these requirements is probably met by the immunoproteasome, which is discussed elsewhere in this volume. Briefly, the immunoproteasome is the same as the 20S proteasome except that the three constitutive catalytic β subunits are replaced by inducible counterparts in response to signaling by INF-γ. Whereas the 20S proteasome cuts after hydrophobic, basic, and acidic amino acid residues, the immunoproteasome is relatively inefficient at cleaving after acidic side chains (DRISCOLL et al. 1993; ELEUTERI et al. 1997; GACZYNSKA et al. 1993, 1996; USTRELL et al. 1995). Thus, immunoproteasomes preferentially produce peptides that have an appropriate C-terminal residue for binding to MHC class I molecules.

Although the proteasome/immunoproteasome seems to make the first cut in the production of ligands for MHC class I molecules, and thereby defines the C-terminal residue, the N terminus can be defined by cytosolic (BENINGA et al. 1998) or luminal (CRAIU et al. 1997; ELLIOTT et al. 1995; SNYDER et al. 1994) N-terminal endopeptidases. Thus, the role of the proteasome is to produce peptides that have an appropriate C terminus and are at least nine residues long. However, biochemical experiments indicate that proteasomes are highly processive enzymes that inflict multiple cuts upon a substrate before releasing the degradation product (AKOPIAN et al. 1997; DICK et al. 1991). Consequently, although 20S proteasomes are capable of generating MHC class I ligands (NIEDERMANN et al. 1996, 1997), this appears to be an inefficient process, and the majority of product peptides are too short for tight binding to MHC class I molecules (KISSELEV et al. 1998, 1999b).

The high processivity of 20S proteasomes presumably derives from the closed architecture that causes inhibition of product release from the proteasome interior. The conclusion that product size is defined by ease of diffusion out of the proteasome interior is supported by the distribution of product size (KISSELEV et al. 1998, 1999b), and by the observation that the product size distribution is unchanged for mutant proteasomes that have reduced numbers of active sites (DICK et al. 1998; NUSSBAUM et al. 1998). Furthermore, the importance of an open gate conformation for product egress is consistent with studies performed on the hydrolysis of different length peptides by the 20S proteasome from *Thermoplasma acidophilum* (DOLENC et al. 1998). The model that 11S regulators open a proteasome product exit port implies that they will reduce processivity and promote the release of longer (>9 residue) products. Because immunoproteasome-generated peptides will have appropriate C-terminal residues, and longer peptides can be trimmed as necessary from their N terminus, this model suggests how the α/β 11S regulator may promote the production of MHC class I ligands.

8 Summary

Although substantial progress has been made in understanding the biochemical properties of 11S regulators since their discovery in 1992, we still only have a rudimentary understanding of their biological role. As discussed above, we have proposed a model in which the α/β complex promotes the production of antigenic peptides by opening the exit port of the 20S proteasome (WHITBY et al. 2000). There are other possibilities, however, that are not exclusive of the exit port hypothesis. For example the α/β complex may promote assembly of immunoproteasome as suggested by PRECKEL et al. 1999, or it may function as a docking module and conduit for the delivery of peptides to the ER lumen (REALINI et al. 1994b).

There are also unanswered structural and mechanistic questions. Higher resolution data are needed to discern important structural details of the PA26/20S proteasome complex. The models for binding and activation that are suggested from the structural data have to be tested by mutagenesis and biochemical analysis. What is the role of homolog-specific inserts? Will cognate regulator/proteasome complexes show conformational changes that are not apparent in the currently available crystal structures, including perhaps signs of allosteric communication between the regulator and the proteasome active sites?

Acknowledgements. We thank Martin Rechsteiner and members of our laboratory for critical comments on this manuscript. Our work on 11S regulators is funded by the NIH (GM59135).

References

Ahn JY, Tanahashi N, Akiyama K, Hisamatsu H, Noda C, Tanaka K, Chung CH, Shimbara N, Willy PJ, Mott JD, Slaughter CA, DeMartino GN (1995) Primary structures of two homologous subunits of PA28, a γ-interferon-inducible protein activator of the 20S proteasome. FEBS Lett 366:37–42

Ahn K, Erlander M, Leturcq D, Peterson PA, Früh K, Yang Y (1996) In vivo characterization of the proteasome regulator PA28. J Biol Chem 271:18237–18242

Akopian TN, Kisselev AF, Goldberg AL (1997) Processive degradation of proteins and other catalytic properties of the proteasome from *Thermoplasma acidophilum*. J Biol Chem 272:1791–1798

Beninga J, Rock KL, Goldberg AL (1998) Interferon-γ can stimulate post-proteasomal trimming of the N-terminus of an antigenic peptide by inducing leucine aminopeptidase. J Biol Chem 273:18734–18742

Beuron F, Maurizi MR, Belnap DM, Kocsis E, Booy FP, Kessel M, Steven AC (1998) At sixes and sevens: characterization of the symmetry mismatch of the ClpAP chaperone-assisted protease. J Struct Biol 123:248–259

Brooks P, Fuertes G, Murray RZ, Bose S, Knecht E, Rechsteiner MC, Hendil KB, Tanaka K, Dyson J, Rivett J (2000) Subcellular localization of proteasomes and their regulatory complexes in mammalian cells. Biochem J 346:155–161

Craiu A, Akopian T, Goldberg AL, Rock KL (1997) Two distinct proteolytic processes in the generation of a major histocompatibility class I presented peptide. Proc Natl Acad Sci USA 94:10850–10854

DeMartino GN, Slaughter CA (1999) The proteasome, a novel protease regulated by multiple mechanisms. J Biol Chem 274:22123–22126

Dick LR, Moomaw CR, DeMartino GN, Slaughter CA (1991) Degradation of oxidized insulin B chain by the multiproteinase complex macropain (proteasome). Biochemistry 30:2725–2734

Dick TP, Nussbaum AK, Deeg M, Heinemeyer W, Groll M, Schirle M, Keilholz W, Stevanovi S, Wolf DH, Huber R, Rammensee HG, Schild H (1998) Contribution of proteasomal beta-subunits to the cleavage of peptide substrates analyzed with yeast mutants. J Biol Chem 273:25637–25646

Dick TP, Ruppert T, Groettrup M, Kloetzel PM, Kuehn L, Koszinowski UH, Stevanovic S, Schild H, Rammensee HG (1996) Coordinated dual cleavages induced by the proteasome regulator PA28 lead to dominant MHC ligands. Cell 86:253–262

Dolenc I, Seemüller E, Baumeister W (1998) Decelerated degradation of short peptides by the 20S proteasome. FEBS Lett 434:357–361

Driscoll J, Brown MG, Finley D, Monaco JJ (1993) MHC-linked *LMP* gene products specifically alter peptidase activities of the proteasome. Nature 365:262–264

Dubiel W, Pratt G, Ferrell K, Rechsteiner M (1992) Purification of an 11S regulator of the multicatalytic protease. J Biol Chem 267:22369–22377

Eleuteri AM, Kohanski RA, Cardozo C, Orlowski M (1997) Bovine spleen multicatalytic proteinase complex (proteasome). Replacement of X, Y, and Z subunits by LMP7, LMP2, and MECL1 and changes in properties and specificity. J Biol Chem 272:11824–11831

Elliott T, Willis A, Cerundolo V, Townsend A (1995) Processing of major histocompatibility class I-restricted antigens in the endoplasmic reticulum. J Exp Med 181:1481–1491

Engelhard VH (1994) Structure of peptides associated with MHC class I molecules. Curr Opinion Immunology 6:13–23

Fabunmi RP, Wigley WC, Thomas PJ, DeMartino GN (2000) Activity and regulation of the centrosome-associated proteasome. J Biol Chem 275:409–413

Früh K, Yang Y (1999) Antigen presentation by MHC class I and its regulation by interferon gamma. Curr Opin Immunol 11:76–81

Gaczynska M, Goldberg AL, Tanaka K, Hendil KB, Rock KL (1996) Proteasome subunits X and Y alter peptidase activities in opposite ways to the interferon-γ-induced subunits LMP2 and LMP7. J Biol Chem 271:17275–17280

Gaczynska M, Rock KL, Goldberg AL (1993) γ-interferon and expression of MHC genes regulate peptide hydrolysis by proteasomes. Nature 365:264–267

Gray CW, Slaughter CA, DeMartino GN (1994) PA28 activator protein forms regulatory caps on proteasome stacked rings. J Mol Biol 236:7-15

Groettrup M, Soza A, Eggers M, Kuehn L, Dick TP, Schild H, Rammensee HG, Koszinowski HH, Kloetzel PM (1996a) A role for the proteasome regulator PA28α in antigen presentation. Nature 381:166-168

Groettrup M, Soza A, Kuckelkorn U, Kloetzel PM (1996b) Peptide antigen production by the proteasome: complexity provides efficiency. Immunology Today 17:429-435

Groll M, Bajorek M, Köhler A, Moroder L, Rubin DM, Huber R, Glickman MH, Finley D (2000) A gated channel into the proteasome core particle. Nat Struct Biol 7:1062-1067

Groll M, Ditzel L, Löwe J, Stock D, Bochtler M, Bartunik HD, Huber R (1997) Structure of 20S proteasome from yeast at 2.4Å resolution. Nature 386:463-471

Hendil KB, Khan S, Tanaka K (1998) Simultaneous binding of PA28 and PA700 activators to 20S proteasomes. Biochem J 332:749-754

Hoffman L, Rechsteiner M (1996) Regulatory features of multicatalytic and 26S proteases. In: Stadtman ER, Chock PB (eds) Current Topics In Cellular Regulation. Academic Press, San Diego, pp 1-32

Jiang H, Monaco JJ (1997) Sequence and expression of mouse proteasome activator PA28 and the related autoantigen Ki. Immunogenetics 46:93-98

Johnston SC, Whitby FW, Realini C, Rechsteiner M, Hill CP (1997) The proteasome 11S regulator subunit REGα (PA28α) is a heptamer. Protein Science 6:2469-2473

Kalmes A, Hagemann C, Weber CK, Wixler L, Schuster T, Rapp UR (1998) Interaction between the protein kinase B-Raf and the alpha-subunit of the 11S proteasome regulator. Cancer Res 58:2986-2990

Kandil E, Kohda K, Ishibashi T, Tanaka K, Kasahara M (1997) PA28 subunits of the mouse proteasome: primary structures and chromosomal localization of the genes. Immunogenetics 46:337-344

Kisselev AF, Akopian TN, Castillo V, Goldberg AL (1999a) Proteasome active sites allosterically regulate each other, suggesting a cyclical bite-chew mechanism for protein breakdown. Mol Cell 4:395-402

Kisselev AF, Akopian TN, Goldberg AL (1998) Range of sizes of peptide products generated during degradation of different proteins by archaeal proteasomes. J Biol Chem 273:1982-1989

Kisselev AF, Akopian TN, Woo KM, Goldberg AL (1999b) The sizes of peptides generated from protein by mammalian 26 and 20S proteasomes. Implications for understanding the degradative mechanism and antigen presentation. J Biol Chem 274:3363-3371

Kloetzel PM, Soza A, Stohwasser R (1999) The role of the proteasome system and the proteasome activator PA28 complex in the cellular immune response. Biol Chem 380:293-297

Knowlton JR, Johnston SC, Whitby FG, Realini CR, Zhang Z, Rechsteiner MC, Hill CP (1997) Structure of the proteasome activator REGa (PA28a). Nature 390:639-643

Kuehn L, Dahlmann B (1996a) Proteasome activator PA28 and its interaction with 20S proteasomes. Arch Biochem Biophys 329:87-96

Kuehn L, Dahlmann B (1996b) Reconstitution of proteasome activator PA28 from isolated subunits: optimal activity is associated with an alpha,beta-heteromultimer. FEBS Lett 394:183-186

Kuehn L, Dahlmann B (1997) Structural and functional properties of proteasome activator PA28. Mol Biol Rep 24:89-93

Li J, Gao X, Joss L, Rechsteiner M (2000) The proteasome activator 11S REG or PA28: chimeras implicate carboxyl-terminal sequences in oligomerization and proteasome binding but not in the activation of specific proteasome catalytic subunits. J Mol Biol 299:641-654

Li N, Lerea KM, Etlinger JD (1996) Phosphorylation of the proteasome activator PA28 is required for proteasome activation. Biochem Biophys Res Commun 225:855-860

Loidl G, Groll M, Musiol HJ, Ditzel L, Huber R, Moroder L (1999) Bifunctional inhibitors of the trypsin-like activity of eukaryotic proteasomes. Chem Biol 6:197-204

Löwe J, Stock D, Jap B, Zwickl P, Baumeister W, Huber R (1995) Crystal structure of the 20S proteasome from the archaeon *T. acidophilum* at 3.4Å resolution. Science 268:533-539

Ma CP, Slaughter CA, DeMartino GN (1992) Identification, purification, and characterization of a protein activator (PA28) of the 20S proteasome (macropain). J Biol Chem 267:10515-10523

Ma CP, Willy PJ, Slaughter CA, DeMartino GN (1993) PA28, an activator of the 20S proteasome, is inactivated by proteolytic modification at its carboxyl terminus. J Biol Chem 268:22514-22519

Macagno A, Gilliet M, Sallusto F, Lanzavecchia A, Nestle FO, Groettrup M (1999) Dendritic cells upregulate immunoproteasomes and the proteasome regulator PA28 during maturation. Eur J Immunol 29:4037-4042

Maksymowych WP, Ikawa T, Yamaguchi A, Ikeda M, McDonald D, Laouar L, Lahesmaa R, Tamura N, Khuong A, Yu DTY, Kane KP (1998) Invasion by Salmonella typhimurium induces increased expression of the LMP, MECL, and PA28 proteasome genes and changes in the peptide repertoire of HLA-B27. Infect Immun 66:4624–4632

Minami Y, Kawasaki H, Minami M, Tanahashi N, Tanaka K, Yahara I (2000) A critical role for the proteasome activator PA28 in the Hsp90-dependent protein refolding. J Biol Chem 275:9055–9061

Mott JD, Pramanik BC, Moomaw CR, Afendis SJ, DeMartino GN, Slaughter CA (1994) PA28, an activator of the 20S proteasome, is composed of two nonidentical but homologous subunits. J Biol Chem 269:31466–31471

Murata S, Kawahara H, Tohma S, Yamamoto K, Kasahara M, Nabeshima Y, Tanaka K, Chiba T (1999) Growth retardation in mice lacking the proteasome activator PA28γ. J Biol Chem 274:38211–38215

Mykles DL (1996) Differential effects of bovine PA28 on six peptidase activities of the lobster muscle proteasome (multicatalytic proteinase). Arch Biochem Biophys 325:77–81

Niedermann G, Grimm R, Geier E, Maurer M, Realini C, Gartmann C, Soll J, Omura S, Rechsteiner MC, Baumeister W, Eichmann K (1997) Potential immunocompetence of proteolytic fragments produced by proteasomes before evolution of the vertebrate immune system. J Exp Med 186:209–220

Niedermann G, King G, Butz S, Birsner U, Grimm R, Shabanowitz J, Hunt DF, Eichmann K (1996) The proteolytic fragments generated by vertebrate proteasomes: Structural relationships to major histocompatibility complex class I binding peptides. Proc Natl Acad Sci USA 93:8572–8577

Nikaido T, Shimada K, Shibata M, Hata M, Sakamoto M, Takasaki Y, Sato C, Takahashi T, Nishida Y (1990) Cloning and nucleotide sequence of cDNA for Ki antigen, a highly conserved nuclear protein detected with sera from patients with systemic lupus erythematosus. Clin Exp Immunol 79:209–214

Nussbaum AK, Dick TP, Keilholz W, Schirle M, Stevanovic S, Dietz K, Heinemeyer W, Groll M, Wolf DH, Huber R, Rammensee HG, Schild H (1998) Cleavage motifs of the yeast 20S proteasome beta subunits deduced from digests of enolase 1. Proc Natl Acad Sci USA 95:12502A–12509

Ordway GA, Neufer PD, Chin ER, DeMartino GN (2000) Chronic contractile activity upregulates the proteasome system in rabbit skeletal muscle. J Appl Physiol 88:1134–1141

Paesen GC, Nuttall PA (1996) A tick homolog of the human Ki nuclear autoantigen. Biochim Biophys Acta 1309:9–13

Pamer E, Cresswell P (1998) Mechanisms of MHC class I – restricted antigen processing. Annu Rev Immunol 16:323–358

Preckel T, Fung-Leung WP, Cai Z, Vitiello A, Slater-Cid L, Winqvist O, Wolfe TG, Von Herrath M, Angulo A, Ghazal P, Lee JD, Fourie AM, Wu Y, Pang J, Ngo K, Peterson PA, Früh K, Yang Y (1999) Impaired immunoproteasome assembly and immune responses in $PA28-/-$ mice. Science 286:2162–2165

Rammensee HG, Falk K, Rotzschke O (1993) Peptides naturally presented by MHC class I molecules. Annu Rev Immunol 11:213–244

Realini C, Dubiel W, Pratt G, Ferrell K, Rechsteiner M (1994a) Molecular cloning and expression of a gamma-interferon inducible activator of the multicatalytic protease. J Biol Chem 269:20727–20732

Realini C, Jensen CC, Zhang Z, Johnston SC, Knowlton JR, Hill CP, Rechsteiner M (1997) Characterization of recombinant REGα, REGβ and REGγ proteasome activators. J Biol Chem 272:25483–25492

Realini C, Rechsteiner M (1995) A proteasome activator subunit binds calcium. J Biol Chem 270:29664–29667

Realini C, Rogers SW, Rechsteiner M (1994b) KEKE motifs: proposed roles in protein-protein association and presentation of peptides by MHC Class I receptors. FEBS Lett 348:109–113

Rechsteiner M, Realini C, Ustrell V (2000) The proteasome activator 11S REG (PA28) and class I antigen presentation. Biochem J 345:1–15

Rock KL, Goldberg AL (1999) Degradation of cell proteins and the generation of MHC class I-presented peptides. Annu Rev Immunol 17:739–779

Schmidtke G, Emch S, Groettrup M, Holzhutter HG (2000) Evidence for the existence of a non-catalytic modifier site of peptide hydrolysis by the 20S proteasome. J Biol Chem 275:22056–22063

Schwarz K, van Den Broek M, Kostka S, Kraft R, Soza A, Schmidtke G, Kloetzel PM, Groettrup M (2000) Overexpression of the proteasome subunits LMP2, LMP7, and MECL-1, but not PA28α/β, enhances the presentation of an immunodominant lymphocytic choriomeningitis virus T cell epitope. J Immunol 165:768–778

Shimbara N, Nakajima H, Tanahashi N, Ogawa K, Niwa S, Uenaka A, Nakayama E, Tanaka K (1997) Double-cleavage production of the CTL epitope by proteasomes and PA28: role of the flanking region. Genes Cells 2:785–800

Snyder HL, Yewdell JW, Bennink JR (1994) Trimming of antigenic peptides in an early secretory compartment. J Exp Med 180:2389–2394

Song X, Mott JD, von Kampen J, Pramanik B, Tanaka K, Slaughter CA, DeMartino GN (1996) A model for the quaternary structure of the proteasome activator PA28. J Biol Chem 271:26410–26417

Song X, von Kampen J, Slaughter CA, DeMartino GN (1997) Relative functions of the α and β subunits of the proteasome activator, PA28. J Biol Chem 272:27994–28000

Soza A, Knuehl C, Groettrup M, Henklein P, Tanaka K, Kloetzel PM (1997) Expression and subcellular localization of mouse 20S proteasome activator complex PA28. FEBS Lett 413:27–34

Stern LJ, Wiley DC (1994) Antigenic peptide binding by class I and class II histocompatibility proteins. Structure 2:245–251

Stohwasser R, Salzmann U, Giesebrecht J, Kloetzel PM, Holzhutter HG (2000) Kinetic evidences for facilitation of peptide channelling by the proteasome activator PA28. Eur J Biochem 267:6221–6230

Tanahashi N, Murakami Y, Minami Y, Shimbara N, Hendil KB, Tanaka K (2000) Hybrid proteasomes. Induction by interferon-γ and contribution to ATP-dependent proteolysis. J Biol Chem 275:14336–14345

Tanahashi N, Yokota K, Ahn JY, Chung CH, Fujiwara T, Takahashi E, DeMartino GN, Slaughter CA, Toyonaga T, Yamamura K, Shimbara N, Tanaka K (1997) Molecular properties of the proteasome activator PA28 family proteins and gamma-interferon regulation. Genes Cells 2:195–211

Tanaka K, Kasahara M (1998) The MHC class I ligand-generating system: roles of immunoproteasomes and the interferon-gamma-inducible proteasome activator PA28. Immunol Rev 163:161–176

To WY, Wang CC (1997) Identification and characterization of an activated 20S proteasome in Trypanosoma brucei. FEBS Lett 404:253–262

Tojo T, Kaburaki J, Hayakawa M, Okamoto T, Tomii M, Homma M (1981) Precipitating antibody to a soluble nuclear antigen 'Ki' with specificity for systemic lupus erythematosus. Ryumachi 21:Suppl:129–140

Ustrell V, Pratt G, Rechsteiner M (1995) Effects of interferon γ and major histocompatibility complex-encoded subunits on peptidase activities of human multicatalytic proteases. Proc Natl Acad Sci USA 92:584–588

Voges D, Zwickl P, Baumeister W (1999) The 26S proteasome: a molecular machine designed for controlled proteolysis. Annu Rev Biochem:1015–1068

Wenzel T, Baumeister W (1995) Conformational constraints in protein degradation by the 20S proteasome. Nature Structural Biology 2:199–204

Whitby FG, Masters EI, Kramer L, Knowlton JR, Yao Y, Wang CC, Hill CP (2000) Structural basis for the activation of 20S proteasomes by 11S regulators. Nature 408:115–120

Wigley WC, Fabunmi RP, Lee MG, Marino CR, Muallem S, DeMartino GN, Thomas PJ (1999) Dynamic association of proteasomal machinery with the centrosome. J Cell Biol 145:481–490

Wojcik C, Tanaka K, Paweletz N, Naab U, Wilk S (1998) Proteasome activator (PA28) subunits, alpha, beta and gamma (Ki antigen) in NT2 neuronal precursor cells and HeLa S3 cells. Eur J Cell Biol 77:151–160

Yao Y, Huang L, Krutchinsky A, Wong ML, Standing KG, Burlingame AL, Wang CC (1999) Structural and functional characterization of the proteasome-activating protein PA26 from Trypanosoma brucei. J Biol Chem 274:33921–33930

Yewdell JW, Bennink JR (1992) Cell biology of antigen processing and presentation to major histocompatibility complex class I molecule-restricted T lymphocytes. Adv Immunol 52:1–123

Zhang Z, Clawson A, Realini C, Jensen CC, Knowlton JR, Hill CP, Rechsteiner M (1998a) Identification of an activation region in the proteasome activator REGα. Proc Natl Acad Sci USA 95:2807–2811

Zhang Z, Clawson A, Rechsteiner M (1998b) The proteasome activator or PA28. Contribution by both α and β subunits to proteasome activation. J Biol Chem 273:30660–30668

Zhang Z, Kruchinsky A, Endicott S, Realini C, Rechsteiner M, Standing KG (1999) Proteasome activator 11S REG or PA28. Recombinant REGα/REGβ hetero-oligomers are heptamers. Biochemistry 38:5651–5658

Zhang Z, Realini C, Clawson A, Endicott S, Rechsteiner M (1998c) Proteasome activation by REG molecules lacking homolog-specific inserts. J Biol Chem 273:9501–9509

Zuhl F, Tamura T, Dolenc I, Cejka Z, Nagy I, De Mot R, Baumeister W (1997) Subunit topology of the Rhodococcus proteasome. FEBS Lett 400:83–90

Immunological Functions of the Proteasome

G. NIEDERMANN

1	Introduction	91
2	Properties of Peptides in the MHC Class I-Antigen Processing Pathway	93
2.1	Peptides that Bind MHC Class I Molecules	93
2.2	Peptides Transported by TAP	94
3	Proteolytic Properties of Proteasomes Influencing Their Immune Functions	95
3.1	Substrates of Proteasomes	95
3.2	Specificities Defined by Small Reporter Substrates	96
3.3	Proteasomes Cleave Polypeptides with Degenerate Specificity	97
3.4	Length of Degradation Products	101
3.5	Mechanisms Underlying the Product-Size Distributions	102
4	Immunoproteasomes	104
4.1	The IFN-γ-Inducible β-Type Subunits	104
4.2	Observations on LMP-Deficient Cells and Knockout Mice	105
4.3	Alterations of Endopeptidase Activities as Observed with Reporter Peptides	106
4.4	Influence on Cleavage Site Usage within Polypeptides	107
4.5	Comparisons of Effects on Epitope Production In Vitro and In Vivo	108
4.6	Ambiguous Consequences of the Dichotomy of Proteasomes for Self-Tolerance and Immune Responsiveness of CTLs	110
4.6.1	Evolution of Antigen Presentation	110
4.6.2	Elimination of Virus-Infected or Malignant Cells	111
4.6.3	Self-Tolerance	112
5	11S REG/PA28: the IFN-γ-Inducible ATP-Independent Activator	114
5.1	Effects on the Processing of Short Polypeptides	114
5.2	Are There Effects on the Processing of Protein Substrates?	116
5.3	Relationships of REG/PA28 with the Immunoproteasome	118
6	Epitope Generation by Purified Proteasomes In Vitro	119
6.1	Epitopes and/or Epitope Precursors	119
6.2	Factors Influencing the Magnitude of Epitope Generation by Proteasomes	126
7	Summary	128
References		128

1 Introduction

The adaptive immune system has evolved in higher vertebrates to defend them efficiently against infections and perhaps also against cancer. Cytotoxic T

Max Planck Institute of Immunobiology, Stübeweg 51, 79108 Freiburg, Germany

lymphocytes (CTLs) represent one major type of antigen-specific immunocyte. CTL recognize peptides from infectious agents and tumor antigens bound to major histocompatibility complex (MHC) class I molecules, and lyse cells that display such MHC class I/peptide complexes at the cell surface. MHC class I-binding peptides are derived through the continuous proteolysis of polypeptides that are synthesized in the cytosol or reach the cytosol in other ways. Suitable peptides bind to the transporter associated with antigen processing (TAP) for translocation into the endoplasmic reticulum (ER), and peptides with an appropriate motif bind to newly synthesized MHC class I molecules. The newly formed MHC class I–peptide complexes then travel to the cell surface for recognition by CTLs (Fig. 1). Because the peptides preferred by TAP are the same size as those presented by the MHC class I molecules or are somewhat longer, the major steps in MHC class I antigen processing can be expected to occur in the cytosol. Proteasomes are highly abundant cytosolic and nuclear protease complexes that degrade most intracellular proteins in higher eukaryotes and appear to play the major role in the cytosolic steps of MHC class I antigen processing. This review summarizes the present knowledge of the role of proteasomes in antigen processing and the impact of proteasomal proteolysis on T cell-mediated immunity.

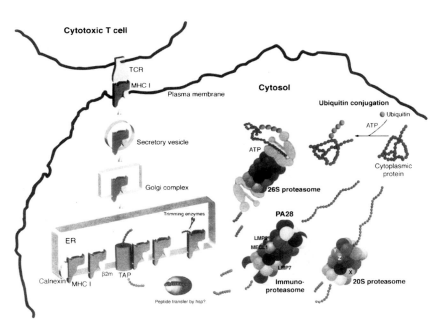

Fig. 1. Schematic representation of the MHC class I antigen processing and presentation pathway. *TAP*, transporter associated with antigen processing; *hsp*, heat shock protein; $\beta_2 m$, β_2-microglobulin; *X*, *Y* and *Z*, the proteolytically active β-type subunits, now designated β5, β1 and β2, respectively

2 Properties of Peptides in the MHC Class I-Antigen Processing Pathway

2.1 Peptides that Bind MHC Class I Molecules

The MHC antigen processing and presentation system has been extensively reviewed (PAMER and CRESSWELL 1998; PIETERS 2000). The two major types of MHC molecules, class I and class II, possess binding grooves that bind small peptides in a more or less extended conformation and present them at the cell surface. Class II presents peptides derived from the endocytic compartment (VILLADANGOS and PLOEGH 2000) and will not be discussed further in this article. Class I presents peptides generated in the cytosol by proteolytic mechanisms that are the subject of this paper. One important feature of the peptide binding groove of MHC class I molecules is that the ends of the groove are 'closed' (MADDEN 1995). MHC class I molecules therefore prefer peptide ligands of relatively restricted size which are usually tightly fixed in the ends of the groove. Almost all ligands characterized have between 8 and 13 amino acids (aa), more than 90% have between 8 and 11, and the most frequent size is nonamers (RAMMENSEE et al. 1997). Classical MHC class I molecules, i.e., those presenting peptides to CTLs, are highly polymorphic. In most species, several isotypes are known, each with a large number of alleles. There are three isotypes of classical class I molecules in humans, with a total number of allelic variants exceeding 300 (BODMER et al. 1997). The polymorphism occurs mainly in the peptide-binding groove, affecting the specificity of the aa side chain-binding subsites (pockets A–F). The polymorphism therefore affects the peptide-binding specificity of the class I molecules. Of the complementary pockets only two are usually relatively restrictive and allow binding of one or a few similar aa side chains (RAMMENSEE et al. 1997). One of these 'anchor pockets' is always the F-pocket, which usually binds the C terminus of the peptide; the other is often, but not always, the B-pocket which harbors the side chain of the second aa from the N terminus. The F-pockets of the class I molecules are of particular interest here as they bind aa directly adjacent to a proteolytic cleavage site within the source protein of the peptide. The F-pockets of the mammalian class I molecules described to date prefer either hydrophobic or basic residues. In chicken, however, one class I allele was discovered that prefers peptides with an acidic aa at the C terminus (KAUFMAN and WALLNY 1996).

Amino acids other than those at the anchor positions are usually less important for peptide binding. A given class I molecule can therefore bind a relatively large number of different peptide ligands. The number of different peptide ligands/allele per cell was estimated to exceed 2000 (HUCZKO et al. 1993). Individual copy numbers of presented peptides differ widely, from fewer than 10 to approximately 10,000 (STEVANOVIC and SCHILD 1999; YEWDELL and BENNINK 1999a). A human individual may express up to six different MHC class I allelic products, supplied by the class I antigen processing machinery with several thousands (perhaps more than 10,000) of different peptides derived from cellular proteins. Considering the entire diversity of class I allelic products, many of which differ in their 'peptide-motif', the number of peptide ligands presented in the human species may be far greater. A typical mammalian cell expresses approximately 10,000 distinct genes. MHC class I-binding peptides generated in the cytosol primarily originate from these endogenous gene products which are, with some exceptions, nonimmunogenic because of self-tolerance of the CTL system (see Sect. 4.6.3). Immunogenic peptides of foreign origin are presented merely by infected cells.

Although the MHC class I peptide binding specificity is broad, a given allelic product will usually present, if any at all, only one or a few peptides of a given protein. For example, anti-viral CTL responses are often directed predominantly against one or a few peptides of all viral gene products, particularly in inbred mice in which MHC diversity is limited by homozygosity (YEWDELL and BENNINK 1999a). However, studies of responses of outbred humans to human immunodeficiency virus (HIV), hepatitis B virus (HBV) and hepatitis C virus revealed that relatively large parts of several viral gene products investigated are covered by different CTL epitopes (CHISARI and FERRARI 1995; ROWLAND-JONES et al. 1997). For example, although the HIV Nef protein is only about 200aa long, more than 50 CTL epitopes have already been identified for only about 20 different human leukocyte antigens (HLAs; i.e., human MHC) class I molecules (KORBER et al. 1997). Whereas some of the epitopes mapped appear to lie directly adjacent to one another, many overlap with others. With a few exceptions, most epitopes are found in one of several epitope clusters. These clusters nevertheless cover over half of the primary Nef sequence.

2.2 Peptides Transported by TAP

The optimal length of TAP-transported peptides is between 8 and 12aa (MOMBURG and HÄMMERLING 1998), and peptides of up to 15 residues bind with similar affinities (VAN ENDERT 1996). Peptides of 6 or up to 40aa have also been reported to be transported, but at low efficiencies (KOOPMANN et al. 1996). TAP is functionally monomorphic in humans and mice, and of limited functional polymorphism in rats. To provide the different MHC class I alleles with the large variety of peptides, TAP has a rather broad peptide sequence specificity. Nevertheless, TAP is not unrestrictive. The strongest influence on peptide binding is exerted by the peptide C terminus. Residues which are typical peptide anchors for class I binding (hydrophobic in mice, hydrophobic and basic in humans and rats) are usually also preferred by TAP (MOMBURG et al. 1994; HOWARD 1995). As demonstrated for human TAP, preferences exist for the three N-terminal residues as well (VAN ENDERT et al. 1995; VAN ENDERT 1996; UEBEL et al. 1997). Among these, the most important seems to be position 3, where hydrophobic residues are preferred. Interestingly, this preference serves as an auxiliary anchor in many class I allele-specific peptide motifs. Position 1 of a peptide also influences TAP specificity. Here, basic, small and polar residues seem to be preferred, while large hydrophobic and acidic aa as well as proline weaken peptide binding. Generally, there seems to be some degree of functional coordination between TAP and MHC class I, but depending on the class I allele, this coordination is more or less pronounced (DANIEL et al. 1998). HLA-B27, for example, is an allele whose binding motif coincides with TAP preferences markedly well. This may explain why HLA-B27 appears to confer advantages in infections with HIV and other viruses and is associated with autoimmune diseases such as Reiter syndrome and ankylosing spondylarthritis (GOMARD et al. 1984; KASLOW et al. 1996; GOULDER et al. 1997). Conversely, there are also class I alleles whose N-terminal binding preferences do not appear to match very well with the transporter. For example, class I alleles belonging to the B7-like superfamily usually use proline as an anchor in position 2 of bound peptides. Peptides with proline in position 2 or 3, however, are strongly disfavored by TAP (NEEFJES et al. 1995; NEISIG et al. 1995). The products of these alleles may therefore derive many ligands from TAP-independent sources such as signal sequence-derived peptides. In addition, a relatively large proportion of ligands for these alleles may be generated from N-terminally extended precursor peptides transported via TAP and subsequently shortened by ER-resident aminopeptidases (see Sect. 6.1).

3 Proteolytic Properties of Proteasomes Influencing Their Immune Functions

3.1 Substrates of Proteasomes

What role do proteasomes play in the production of the large, but nevertheless limited variety of peptides loaded onto class I molecules? Proteasomes are endoproteases with broad cleavage site specificity that degrade proteins and polypeptides into heterogeneous collections of oligopeptides, many of which nonetheless show some characteristic structural features (see below). Proteolysis requires unfolded substrates, as it occurs within the central chamber of the 20S proteasome, accessible only through a narrow channel (BAUMEISTER et al. 1998). Proteasomes occur both in the cytosol and nucleus and are remarkably abundant, representing approximately 0.5%–1% of cellular protein. The importance of proteasomal proteolysis in particular in conjunction with the ubiquitin system for almost every aspect of cell physiology has been increasingly appreciated in recent years (COUX et al. 1996; CIECHANOVER et al. 2000). Inhibitor studies have demonstrated that proteasomes degrade not only short-lived and abnormal proteins, as initially thought, but also the bulk of long-lived cellular proteins (ROCK et al. 1994). Recent studies have shown that not only cytosolic and nuclear proteins are proteasome substrates: integral membrane and even ER-luminal proteins have been identified that are degraded by proteasomes, the latter being accessible only after re-export to the cytosol (KOPITO 1997; LORD et al. 2000). In addition, a recent study has determined that the proportion of newly biosynthesized proteins that never attain their native state – so-called defective ribosomal products (DRIPs) – is quite high,

30% or above, and that these are rapidly destroyed by proteasomes (SCHUBERT et al. 2000). All in all, proteasomes appear to have a very broad substrate specificity and may degrade most polypeptides synthesized within a cell. As a consequence, substrates degraded by proteasomes include endogenous proteins as well as proteins foreign to the cell. Thus, proteasomes are uniquely suitable as processing enzymes for the MHC class I-peptide presentation system, the main function of which is the unmasking of intracellular viral and tumor antigens.

Proteasome inhibitors were originally used to study the dependence of class I peptide loading on proteasomes. While initially a pronounced blockade of peptide supply was reported for several class I molecules (ROCK et al. 1994; BAI and FORMAN 1997; CERUNDOLO et al. 1997; CRAIU et al. 1997b), several recent reports have revealed more complex effects for others (VINITSKY et al. 1997; BENHAM et al. 1998). Most proteasome inhibitors suppress the different proteasomal peptidases with different efficiencies or even enhance selected activities (BOGYO et al. 1997; VALMORI et al. 1999). Therefore, lack of inhibition per se does not rule out a role for proteasomes in peptide supply for these alleles. A few cytosolic or nuclear proteins are not degraded by the ubiquitin–proteasome pathway. The classic example is the Epstein–Barr virus (EBV) nuclear antigen 1, which is not presented by MHC class I molecules, leading to latency of the oncogenic virus (LEVITSKAYA et al. 1997). This and other examples are consistent with a pivotal role of the proteasome in providing the peptide ligands for most class I molecules.

3.2 Specificities Defined by Small Reporter Substrates

For information on the structure of the different types of proteasomes, the reader is referred to recent reviews (BAUMEISTER et al. 1998; BOCHTLER et al. 1999, and other chapters in this volume). Regarding their peptide products and cleavage specificities, 20S proteasomes have been investigated in more detail than the other types of proteasomes, but it is not clear whether the 20S proteasome itself contributes significantly to intracellular proteolysis or functions only as the proteolytic core of the larger proteasome particles. It is generally thought that intracellular proteolysis is mediated largely by the ATP-dependent 26S proteasome, which degrades mainly ubiquitin-tagged proteins. However, the 20S proteasome is able to degrade denatured proteins in vitro (see Sects. 3.3 and 6) and there is evidence for the degradation of oxidatively damaged proteins by 20S proteasomes in vivo (GRUNE et al. 1997).

Early work with chromogenic tri- and tetrapeptide reporter substrates had already shown that mammalian 20S proteasomes are multicatalytic (this is true for all eukaryotic proteasomes) and have broad cleavage specificity. Classical activities responsible for efficient cleavage on the carboxyl side of acidic (glutamate, aspartate), basic (usually arginine), and hydrophobic (e.g., leucine, tyrosine and phenylalanine) P1 residues have been respectively referred to as peptidylglutamyl–peptide bond hydrolase (also called postglutamyl, postacidic or caspase-like), trypsin-like, and chymotrypsin-like activities (WILK and ORLOWSKI 1980). By mutational analysis in yeast, these three activities have been assigned to the three different active β-type subunits, β1, β2, and β5, respectively (HEINEMEYER et al. 1993; ENENKEL et al. 1994). Analyses of cleavages in the peptidyl portion of reporter substrates even revealed cleavages on the carboxyl side of very small P1 aa such as glycine and alanine when the P1'-residue was also small. This activity was called small neutral aa-preferring (SNAAP) activity. Another activity, defined with the help of reporter substrates and inhibitors, is the branched chain aa-preferring (BrAAP) activity (ORLOWSKI et al. 1993). This activity has P1 preferences similar to those of the chymotrypsin-like activity. Whereas the three classical peptidase activities explained above cleave the peptidyl-arylamide bonds between the P1 residue and the chromo- or fluorogenic leaving group in reporter substrates, the BrAAP and SNAAP activities are not able to do so and are defined by cleavage of peptide bonds in the peptidyl portion of reporter substrates.

3.3 Proteasomes Cleave Polypeptides with Degenerate Specificity

The cleavage preferences for hydrophobic and charged residues, as determined with small reporter substrates, have been a major reason to consider proteasomes as candidate enzymes for the generation of the C termini of class I ligands (DRISCOLL et al. 1993; GACZYNSKA et al. 1993). However, other studies had created the impression that the specificities defined by short reporter substrates had only limited predictive value for the proteasomal cleavage of longer polypeptides. By digesting polypeptides such as the 30-aa insulin B-chain, it was shown that the proteasome can cleave after additional residues including glutamine and cysteine (WILK and ORLOWSKI 1980; RIVETT 1985). Later studies suggested that the proteasome may be able to cut polypeptides between almost any two aa and that the P1 residue plays a minor role as specificity determinant (DICK et al. 1991, 1994; EHRING et al. 1996; STEIN et al. 1996). The complexity in the discussion on proteasomal cleavage of polypeptides was augmented further by the discovery of the length restriction of proteasomal fragments (see next section). Together, these studies were interpreted by many investigators to suggest that cleavage site usage by the proteasome was determined primarily by the mechanism responsible for length restriction, and scarcely by proteolytic specificity.

We reasoned that the question of specificity could only be answered if cleavage site usage in polypeptides was analyzed comprehensively and quantitatively. To approach this aim, we first performed a comprehensive analysis of all detectable 20S proteasomal cleavage sites in polypeptides of around 20 to 25aa by quantitative Edman degradation pool-sequencing (NIEDERMANN et al. 1995). P1 preferences similar to those found with reporter peptides were observed, although some cleavages after other aa also occurred. Moreover, not all of the sites predicted by reporter peptide hydrolysis were cleaved. Because the complete identification of all products from a full-length protein was not possible, we then decided to comprehensively characterize digests of several polypeptides of up to 44aa by tandem mass spectrometry, using murine 20S proteasomes isolated from EL4-thymoma cells

(NIEDERMANN et al. 1996, 1999). The results clearly revealed a preference for most hydrophobic and charged (positively and negatively charged) residues in the P1 position of cleavage sites (the C terminus of products), although some cleavages after other residues were also detected; 58% of cleavages were after hydrophobic aliphatic (i.e., branched-chain; 22% after leucine) or aromatic residues, 31% after charged aa and the remaining cleavages (11%) were after other residues. We also observed preferences in the P1' position of cleavage sites (the N terminus of product), although not as pronounced as for P1. In P1', small and/or polar aa were enriched, particularly serine. On the basis of these results we concluded that, in addition to the mechanism restricting the distance between cleavages (see below), both the P1 and P1' positions influenced cleavage site selection, the former more strongly than the latter. Specific flanking sequence preferences appeared to exist for the P2 and P3 positions, e.g., cleavages after negatively charged P1 residues were favored when a hydrophobic aa was in P2 and/or P3 upstream of the cleavage site. This agrees with reporter peptide/inhibitor studies that have shown a preference of the postglutamyl activity for hydrophobic residues in P2 and P3 (ORLOWSKI and MICHAUD 1989).

For comparison with proteasomal fragments, we analyzed the C- and N-terminal positions of more than 300 and the immediate flanking (C + 1 and N − 1) positions of 134 acid-eluted class I ligands (NIEDERMANN et al. 1996, 1999). With the exception of acidic aa, we noted a good correlation of aa frequencies at the C-terminal position between proteasomal products (see above) and class I ligands. Of the latter, 73% had a hydrophobic aliphatic or aromatic aa, with leucine constituting about one-fourth of all ligand C termini; 20% had a charged (almost exclusively positively charged) aa; and 7% had other aa at the C terminus. This strong correlation suggested that proteasomes generate the C termini of many MHC class I ligands. This conclusion was further supported by the finding that, similar to the P1'-position of proteasomal cleavage sites (the N terminus of products), small and polar aa were also found to be enriched in the immediate flanking position (C + 1) of class I ligands. Besides the three small aa serine, alanine and glycine, the basic aa lysine was found to be over-represented at the C + 1 position. Conspicuously, the same four aa were found to be enriched to a very similar extent in the N-terminal position of the class I ligands analyzed. In a recent analysis of 286 acid-eluted class I ligands (ALTUVIA and MARGALIT 2000), similar residues were found to be enriched in the immediate flanking position C + 1; besides serine, alanine and lysine, arginine was also found to be over-represented in this position. At the N terminus of the 286 ligands, alanine, glycine, arginine and lysine, but not serine, were found to be over-represented. We have now analyzed all 543 acid-eluted class I ligands listed in a compilation by (RAMMENSEE et al. 1997). At the N terminus of these 543 ligands, lysine, alanine, serine, arginine and glycine were found to be enriched; their appearance relative to their natural occurrence in proteins was 1.76, 1.74, 1.67, 1.64, and 1.35-fold, respectively. Together, these five small or basic aa constitute the N terminus of about 55% of the class I ligands analyzed. The resemblance of the frequencies of small and basic aa in C + 1 and at the N termini of class I ligands to those at the N termini of proteasomal products suggested to us that proteasomes may also liberate a proportion of ligand N termini (see also Sect. 6).

CARDOZO and KOHANSKI (1998) analyzed cleavages made by bovine 20S proteasomes in the small protein lysozyme (129aa). Although some differences between pituitary (mainly housekeeping) and spleen (mainly immuno-) proteasomes were found (see below), also here most major cleavages were located after hydrophobic aliphatic (branched chain), aromatic or charged P1 residues. Cleavages after other residues (e.g., alanine, serine, asparagine and cysteine) were also detected, but these were uniformly present in small amounts. Similar to our results, this study also showed that 20S proteasomes prefer small neutral or polar residues (alanine, serine, glycine, cysteine, and asparagine) in the P1' position of cleavage sites; this was true for both pituitary and spleen proteasomes. Furthermore, the results suggested that cleavage sites with basic P1 residues often also have a basic P2 or P3 residue. Whereas the BrAAP activity of the spleen proteasome appeared to cleave more efficiently after branched chain P1 residues when a hydrophobic residue was present in P3 and a small neutral or polar residue in P2, that of the pituitary proteasome appeared to prefer a small neutral residue in P3 (including proline) and a hydrophobic residue in P2.

NUSSBAUM et al. (1998) analyzed cleavages by yeast 20S proteasomes in enolase. Although, cleavages after almost every type of aa were detected (except proline and glutamine), the P1 position was found to exert a significant influence on cleavage site selection. Most hydrophobic and charged residues (except lysine) were favored as P1 residue. Peptide bonds with leucine and arginine in P1 were cleaved every second time they appeared in enolase. Disfavored in P1 were proline, glycine, serine, threonine, asparagine, glutamine, and lysine (i.e., mainly small and/or polar residues). Also the yeast proteasome was found to prefer small residues in P1'. Furthermore, a preference for proline in P4 also appeared to be a general feature of the yeast 20S proteasome. Besides these preferences which were generally significant for the yeast proteasome, some subunit-specific preferences were also observed, including a weak one for basic residues in P1' for subunit β5/Pre2.

EMMERICH et al. (2000) analyzed cleavages made by human erythrocyte proteasomes [which did not contain low molecular mass protein (LMP)7 and therefore probably are housekeeping proteasomes] in casein. Most prominent for both 20S and 26S proteasomes was a preference for hydrophobic aa, especially leucine, in P1: the 20S proteasome cleaved after 11 of the 22 available leucine residues, and the 26S proteasome cleaved after 14 of the 22. The 26S, but not the 20S, also showed significant preferences for proline in P4 and β-turn-promoting aa, especially proline in P1'; β-turn promoting are aa with small, polar or no side chain.

Taken together, recent surveys of cleavages in polypeptides show that proteasomes (in most experiments housekeeping proteasomes or mixtures of housekeeping- and immunoproteasomes), despite their broad specificity, do not cleave polypeptides completely nonspecifically. Although almost every aa has been observed as a P1 residue, nearly all studies have also shown a clear preference for most hydrophobic and charged residues in polypeptide processing. Furthermore, all recent studies show that the P1' position also influences cleavage site selection, although the proteasome's P1' specificity is even broader than its P1 specificity. Every aa can be found in P1', but especially small aa seem to facilitate hydrolysis at the P1–P1' site. The results on cleavage preferences are largely in agreement with crystal structure analyses of the yeast proteasome. The S1 pockets of the β1, β2 and β5-subunits are the major specificity determinants and are appropriately polar and sized to accommodate acidic, basic, and apolar P1 side chains, respectively, but can also bind noncomplementary residues. For example, all subunits bind the hydrophobic norleucine of peptide aldehyde inhibitors (GROLL et al. 1997; BOCHTLER et al. 1999). There is also evidence that the peptide backbone has to be bent sharply for the hydrolysis to occur, perhaps explaining the beneficial effect of small aa in the P1' position (DITZEL et al. 1997, 1998).

Up- and downstream flanking residues also influence cleavage site selection and cleavage rate. Among the upstream residues, a preference for proline in P4 has been found in several analyses, as well as hydrophobic residues in P3 of acidic P1 residues, or for basic aa in P2/P3 of basic P1 residues. Several of the findings made with reporter peptides and inhibitors are consistent with the flanking sequence effects which have been observed in the analysis of polypeptide degradation products and cleavage sites (ORLOWSKI and MICHAUD 1989; CARDOZO et al. 1994; VINITSKY et al. 1994; BOGYO et al. 1998). Among the downstream residues, the P1' residue seems to have the most prominent influence. This conclusion is also supported by reports that mutations in the immediate flanking position of the C terminus of CTL epitopes (C + 1) can affect epitope generation (NIEDERMANN et al. 1995; YELLEN-SHAW et al. 1997; THEOBALD et al. 1998; BEEKMAN et al. 2000). Studies on a greater number of polypeptide/protein substrates would reveal which of the specific flanking sequence effects consistently occur in polypeptide cleavages.

Studies comparing cleavage site specificities of the different types of proteasomes have not yet been extensively performed. Apparently, different types of proteasomes generate partially differing peptide spectra. The nature and extent of cleavage site specificity differences between housekeeping- and immunoproteasomes have been investigated and the results are under debate, as will be discussed in more detail below. There also appear to be certain differences between 20S and 26S proteasomes. The reverse-phase high performance liquid chromatography spectra of peptides produced from casein by 20S and 26S rabbit muscle proteasomes were not identical (KISSELEV et al. 1999). In another study on casein degradation (EMMERICH et al. 2000), about 50% of cleavage sites used by 20S and 26S human erythrocyte proteasomes were identical, although only about 25% of the products were identical, with those of the 26S being on average only half as long than those of the 20S. Nevertheless, both 20S and 26S erythrocyte proteasomes shared a strong preference for hydrophobic P1 residues, especially leucine, which is particularly frequent at the C terminus of class I ligands. Therefore, regarding the C termini of products, both types of proteasomes theoretically can deliver immunocompetent peptides.

3.4 Length of Degradation Products

As mentioned above, proteasomes do not always cleave even optimal sites in a polypeptide. Puzzling as this may be, it may be related to some extent to the characteristic length distribution of proteasomal products, suggesting a mechanism that determines the distance between proteasomal cleavages. This was first described for the archaebacterial 20S proteasome (WENZEL et al. 1994). Archaebacterial proteasomes degraded the insulin B-chain (30aa) and hemoglobin (140aa) into a large number of different peptides; virtually all fell into the size range of 3 to 13aa. With both substrates, hepta-, octa- and nonapeptides constituted about 45%, and fewer than 14% were longer than decapeptides. These observations led Baumeister's group (WENZEL et al. 1994) to propose that proteasomes possess an intrinsic 'molecular ruler'. They further proposed that the ruler may be provided by the distance between the active sites or by properties of the substrate binding groove (see below).

We studied the product length of mammalian proteasomes for comparison with that of MHC class I ligands. Digesting a 44-aa ovalbumin (OVA) fragment and the small subunit of ribulose 1,5 bisphosphate carboxylase (123aa), we found a very similar length distribution for the products of murine 20S proteasomes (NIEDERMANN et al. 1996, 1997). For both substrates, almost all internal fragments were between 3 and 14aa. At all time points, including early during digestion, about 45% of the internal products had between 8 and 11aa, i.e., roughly the preferred size of class I ligands; 15% were longer, and the rest ranged from 3 to 7aa (i.e., too short for class I binding). With both substrates, 7- to 11-mers appeared to be produced most efficiently. Digestion of the two substrates with yeast or *Drosophila* proteasomes yielded a very similar length distribution of the products (NIEDERMANN et al. 1997). Recently, we digested the 207-aa HIV-1 Nef protein with human 20S proteasomes and again found 7- to 11-mers to be most abundant (about 45%). Only a few products were shorter than 8aa and about 15% were longer than 14aa (up to 22aa). Most long peptides were either proline-rich or from hydrophilic regions of the protein that do not contain CTL epitopes (M. Lucchiari-Hartz, V. Lindo, F. Greer, P.M. van Endert, K. Eichmann and G. Niedermann, unpublished data).

The 66 fragments identified by NUSSBAUM et al. (1998) from digestion of the 436-aa protein enolase by yeast 20S proteasomes were between 3 and 23aa; 4- to 9-mers dominated and only two peptides were longer than 13aa. The sizes of peptide products of insulin-like growth factor (IGF) (70aa), β-casein (209aa) and OVA (385aa) produced by rabbit proteasomes were studied by KISSELEV et al. (1999) using size-exclusion chromatography. Also here, fragments between 3 and 22 residues were found. Casein also yielded a few longer fragments of up to 30 residues, perhaps because of its unusual primary structure (16.7% prolyl residues) and because it is phosphorylated at multiple sites, both of which may retard cleavages. With all three substrates, two-thirds of the products appeared to be too short for presentation (< 8aa), 15% had 8 or 9 residues, and 15 to 20% were longer. The product size distribution appeared to follow a log-normal distribution, i.e., the abundance of the products tended to decrease as their size increased. Digesting casein with human erythrocyte 20S proteasomes, EMMERICH et al. (2000) also found products ranging from 4 to up to 35aa, but here only few products were shorter than 8aa. While qualitatively, the 20S' products appeared to be almost evenly distributed over the whole product size range, quantitatively, nonamers appeared to be most prevalent.

Only a few reports describe comparisons of fragment length among 20S and 26S proteasomes. Some of these data suggest that the 26S proteasomal products may be somewhat shorter than those of the 20S proteasome. The 12 major fragments identified in a 26S proteasomal digest of ornithine decarboxylase were between 5 and 11aa (TOKUNAGA et al. 1994). KISSELEV et al. (1999) found that the mean number of product residues was slightly (on average one or two residues) shorter for IGF (6.5 vs. 8.3) and OVA (7.4 vs. 9.5) and almost identical for β-casein (9.0 vs. 8.9). In contrast, EMMERICH et al. (2000), analyzing digests of β-casein produced by human erythrocyte proteasomes, found that the 26S proteasomal products were much shorter than those of the 20S proteasome (mean fragment sizes 18.3 for the 20S vs. 10.1 for the 26S). Both in number and particularly in amount, many 26S proteasomal casein products were shorter than 8aa (i.e., too short for MHC binding), whereas only very few 20S proteasomal products were shorter than 8aa (see above). However, as already mentioned, casein may not be a typical proteasome substrate owing to its unusual primary structure, and the difference in the product sizes between 26S and 20S proteasomes will need to be examined with additional, more typical proteasome substrates.

3.5 Mechanisms Underlying the Product-Size Distributions

The degradation of most protein substrates into a mixture of short oligopeptides is a consequence of the fact that proteolysis is confined to the proteolytic chamber in the proteasome. This complete degradation by multiple site cleavage without release of intermediates is called 'processive protein degradation', and is a feature of all self-compartmentalizing proteases (DICK et al. 1991; THOMPSON et al. 1994; BAUMEISTER et al. 1998). Both 20S and 26S proteasomes degrade most substrate proteins completely in a processive manner (AKOPIAN et al. 1997; NUSSBAUM et al. 1998; KISSELEV et al. 1999). A few exceptions are known, for example the nuclear factor (NF)-κβ-precursor protein p105, which contains a glycine-rich region that halts digestion and leads to the release of the p50 active subunit (LIN and GHOSH 1996).

For in vitro studies short polypeptides have often been used. However, with short polypeptides, processivity is usually less pronounced. 20S Proteasomes often cut short substrates first only at one site, so that mainly so-called 'single cleavage' intermediates are formed. However, analysis of 20S proteasome-mediated degradation of the 30-aa insulin B-chain revealed that, besides single cleavage products, some 'dual-cleavage products' are also made already at early times of substrate turnover (DICK et al. 1991). Careful analysis of the accumulation of a dominant dual-cleavage product (a nonamer) revealed that this peptide was produced via two pathways. One of them required the concerted action of two different active sites, leading to the immediate generation of this product by processive substrate degradation. This pathway dominated at early times of substrate consumption. The other one dominated later, and here the nonamer was generated from single cleavage intermediates that had been released into solution and recaptured.

Despite the fact that only two active subunits of the eukaryotic proteasome are direct neighbors (SEEMÜLLER et al. 1995; GROLL et al. 1997; see also Fig. 1), a proportion of final products may indeed be directly produced by 'simultaneous' dual cleavages on $\beta 1$ and $\beta 2$, or $i\beta 1$ and $i\beta 2$, respectively. This may, for instance, be true for the above mentioned nonamer derived from the insulin B-chain. Both the postglutamyl- and trypsin-like activities appeared to be responsible for its production (DICK et al. 1991), and it is known that the juxtaposed $\beta 1$- and $\beta 2$-subunits are responsible for these two activities (ENENKEL et al. 1994; ARENDT and HOCHSTRASSER 1997; HEINEMEYER et al. 1997).

It may be that short polypeptides are not preferentially degraded in a processive manner because the flanking regions of most cleavage sites are too short, i.e., have too low binding affinity to the substrate-binding regions of the proteasome. SHIMBARA et al. (1997), for example, found that 20S proteasomes generated a decamer precursor of an octamer CTL epitope derived from the c-Akt proto-oncogene by processive dual-site cleavage only from c-Akt fragments of at least 29aa. In this substrate, the N-terminal flank of the N-terminal cleavage site had 7aa and the C-terminal flank of the C-terminal cleavage site was 12aa long. In c-Akt fragments shortened N- or C-terminally, either only the N- or the C-terminal site was used by the 20S proteasome. Efficient processive dual-cleavage excision from the shorter polypeptides was possible in the presence of the interferon (IFN)-γ-inducible proteasome activator PA28, as had first been described by DICK et al. (1996) (see Sect. 5). The dual-cleavage products whose excision from short polypeptides was found to be accelerated by PA28 had between 7 and 11aa (DICK et al. 1996; NIEDERMANN et al. 1997; SHIMBARA et al. 1997).

Proteasomes cleave polypeptides specifically (see Sect. 3.3) but, as explained, a certain length of flanking residues is required for cleavage at a specific site (SHIMBARA et al. 1997; DOLENC et al. 1998). It is conceivable that there is no strict length minimum for the substrate anchoring residues. The substrate binding affinity may be determined not only by the number of residues, but may also be co-determined to some extent by the nature of the flanking residues and the nature of the specific P1–P1'-site. In addition, binding of regulatory/activating cap complexes may alter (reduce) the length required for efficient cleavage. In agreement with the importance of the binding affinity, DOLENC et al. (1998) have shown for the archaebacterial 20S that the substrate length critically determines its rate of degradation; e.g., a 30-mer was degraded 70 times faster than a 12-mer, and a sharp decrease in the velocity of degradation was observed for peptides shorter than 14 residues. They concluded that these peptides are not efficiently retained by the proteasome and have a high probability to exit, and that this is equivalent to a low affinity for peptides shorter than 14 residues. This interpretation is compatible with the observation made for the murine 20S that products of up to 14aa are remarkably stable in the presence of active proteasomes (NIEDERMANN et al. 1996).

To summarize, all types of proteasomes investigated degrade polypeptides efficiently into a mixture of mostly short oligopeptides. With most substrates, the vast majority of products are shorter than 14aa, and in several studies, a considerable proportion of products had roughly the preferred size of MHC class I ligands (i.e., 8 to 11 residues). Concerning class I presentation, the data imply that proteasomes produce peptides which fit directly into the groove, as well as peptides which are slightly longer, and which may often become presented after having been trimmed. In some studies, the percentage of peptides which would be too short for MHC binding appeared to be quite high. This is particularly true for the 26S proteasome. For the 20S, the data are less consistent, as in some 20S studies the percentage of products shorter than 8aa was low. In some studies, the percentage of peptides larger than 14aa was higher, but this may be due to unusual primary sequences of the substrates used.

4 Immunoproteasomes

Some elements of the vertebrate proteasomal system are upregulated by IFN-γ, an immunomodulatory cytokine secreted mainly by T helper 1, T cytotoxic 1 and natural killer (NK) cells in response to immune activation. In most cell types, its function is focused mainly on inhibition of viral replication and enhancement of antigen presentation (BOEHM et al. 1997). In addition to other components of the class I antigen processing and presentation machinery such as MHC class I, β2-microglobulin and TAP, IFN-γ-inducible elements include homologs of the three proteolytically active β-subunits of the proteasome and the 11S regulator (REG or PA28). 11S REG/PA28 was originally identified as a strong activator of the peptidase activities of the 20S proteasome from mammalian cells and forms a cap at one or both ends of the 20S proteasome (TANAKA and KASAHARA 1998; RECHSTEINER et al. 2000). Proteasomes containing the IFN-γ-inducible β-subunits are commonly referred to as 'immunoproteasomes' (see also Fig. 1). The 11S regulator will be discussed in a separate chapter.

4.1 The IFN-γ-Inducible β-Type Subunits

The inducible homologs of the three constitutive proteolytically active subunits β1, β2, and β5, are referred to as inducible β1 (iβ1) or LMP2, iβ2 or multicatalytic endopeptidase complex-like (MECL1), and iβ5 (LMP7). Upon induction, these subunits are incorporated into newly synthesized proteasomes in place of their constitutive counterparts (TANAKA and KASAHARA 1998). Incorporation of the three subunits occurs interdependently, so that under physiological conditions formation of homogeneous 'immunoproteasomes' appears to be favored (GROETTRUP et al. 1997; GRIFFIN et al. 1998). Thus, there seem to be mainly two types of 20S proteasomes, those with the constitutively expressed catalytic subunits and 'immunoproteasomes' incorporating the inducible subunits. In some normal tissues, particularly in hematopoietic cells (including professional antigen-presenting cells), there is also considerable constitutive expression of the inducible subunits. The *lmp2* and *lmp7* genes are located in the mammalian MHC, and immunoprecipitation of LMP-containing protein complexes with an anti-MHC alloantiserum was a key finding in defining the role of proteasomes in antigen processing (MONACO and MCDEVITT 1986; NANDI et al. 1998).

Two nonmutually exclusive hypotheses were put forward after the discovery of the close linkage of the *lmp* and *tap* genes in the mammalian MHC: (a) the LMPs may have a docking function to physically link proteasomes to TAP, thus facilitating the direct transfer of antigenic peptides, and (b) they may tailor proteasomal specificity for production of peptides better suited for binding to class I (MONACO 1992). In support of the former hypothesis, electron microscopic studies have shown that a small proportion of proteasomes is associated with the ER in different cell types in higher vertebrates (RIVETT et al. 1992), and subcellular fractionation studies have demonstrated an enrichment of the LMPs in the microsomal fraction (YANG et al. 1992; PALMER et al. 1996). However, proteasomes containing an LMP2–green fluorescent protein fusion appeared to be more or less evenly distributed throughout the cytoplasm and nucleus even after treatment of the cells with IFN-γ (REITS et al. 1997). Furthermore, association of proteasomes to the ER appears much more pronounced in yeasts than in vertebrates (ENENKEL et al. 1998). Many proteasomes may simply reside in the vicinity of the ER through physical engagement of ER substrates (HIRSCH and PLOEGH 2000) and perhaps also through engagement of misfolded nascent polypeptides which are continuously and rapidly eliminated by proteasomes (see Sect. 3.1). A docking function of LMP subunits therefore remains uncertain. In contrast, evidence has been accumulating that immunoproteasomes have altered peptidolytic properties, consistent with the fact that of the 14 different proteasomal subunits only the three proteolytically active have inducible homologs (SEEMÜLLER et al. 1995; GROLL et al. 1997). While differences in enzymatic activities have been observed by several groups, the data have not always been consistent (see below). Therefore, the nature and magnitude of these alterations – as well as the consequences for the immune system – are not yet fully understood. Both a closer association of immunoproteasomes with TAP as well as altered specificities could affect peptide presentation.

4.2 Observations on LMP-Deficient Cells and Knockout Mice

Initial studies were performed with lymphoblastoid cell lines with a large homozygous deletion in the MHC class II region encompassing both *tap1/2* and *lmp2/7*, which were transfected with *tap*. Such transfectants showed neither reduced class I cell surface expression, which is found if stabilizing peptide is lacking, nor significant defects in CTL recognition of a relatively large panel of TAP-dependent viral epitopes (ARNOLD et al. 1992; MOMBURG et al. 1992; YEWDELL et al. 1994; ZHOU et al. 1994). Although CTL lysis assays may not always be sensitive to subtle or even modest changes in epitope densities, for many epitopes the MHC-encoded proteasome subunits did not appear to be essential for presentation. However, subsequent work with such cells demonstrated selective presentation defects. In one study, presentation of three viral epitopes was severely compromised although two of them had been found in a previous study to be presented in the absence of the LMPs. Presentation of four other viral epitopes remained unimpaired (CERUNDOLO et al. 1995). For one of the compromised epitopes it was demonstrated that it could be presented when LMP7 was expressed (CERUNDOLO et al. 1995; GILEADI et al. 1999). Also, presentation of a subdominant HIV-1 reverse transcriptase epitope was improved when LMP7 was expressed, whereas that of an immunodominant (ID) reverse transcriptase epitope was independent of the LMPs, but was still inhibitable by proteasome inhibitors and therefore probably depended on the constitutive proteasome (SEWELL et al. 1999). An LMP2-deficient mouse cell line presented three out of four tested H-2K^k-restricted influenza virus epitopes inefficiently; although basal class I cell surface expression was normal, full upregulation of H-2K^k and D^k upon IFN-γ treatment could not be achieved (SIBILLE et al. 1995). In view of the cleavage specificity differences between housekeeping- and immunoproteasomes discussed below, it is of interest that all epitopes investigated in this study contained acidic residues and that H-2K^k uses glutamic acid as an anchor in position 2.

Partial presentation defects had been revealed before with knockout mice. LMP7-deficient mice had a slight reduction of MHC class I molecules on lymphocytes and macrophages and presented the HY male antigen with reduced efficiency (FEHLING et al. 1994). LMP2-deficient mice, in contrast, had unimpaired class I cell surface expression, including on spleen cells which have constitutively high expression of the inducible β-subunits, but had slightly reduced numbers of $CD8^+$ T lymphocytes and generated fewer CTL precursors to an influenza virus nucleoprotein (NP) epitope. However, the response to Sendai virus and recognition of the OVA-derived SIINFEKL epitope were normal (VAN KAER et al. 1994). Together, the studies on LMP2 and/or LMP7-deficient cells and mice have suggested that the MHC-encoded subunits are important for the presentation of some, but not all class I-binding peptides.

4.3 Alterations of Endopeptidase Activities as Observed with Reporter Peptides

Several groups have studied the influence of the IFN-γ-inducible β-type subunits on proteasomal endopeptidase activities with fluoro- or chromogenic reporter peptides. The results have not always been consistent and the reasons for the inconsistencies are not clear. Initially, proteasomes isolated from IFN-γ-treated cells were reported to have about twofold higher chymotrypsin- (Suc-LLVY-MCA-hydrolyzing) and trypsin-like (Boc-LLR-MCA-hydrolyzing) activities (DRISCOLL et al. 1993; GACZYNSKA et al. 1993; AKI et al. 1994), but lost about half of the postglutamyl (Cbz-LLE-MNA-hydrolyzing) activity (GACZYNSKA et al. 1993; AKI et al. 1994). It was suggested that these functional alterations favor the degradation of proteins to peptides that terminate in hydrophobic or basic residues that are usually found on mammalian MHC class I molecules. However, USTRELL et al. (1995) found no significant impact of LMP subunits on the peptidase activities of purified proteasomes. Kloetzel and colleagues (BOES et al. 1994; KUCKELKORN et al. 1995) found reduced chymotrypsin-like and postglutamyl activities and unaffected trypsin-like activity. The same was described by ELEUTERI et al. (1997), who, in addition, found strongly enhanced cleavages of the BrAAP activity after branched chain or aromatic P1 residues in the peptidyl portion of reporter substrates. In transfection experiments, GACZYNSKA et al. (1996) found that overexpression of LMP7 (iβ5) enhanced the chymotrypsin- and trypsin-like activities, while that of LMP2 (iβ1) led to reduced postglutamyl activity. In another study (KUCKELKORN et al. 1995) the chymotrypsin-like activity was unaffected when either of the MHC-encoded subunits was expressed alone, but was strongly reduced when the two subunits were expressed together. In addition, the postglutamyl activity was strongly suppressed. The only relatively consistent finding of the reporter peptide studies is a reduced post-acidic activity of the immunoproteasome. Both the chymotrypsin-like and BrAAP-activities cleave after hydrophobic residues (see Sect. 3.2). Whereas the data on the chymotrypsin-like activity are controversial, the BrAAp activity appears to be strongly increased.

4.4 Influence on Cleavage Site Usage within Polypeptides

Upon digestion of 20- to 30-mer polypeptides different effects of the IFN-γ-inducible subunits have been reported as well. Kloetzel and colleagues, while observing changes in the cleavage profiles, could not correlate these changes with those observed with fluorogenic substrates measuring the three classical peptidase activities. Proteasomes from cytokine-induced cells showed only quantitative differences in cleavage site usage within a pp89 25-mer fragment (BOES et al. 1994), whereas proteasomes from cells transfected with either LMP2 or LMP7 or both also showed some qualitative alterations (GROETTRUP et al. 1995; KUCKELKORN et al. 1995). EHRING et al. (1996) did not observe any differences in the cleavage pattern of the oxidized insulin B-chain depending on the immunosubunits. In two polypeptide substrates, we found enhanced cleavage after hydrophobic aliphatic (leucine and valine) P1 residues (i.e., branched chain aa) and reduced cleavage after acidic (glutamate) P1 residues (NIEDERMANN et al. 1997). Reduced cleavages after acidic and enhanced cleavages after branched chain aa were also found by CARDOZO and KOHANSKI (1998) for the small protein lysozyme. However, cleavages were decreased after aromatic aa, which along with branched chain aa are also frequent in the C-terminal position of the ligands of several class I alleles. Both quantitative and qualitative changes in cleavage site usage were observed, although some of the peptides exclusively found in 'immunoproteasome' digests were present only in small amounts.

The X-ray structure of the yeast proteasome (GROLL et al. 1997; BOCHTLER et al. 1999) predicts that the binding sites for the P1 residue (S1 pocket) of the IFN-γ-inducible subunits LMP7 (iβ5) and MECL1 (iβ2) do not differ substantially from those of their housekeeping counterparts. In contrast, two aa in the predicted S1 pocket of LMP2 (iβ1) differ substantially from those of the constitutive subunit Y (Arg45 → Leu at the base of the pocket, and Thr31 → Phe), making the S1 pocket of LMP2 more apolar and abolishing its preference for acidic residues. Thus, the crystal structure data are consistent with reduced capability of the immunoproteasome to produce peptides with an acidic residue at the C-terminus and enhanced capability to produce peptides with a hydrophobic end, although it should be emphasized that the constitutive proteasome already has a high capacity to cleave after hydrophobic residues (NIEDERMANN et al. 1997; CARDOZO and KOHANSKI 1998; NUSSBAUM et al. 1998). Interestingly, both inducible subunits encoded in the MHC have a preference for hydrophobic P1 residues. The shift in preference of the S1-pocket of LMP2 from acidic to hydrophobic P1 residues may indeed represent an adaptation to the binding requirements of TAP and MHC class I.

4.5 Comparisons of Effects on Epitope Production In Vitro and In Vivo

Several studies have addressed the question whether the IFN-γ-inducible β-type subunits affect epitope generation by purified proteasomes in vitro. Although purified immunoproteasomes produce more epitope precursors of the ID L^d-binding pp89 epitope of murine cytomegalovirus (MCMV) than housekeeping proteasomes (GROETTRUP et al. 1995; KUCKELKORN et al. 1995), overexpression of the LMPs did not improve recognition of pp89-transfected cells by CTLs (GROETTRUP et al. 1996a). Also, the quantitative changes in the ID epitope SIINFEKL and a precursor thereof in OVA fragment digests (NIEDERMANN et al. 1997) were not reflected in significant changes in CTL epitope recognition of OVA transfectants upon pretreatment with IFN-γ (NIEDERMANN et al. 1995). In contrast, several recent studies have demonstrated correlations between improved in vitro production by immunoproteasomes and CTL recognition of viral epitopes. Most of these studies were done with fibroblast cell lines that express all three immunosubunits together upon induction. Given that interdependent immunosubunit incorporation seems to be favored, this triple transfection system may generate more significant results than previous transfections of only single subunits (see above). Furthermore, the inducible system makes it possible to investigate the immunosubunit effects independently of those of the many other IFN-γ-inducible gene products and avoids clonal effects.

In one of the studies using this system, an E1B-derived CTL epitope of adenovirus type 5 was shown to be recognized more efficiently upon expression of the three inducible β-type subunits. Although both types of 20S precisely excised the epitope in vitro, immunoproteasomes cleaved the epitope boundaries (Leu–Val and Ile–Ser bonds) faster than the constitutive proteasomes; an E1A-derived epitope was not affected by the inducible subunits (SIJTS et al. 2000b). Using the same overexpression system, CTL recognition of an ID H2-L^d-binding epitope derived from the NP of lymphocytic choriomeningitis virus (LCMV) was enhanced two- to threefold. Here, purified immunoproteasomes generated higher amounts of N-terminally elongated precursors of the epitope, which is probably not a good TAP substrate owing to a proline in position 2 (SCHWARZ et al. 2000b). VAN HALL et al. (2000) used the triple-transfectant system to study the effect on presentation of two Moloney murine leukemia virus (MuLV)-derived CTL epitopes, an ID Gag-derived and a subdominant Env-derived epitope. Presentation of the ID epitope was markedly enhanced when expression of the immunosubunits was induced. Basal presentation by uninduced fibroblasts was inhibitable by several proteasome inhibitors and was therefore probably dependent on housekeeping proteasomes. Presentation of the Env-derived epitope was not improved upon immunosubunit expression and appeared to be entirely dependent on the housekeeping proteasome.

A very pronounced effect was seen for an 11-aa ID epitope of the HBV core protein (SIJTS et al. 2000a). CTLs recognized the epitope only on IFN-γ-treated, but not on untreated HeLa cells. This may have been due to low expression of the LMPs and MECL-1, as in vitro only immunoproteasomes produced the 11-mer. Both constitutive and immunoproteasomes precisely cleaved at the epitope N terminus, but only the latter efficiently liberated the C terminus. Since hepatocytes contain mostly constitutive proteasomes, the epitope may be efficiently presented by HBV-infected hepatocytes only during the acute phase of infection when sufficient IFN-γ is generated. The authors therefore proposed that the failure of hepatocytes to generate certain viral epitopes could contribute to persistence of HBV in chronically infected patients.

Another study (MOREL et al. 2000) demonstrated that immunoproteasomes are not always better than housekeeping proteasomes in the production of antigenic peptides. On the contrary, the study has identified epitopes which are made by housekeeping proteasomes, but not by immunoproteasomes. By stimulating lymphocytes with an autologous renal carcinoma, the authors obtained CTLs specific for a ubiquitously expressed protein, RU1. These CTLs recognized tumor cells and other autologous cells in a proteasome-dependent manner, but failed to lyse EBV-transformed B cells although the latter expressed the protein and could be lysed if pulsed with antigenic peptide. EBV-B cells constitutively express mostly immunoproteasomes (FRISAN et al. 1998) and it was therefore conceivable that inadequate immunoproteasomal processing was responsible for lack of presentation. In vitro digestion of an RU1 fragment covering the epitope provided evidence for this scenario. Renal carcinoma cell proteasomes, which were shown to be mostly housekeeping proteasomes, generated epitope-containing fragments by cleavage after the C-terminal valine. In contrast, proteasomes from EBV-B cells and IFN-γ-treated carcinoma cells cleaved at the C terminus only poorly or not at all and destroyed the epitope mainly by internal cleavage after a phenylalanine.

Together, the above cited in vivo and in vitro data allow the following conclusions:

a. Both types of (20S) proteasomes, housekeeping and immunoproteasomes, appear to be involved in generation of CTL epitopes and MHC class I ligands in general. This is suggested by inhibitor experiments on cells exclusively expressing one or the other type as well as by in vitro digestions of epitope-containing substrates.
b. The cleavage specificity of the immunoproteasome is altered compared to that of the housekeeping proteasome. There appears to be a consensus now that the most evident alteration is the increased production of peptides with hydrophobic C termini, at the expense of peptides with acidic ends. A proportion of peptide products, including epitopes or epitope precursors, is qualitatively identical although there can be quantitative differences. A certain proportion of peptides, including epitopes and epitope precursors, is produced only by one or the other form.

4.6 Ambiguous Consequences of the Dichotomy of Proteasomes for Self-Tolerance and Immune Responsiveness of CTLs

We have previously shown that the proteolytic fragments generated by housekeeping proteasomes of lower eukaryotes, i.e., before evolution of the adaptive immune system, comprise many peptides perfectly suitable for binding to MHC class I (NIEDERMANN et al. 1997). The appearance of the immunoproteasome is therefore most probably the result of an evolutionary pressure on the preexisting housekeeping proteasome towards the optimization of certain functional aspects of the adaptive immune system. Here we consider three such aspects, the evolution of antigen presentation from a putative primordial system with a different function, the cytolytic elimination of virus-infected or malignant cells, and self-tolerance.

4.6.1 Evolution of Antigen Presentation

The evolution of antigen presentation itself may have been a driving force that led to the appearance of the immunoproteasome. There is evidence that class I-like molecules existed in evolution before the appearance of the adaptive immune system (KASAHARA 1998). They arose later in evolution than the proteasome and we previously suggested that their peptide binding groove may have adapted to the preexisting products of proteasomal proteolysis (NIEDERMANN et al. 1997). Accordingly, their peptide supply came from housekeeping proteasomes. The function of this system may have been associated with innate immunity, most probably to provide cell surface ligands for receptors of NK cells (KÄRRE 1997). These primordial MHC molecules may not have been highly polymorphic, and the nature of the repertoire of self-peptides presented to NK cells may not have been of critical consequence. The pronounced MHC polymorphism is likely to have arisen together with the appearance of the elements of the adaptive immune system and may have required alterations and specialization of peptide repertoires. For recognition of cells by antigen-specific CTLs, and in particular for induction of specific CTLs, individual peptides had to associate in sufficient copy numbers with at least one of the several different class I molecules expressed by the cell. Moreover, species highly polymorphic for MHC class I require a functionally monomorphic TAP to service all possible class I molecules (LOBIGS and MULLBACHER 1993; YEWDELL et al. 1993). For a monomorphic transporter it may be difficult to efficiently transport all of the peptide species produced by the housekeeping proteasome, especially peptides with oppositely (positively or negatively) charged C terminus in addition to the bulk of peptides with a hydrophobic one. Together these factors may have provided the evolutionary pressure to generate a specialized peptide repertoire adjusted to interact with TAP and MHC class I molecules. As discussed above, the immunoproteasome differs from the housekeeping proteasome primarily in the LMP2 (iβ1) subunit which prefers hydrophobic P1 residues in contrast to the preference of the constitutive β1 subunit for acidic P1 residues. As a result, the immunoproteasome, compared to the housekeeping proteasome, exhibits a further enhanced preference for hydrophobic P1 residues and thus appears functionally more restricted. As has already been often speculated, this may reflect adaptation to MHC class I and TAP, and may be the result of co-evolution of the three systems. Most proteins contain more hydrophobic than either positively or negatively charged residues, and housekeeping proteasomes (even from invertebrates) already have a pronounced preference for hydrophobic P1 residues. Thus, proteasomal products with hydrophobic C termini appear to be well suited for representation of the cytosolic protein content (by the many different MHC class I allelic products), and a primordial class I–peptide presentation system may have already evolved to present such peptides. Previous hypotheses had invoked the need for a greater peptide diversity as the main evolutionary reason for the advent of the immunoproteasome (GROETTRUP et al. 1996b; NANDI et al. 1996). While a greater diversity of peptides may indeed be generated with two types of proteasomes, we feel that the most plausible purpose of the immunoproteasome may be the efficient production of a specialized and restricted peptide repertoire, compatible with TAP and MHC and therefore efficiently represented at the cell surface.

4.6.2 Elimination of Virus-Infected or Malignant Cells

Proteasomal peptides appear to be divided into three repertoires, those exclusively produced by housekeeping proteasomes, those exclusively produced by immunoproteasomes, and the common repertoire produced by both. Effective immune defenses against viruses and tumors depend on specific CTLs that lyse the cells that they recognize by antigenic peptides presented by class I molecules. Antiviral CTLs are induced predominantly by dendritic cells (DCs) processing viral proteins, either synthesized within the DCs or by cross-presentation, i.e., after phagocytosis of apoptotic virus-infected cells. Mature DCs, i.e., after migration to the lymph node and at the time of CTL stimulation, express predominantly immunoproteasomes and possess hardly any housekeeping proteasomes (MACAGNO et al. 1999; MOREL et al. 2000). Accordingly, antiviral CTLs recognize predominantly viral peptides generated by immunoproteasomes. Most other cells express housekeeping proteasomes, whereas expression of immunoproteasomes depends on IFN-γ, i.e., on an ongoing immune response. As most viruses infect cells other than DCs, effective cytolysis of infected cells by antiviral CTLs may often depend on the induction of immunoproteasomes by IFN-γ. IFN-γ is amply produced during acute infection, but may fall below critical levels thereafter. Virus-infected cells surviving acute clearance present viral peptides produced by housekeeping proteasomes and may therefore be ignored by the proportion of CTLs which recognize peptides exclusively produced by immunoproteasomes (SIJTS et al. 2000a). Depending on how great this proportion is, this may lead to efficient viral clearance or to persisting virus infections and cases of virus latency. For example, noncytopathic or poorly cytopathic viruses are often not completely eradicated by the initial immune response (ZINKERNAGEL et al. 1997). Thus, the strategy of restricting immune induction to peptide repertoires produced by immunoproteasomes may have disadvantages with respect to viral clearance in chronic infections. For the sake of effective antiviral immunity, at least in the chronic state, the common repertoire of peptides produced by both types of proteasomes should be as large as possible.

Tumor antigens are for the most part endogenous proteins that are processed by housekeeping proteasomes. Anti-tumor CTLs are induced by DCs processing tumor proteins by cross-presentation. These CTLs recognize tumor antigens processed by immunoproteasomes, derived predominantly from tissue or tumor-specific proteins. Therefore, effective immune tumor destruction by CTLs is restricted mostly to the recognition of the common peptide repertoire. Hence, as for antiviral immunity, effective anti-tumor immunity requires a large common peptide repertoire.

4.6.3 Self-Tolerance

One of the most profound requirements of an adaptive immune system is that the repertoire of immunocytes, expressing an almost infinite number of different receptors, must be capable of mounting an immune response to essentially any foreign structure while exhibiting tolerance to virtually all endogenous structures. Nonresponsiveness to self can be induced and maintained by various mechanisms at several stages of development of a T cell. For the present discussion we focus on two major mechanisms, tolerance by thymic deletion and ignorance (LANGMAN 2000). For T lymphocytes, it is now known that negative selection occurs upon differentiation in the thymus, deleting the T cells with receptors that have high affinity for self-peptides in conjunction with self-MHC molecules. However, even though expression of peripheral proteins in the thymus has been observed (KLEIN and KYEWSKI 2000), it seems hardly possible that all endogenous proteins, including proteins specifically expressed in peripheral tissues, are presented to T cells in the thymus. Moreover, negative selection in the thymus is mediated primarily by cells of hemopoietic origin, predominantly DCs. The overall nonresponsiveness of peripheral T cells to peripheral tissue-specific antigens apparently not presented by thymic DCs has become easier to understand since it became known that the initial activation of naive T cells in the periphery is mediated almost exclusively by DCs as well (BANCHEREAU and STEINMAN 1998). Peripheral activation of CTLs, on the one hand, is restricted to peptides derived from proteins processed in DCs, thus excluding the endogenous proteins to which deletion has been induced in the thymus. On the other hand, endogenous proteins not expressed in DCs usually do not activate CTLs, although CTLs recognizing these proteins are not deleted in the thymus. This latter phenomenon has been termed ignorance, to emphasize the difference from tolerance, i.e., thymic deletion.

The composition of the proteasomes expressed in thymic DCs is not known at present. However, the differences in the peptide repertoires generated by housekeeping- and immunoproteasomes summarized in the previous chapters of this volume potentially adds a novel dimension to present concepts of the roles of tolerance and ignorance in maintenance of nonresponsiveness to self. If thymic DCs express predominantly immunoproteasomes similar to mature peripheral DCs, tolerance would not be restricted only to endogenous proteins processed in DCs, but also to those peptide repertoires generated by immunoproteasomes. Conversely, the peptides generated exclusively by housekeeping proteasomes are ignored, even though the source proteins may be processed by DCs as well. The putative strategy of the immune system may be a limitation of the peptide repertoire to which tolerance is induced, i.e., only that produced by immunoproteasomes. A positive biological consequence of this strategy would be a maximal diversity of T cell specificities surviving thymic deletion. Another potential advantage would be that the peptide repertoire exclusively produced by housekeeping proteasomes would be excluded from potential immune activation in the periphery. Either purpose requires a common repertoire of limited size, in apparent contrast with anti-viral and anti-tumor immunity. With respect to the avoidance of autoimmunity such a strategy would appear at least to be ambiguous. On the one hand, a common repertoire of limited size may minimize the chance of autoaggression by activated peripheral CTLs. On the other hand, a maximally diverse CTL repertoire contains more potentially autoreactive cells that could become autoaggressive upon accidental activation. Two types of experiments seem relevant to answer these questions: assessing the nature of the proteasomes and characterizing the peptide repertoires presented by cells involved in thymic selection and peripheral immune activation. Taken together, the present knowledge suggests that the immunoproteasome may have evolved in the context of the adaptation of the primordial molecular system to the requirements of foreign antigen presentation. However, not all aspects of adaptive immunity profit from a drastically different immunoproteasome, perhaps a reason for the maintenance of a fair degree of functional similarity to the housekeeping proteasome.

5 11S REG/PA28: the IFN-γ-Inducible ATP-Independent Activator

11S REG, also referred to as PA28, is one of two types of regulatory cap complexes of the 20S proteasome. It forms hexa- or heptameric ring-shaped complexes of 28 kDa subunits (RECHSTEINER et al. 2000). The complex has a molecular mass of approximately 200 kDa and sediments at 11S (DUBIEL et al. 1992). Similar to the 19S caps (PA700), the 11S regulator binds to one or both α-rings of the 20S proteasome (GRAY et al. 1994; see also Fig. 1) and stimulates its peptidase activities, although to a much greater extent than the 19S caps, and independently of ATP hydrolysis (YUKAWA et al. 1991; DUBIEL et al. 1992; MA et al. 1992; RECHSTEINER et al. 2000). The regulator does not confer the ability to degrade ubiquitinated proteins to the 20S proteasome (DUBIEL et al. 1992; MA et al. 1992), but it cannot be excluded that 20S proteasome–PA28 complexes degrade proteins if they are provided unfolded. Furthermore, 11S REG/PA28 may affect protein degradation when contained within the more recently identified so-called 'hybrid proteasome' which contains a 19S cap on one side and an 11S cap on the other (HENDIL et al. 1998).

The REG/PA28 family consists of three members: α, β, and γ. Whereas γ forms homo-oligomers found largely in the nucleus, α and β predominantly form hetero-oligomers, which have been found mostly in the cytoplasm. In some studies, the α- and β-polypeptides were also found in the nucleus (DUBIEL and KLOETZEL 2000; RECHSTEINER et al. 2000). REG/PA28 α and β-subunits occur only in vertebrates and are upregulated by IFN-γ. Together with their relatively high expression in cells and organs of the immune system, this has suggested a role in antigen processing. Unlike the LMPs, REG/PA28 α and β are not encoded in the MHC (McCUSKER et al. 1999). One of the first direct hints for a role in antigen processing was the observation that REG/PA28α overexpression on fibroblasts, transfected with pp89 or infected with influenza virus, facilitated CTL recognition of two ID epitopes, the L^d-binding pp89-epitope of MCMV and the K^d-binding influenza virus epitope NP147–155 (GROETTRUP et al. 1996a).

5.1 Effects on the Processing of Short Polypeptides

To understand the positive effect on antigen presentation mechanistically, short synthetic polypeptides have been digested with purified 20S proteasomes in the presence or absence of the activator. In the first of these studies (GROETTRUP et al. 1995), a 25-aa pp89-derived fragment covering the ID L^d-binding epitope was digested and the peptide products were analyzed after complete substrate turnover. Under these conditions, binding of purified cytosolic REG/PA28 to the 20S proteasome markedly changed both the quality and quantity of products, but did not appear to be beneficial for production of the epitope. Subsequent kinetic studies on this substrate revealed that REG/PA28 binding strongly enhanced the production of peptides which required two cleavages, especially at early times of substrate consumption (DICK et al. 1996). These peptides were 7–11aa long and included the nonamer pp89 ID epitope and a precursor thereof. A pronounced effect was also seen for the dominant H2-K^d-ligand SYFPEITHI contained within a 19- or 21-mer fragment of the source protein JAK1. Importantly, PA28/REG binding, while markedly enhancing the excision of dual-cleavage peptides from these short substrates, did not significantly increase the rate of substrate consumption and did not change the proteasome's cleavage site preferences. However, a difference was observed in the cleavage mechanism. In the absence of REG/PA28, the 20S proteasome initially cleaved the short substrates only at one site, and further attack of accumulating single cleavage intermediates was not observed until much later. In the presence of REG/PA28, the very same cleavages appeared to occur 'simultaneously' or in a 'coordinated' fashion, and kinetic data suggested that either the two cleavages occurred in the substrate cooperatively bound to two neighboring β subunits or that the second cleavage occurred on an intermediate still bound to one of the two active sites, i.e., on a long-lived acyl-intermediate.

Similar observations were made in experiments monitoring the production of the OVA epitope SIINFEKL by digesting a 22-mer OVA fragment in the presence and absence of recombinant REG/PAα (NIEDERMANN et al. 1997). Furthermore, SHIMBARA et al. (1997) studied production of an L^d-binding epitope from the c-Akt proto-oncogene by digesting antigenic fragments of variable length. Irrespective of the presence of REG/PA28, 20S proteasomes produced the epitope precursor SIIPGLPLSL from all substrates in which the epitope's flanking regions had the minimally required length. When the N- and C-terminal flank had at least 7 and 12aa, respectively, the 20S proteasome produced this precursor directly by dual cleavages, even in the absence of REG/PA28, and addition of the latter yielded a two- to threefold higher level of dual-cleavage peptides. When the flanking regions were shortened, the 20S proteasome changed to the single cleavage mode, i.e., cleaved only at one or the other end of the precursor peptide. In contrast, 20S–PA28 complexes continued to use the dual-cleavage mode, at least with some substrates with shorter flanking regions. When shortened further, 20S–PA28 complexes also changed over to the single-cleavage mode, but these single cleavages were accelerated compared to those of the free 20S proteasome. These data suggested that PA28/REG binding can accelerate single and double cleavages, perhaps by increasing substrate affinity.

DICK et al. (1991) had already demonstrated for a nonamer peptide produced upon degradation of the 30aa oxidized insulin B-chain that the 20S proteasome per se can excise at least some fragments directly by dual cleavages even from short substrates. The mechanism proposed at that time was channeling of intermediates. This process, which leads to dual- or multiple-site cleavage without release of intermediates, was later termed 'processive degradation' (THOMPSON et al. 1994), as discussed above in the context of fragment length (see Sect. 3.5). In the case of short polypeptide substrates ($\approx$20–30-mers), REG/PA28-binding can enhance the processivity of substrate degradation and thereby improve the 'one-step' liberation of epitopes and/or slightly larger epitope precursors from the substrate.

It has been questioned that REG/PA28 induces 'coordinated' or 'simultaneous' cleavages on two neighboring active sites, arguing that subunit β5 (X)/iβ5 (LMP7), which is flanked by two inactive β-subunits on either side, prefers hydrophobic P1 residues and is thought to be important for the production of the C termini of class I-binding peptides. However, as already mentioned, all three different constitutive active β-subunits bind the hydrophobic norleucine P1 residue of peptide aldehyde inhibitors and the BrAAP activity of the constitutive proteasome has been at least partially assigned to β1 (Y) (CARDOZO et al. 1996; DICK et al. 1998a; MCCORMACK et al. 1998). Furthermore, yeast proteasome mutants lacking β5 still cleave efficiently after hydrophobic residues in proteins, showing that β1 and/or β2 can indeed cleave after such residues (NUSSBAUM et al. 1998). As already mentioned, based on the crystal structure of the yeast 20S, it has been predicted that the S1-pocket of iβ1 (LMP2) is apolar and therefore probably prefers hydrophobic residues anyway. Irrespective of this, it is conceivable that the observed enhanced processivity of substrate degradation need not necessarily be due to 'simultaneous' cleavages at two neighboring active sites. The same effect may be achieved by longer retention of single cleavage intermediates within the catalytic chamber and/or enhanced catalytic activity of individual active sites – the latter being equivalent to a higher affinity of cleavage intermediates and short substrates.

It is not known whether 20S–PA28 complexes are provided with such short substrates intracellularly. For a while it had been assumed that proteins may first be cut by 26S proteasomes into relatively large intermediate pieces that could then serve as substrates for 20S–PA28 complexes, finally leading to small peptide fragments (DICK et al. 1996). This scenario now appears to be rather unlikely, as the 26S proteasome itself has been shown to degrade proteins in a highly processive manner to oligopeptides, most being too short for class I binding (see Sect. 3.4). Nevertheless, it cannot be excluded that the PA28-capped proteasome is provided with relatively short substrates; e.g., cryptic translation products (BOON and VAN PEL 1989; MAYRAND and GREEN 1998; MALARKANNAN et al. 1999) and some of the DRIPs may be relatively short, or nonproteasomal proteases may provide intermediate-sized processing intermediates. The so-called DRIPs (YEWDELL et al. 1996), the result of an apparently high failure rate of proper folding of nascent polypeptides, appear to be particularly abundant (see Sect. 3.1). Two recent studies suggest that MHC class I molecules primarily present peptides that originate from newly synthesized polypeptides (perhaps mainly from DRIPs) to ensure rapid presentation to the immune system (REITS et al. 2000; SCHUBERT et al. 2000).

5.2 Are There Effects on the Processing of Protein Substrates?

In contrast to short polypeptide substrates, proteins have been reported to be degraded by both 20S and 26S proteasomes in a highly processive manner (see Sect. 3.5). Moreover, it has not been demonstrated that PA28 binding improves the 20S proteasome's capacity to degrade proteins. If 20S–PA28 complexes indeed degrade proteins, it would be interesting to see whether, similar to the findings made with short polypeptide substrates, PA28 binding would further improve the direct liberation of epitopes and epitope precursors from the longer substrates. Several possible mechanisms for the effects of PA28 on the processing of short polypeptides have been postulated as discussed above. Activation of the peptidase activities (reducing the length requirement for the substrate anchoring residues), induction of 'simultaneous'/'coordinated' dual cleavages on β1/iβ1 and β2/iβ2, or longer retention of degradation intermediates (enhancing the chance of more cleavages), would all yield more rapidly small oligopeptides (including epitopes and epitope precursors) that can readily diffuse out of the catalytic chamber. Thus, it is conceivable that the time needed for complete digestion of a protein would be reduced. In addition, the proportion of longer protein cleavage products might be reduced.

Although the studies on polypeptides do not indicate that PA28 binding changes the cleavage site specificity of the proteasome, certain alterations cannot be excluded, as studies with reporter substrates have shown dramatic differences in the extent of the increase of hydrolysis depending on the substrate (RECHSTEINER et al. 2000).

It has also been proposed that REG/PA28 binding may facilitate the egress of peptide products from the proteolytic chamber of the proteasome by enlarging putative product exit sites. Crystal structure analysis of recombinant REGα-heptamers together with mutational analyses has provided support for the model that REG binding widens the α-ring channel of the 20S proteasome thus forming a continuous channel leading from the upper surface of REG to the interior of the proteasome. As the surface of the regulator's pore appears to be very hydrophilic, it has been suggested that REG binding may facilitate the entry or egress of small, water-soluble peptides (KNOWLTON et al. 1997). Facilitation of the efflux of products has also been considered by Baumeister and colleagues based on the observation that binding of the regulator weakens the interaction between rings in the 20S proteasome to the extent of partial disassembly (BAUMEISTER et al. 1998). Facilitation of product release would perhaps also speed up substrate turnover but might result in increased rather than decreased product length. In this context, it would be interesting to examine whether the hybrid proteasome, which may degrade many ubiquitinated proteins in infected cells, produces peptides that are on average larger than that of the pure 26S proteasome (see Sects. 3.4 and 3.5).

Recently, evidence for an in vivo association of PA28 with misfolded proteins has been obtained, namely in heat shock protein (Hsp)90-dependent protein refolding (MINAMI et al. 2000). In this pathway, PA28 is believed to take over misfolded proteins which have been kept by Hsp90 in a refolding competent state, and to transfer them to Hsp70 for full reactivation. Given the apparent dual role of other regulatory protease cap-complexes in protein degradation and refolding (ZWICKL and BAUMEISTER 1999), it is conceivable that under conditions where transfer to Hsp70 is unsuccessful, PA28 may reroute the substrate to the 20S proteasome in order to degrade it. In addition, PA28 may bind and transfer to the proteasome unfolded proteins independently of its apparent role in Hsp90-dependent protein remodeling. In this context, it is interesting to note that some proteasome preparations contain chaperones, particularly Hsp90 (TSUBUKI et al.

1994). Moreover, PA28/REG α, β and γ contain subunit-specific inserts that are not required for proteasome binding or activation. These inserts are located at the top edge of the regulator distal to the proteasome-binding surface (KNOWLTON et al. 1997). Because of their location, the inserts have been proposed to couple the proteasome to other cellular components. In particular the PA/REGα-specific KEKE-motif has been proposed to mediate interactions with other proteins containing such motifs (REALINI et al. 1994). Among the cellular proteins containing KEKE motifs are the cytosolic chaperons Hsp90 and Hsp70. Another one is the ER membrane chaperone calnexin which is involved in the folding of MHC class I heavy chains in the ER. Its cytoplasmic tail contains a KEKE motif and it was therefore speculated that PA28 might couple proteasomes via this interaction to the ER membrane (RECHSTEINER et al. 2000).

5.3 Relationships of REG/PA28 with the Immunoproteasome

Further insights into the role of PA28/REG in the immune system had been expected from the analysis of gene-targeted mice. Analysis of both PA28γ- and PA28β-knockout mice have recently been reported. PA28γ-knockout mice did not show any obvious defects in their immune system (MURATA et al. 1999), consistent with previous reports that PA28γ is not upregulated by IFN-γ and is also found in invertebrates lacking an adaptive immune system. The authors did, however, not want to exclude a specialized role in the processing of nuclear antigens. In this context, it is of interest that recent studies have suggested that proteasomes of the nucleus, in particular those of certain nuclear substructures, the promyelocytic leukemia oncogenic domains, may play a role in the production of antigenic peptides (YEWDELL et al. 1999b).

Disruption of the *PA28β*-gene caused a relatively severe immunological phenotype (PRECKEL et al. 1999). Besides PAβ, the PAα-polypeptide was also not found in the $β^{-/-}$ mice, confirming previous findings indicating that cytosolic PA28 exists predominantly as a heteropolymer (AHN et al. 1996; SONG et al. 1996). Although surface class I expression levels and numbers of CD4 and CD8 T cells appeared to be normal, the processing of several epitopes derived from exogenous and endogenous antigens as well as primary in vivo CTL responses were found to be severely impaired. Both a D^b-binding epitope derived from the endogenous male self-antigen HY and the ID K^b-binding epitope SIINFEKL from osmotically loaded OVA, were recognized on lipopolysaccharide blasts made from wild-type but not from knockout mice. However, CTLs specific for the latter were detected in knockout mice upon in vivo priming with OVA, although the response was much lower (around one-fifth as many CTLs according to lytic units) than in wild-type mice. Furthermore, peritoneal macrophages infected with influenza virus were less sensitive to lysis by NP366-374-specific CTLs, and primary in vivo CTL-responses against two viruses were severely affected: an MCMV-specific response was virtually undetectable and in the case of LCMV infection, only one-third of wild-type CD8 T cell activity was generated. Perhaps most surprising was the finding that the assembly of the immuno- but not of the constitutive proteasome was severely impaired in the $PAb^{-/-}$ mice, leading the authors to conclude that, in addition to modulating proteasomal cleavage, PA28 may enhance MHC class I antigen processing indirectly by promoting immunoproteasome assembly. Due to the lack of incorporation of all the three immuno β-subunits and the apparent lack of the PAα-subunit, the $PA28\beta^{-/-}$ mice may actually be equivalent to mice defective in all known inducible proteasomal elements. The strong deficit in the priming phase of immune responses is consistent with the finding that mature DCs, the only antigen presenting cells believed to be capable of inducing primary T-cell responses, normally contain almost exclusively immunoproteasomes and high levels of PA28 (MACAGNO et al. 1999). Both the immunosubunits and PA28α/β are normally also upregulated by LPS (MACAGNO et al. 1999). This may perhaps also explain the strong presentation defect of the LPS blasts of the $\beta^{-/-}$ mice.

The conclusion that PA28 is required for immunoproteasome assembly is not easily reconciled with recent studies on fibroblast cell lines with no basal expression of the immunosubunits and only very low expression of PA28 (no β and only little α). These experiments have shown that PA28 and the immunosubunits can affect antigen presentation independently of one another, suggesting that they may influence epitope presentation by different mechanisms. For example, the presentation of a D^b-binding epitope of the Gag-protein of MuLV was enhanced by expression of either the three immunosubunits or PA28α and β (VAN HALL et al. 2000). That PA28 can act in conjunction with the constitutive proteasome was already suggested by the enhanced recognition of two viral epitopes on PA28α-transfected fibroblasts (GROETTRUP et al. 1996a) and the observation that purified PA28 can stimulate the peptidase activities of both the constitutive and the immunoproteasome (RECHSTEINER et al. 2000). On the other hand, epitopes have been identified which are better presented upon expression of the three immunosubunits, but not of PA28α and β (SCHWARZ et al. 2000b; SIJTS et al. 2000b). However, it has not been tested whether PA28 enhances the presentation of these epitopes in conjunction with the expression of immunoproteasomes. The latter studies also demonstrate that immunoproteasomes can assemble in the absence of substantial levels of PA28, at least when the immunosubunits are overexpressed. Taking these results together, it appears that PA28 improves the presentation of at least some epitopes, but the mechanism(s) which operate under in vivo conditions remain to be clarified.

6 Epitope Generation by Purified Proteasomes In Vitro

6.1 Epitopes and/or Epitope Precursors

Analyses of the generation of MHC class I ligands within intact cells and of the function of individual components involved in this process have been severely hampered by quantitative constraints. Detection of epitopes generated within cells usually requires CTLs that can recognize minute amounts of peptide presented by an appropriate class I molecule. Potential precursors are more difficult to detect, as binding to class I is less avid and detection by specific CTLs is less sensitive. Without protection by the appropriate class I molecule, peptides have short half-lives in vivo and detection was successful only rarely. In some cases, epitopes and epitope precursors have been traced, either in association with cytosolic or ER-resident heat shock proteins which provide some degree of protection (NIELAND et al. 1996; BRELOER et al. 1998; ISHII et al. 1999) or if very large numbers of cells were extracted using acid (UDAKA et al. 1993; UENAKA et al. 1994).

A readily feasible and informative assessment of the role of proteasomes in epitope production was the in vitro digestion of antigenic proteins and epitope-containing protein fragments with purified proteasomes. With a few exceptions, 20S proteasomes have been used in such experiments. While the relevance of this approach has been questioned (ROCK and GOLDBERG 1999), a considerable number of studies demonstrated a good correlation between 20S proteasomal production of epitopes or epitope precursors in vitro and their presentation in vivo (EGGERS et al. 1995; NIEDERMANN et al. 1995; DICK et al. 1996; GROETTRUP et al. 1996a; OSSENDORP et al. 1996; SHIMBARA et al. 1997; THEOBALD et al. 1998; VALMORI et al. 1999; BEEKMAN et al. 2000; LUCCHIARI-HARTZ et al. 2000; MOREL et al. 2000; SCHWARZ et al. 2000b; SIJTS et al. 2000a,b; YAGUE et al. 2000). Some of these studies will be discussed below in more detail.

DICK et al. (1994) were the first to demonstrate that 20S proteasomes can liberate minimal class I ligands from antigenic proteins, as shown for the OVA-derived epitope SIINFEKL and the β-galactosidase-derived epitope TPHPARIGL. Thereafter, in vitro production by 20S proteasomes was demonstrated for a number of CTL epitopes (see Table 1). While it has been argued that proteasomes produce definitive epitopes only inefficiently if at all, several of the epitopes studied to date are produced by proteasomes as major products (see Table 1 and Fig. 2). One of these is the JAK1-derived dominant self-ligand SYFPEITHI, which is presented in about 10,000 copies by K^d molecules of P815 mastocytoma cells (STEVANOVIC and SCHILD 1999). 20S proteasomes excise this dominant class I ligand as the major product from epitope-containing JAK1 fragments (DICK et al. 1996, 1998b; NIEDERMANN et al. 1997) and PA28 binding leads to a further drastic improvement of its production (DICK et al. 1996). These in vitro results appear to be physiologically relevant, as the very same fragments found in such proteasomal digests, namely the epitope SYFPEITHI and fragment FPEITHI, were immunoprecipitated from cell lysates (DICK et al. 1998b). Meanwhile, there are several examples of epitopes or self-ligands produced by purified 20S proteasomes as major products with no or virtually no precursor peptide production (Table 1). In other cases, proteasomes generate both the minimal ligands and potential precursors, or only precursor peptides (Table 1). One example for the latter category is the epitope IPGLPLSL, which is derived from the c-Akt-proto-oncogene. In vitro, this peptide is not made by 20S proteasomes, and only the decamer precursor SIIPGLPLSL is produced (SHIMBARA et al. 1997; see also Fig. 2). Nevertheless, there is again a good correlation with in vivo data, with the decamer precursor being found in cell extracts in addition to the octamer L^d-ligand when 2.5×10^{11} tumor cells were extracted using acid (UENAKA et al. 1994).

Proteasomes can make N- as well as C-terminally elongated forms of a minimal epitope. C-terminal length variants can be presented in the form produced by proteasomes if their C terminus is TAP-compatible. This has been shown for the HIV Nef-derived ligands pairs TPGPGVRY and TPGPGVRYPL as well as PLTFGWCYKL and PLTFGWCYKLV (LUCCHIARI-HARTZ et al. 2000) (Fig. 2). Recent unpublished data from our laboratory show that these cases are not exceptional. Crystal structure analyses have revealed that elongated peptides can be accommodated by bulging the middle part of the peptide; in other cases, the peptide C terminus can extend over the end of the groove (MADDEN 1995).

Various kinds of evidence suggest that proteasomes are essential for liberation of most ligand C-termini and that there is only little if any carboxypeptidase activity present in the cytosol and the ER. The influenza NP epitope TYQRTRALV is presented when expressed from a minigene or from the full-length NP. However, the C-terminally extended variant TYQRTRALVTG was not presented unless the carboxydipeptidase ACE was co-expressed (EISENLOHR et al. 1992). Proteasomes also appear to be essential for the liberation of the C termini of other epitopes such as the OVA epitope SIINFEKL (CRAIU et al. 1997a; Mo et al. 1999). Peptides extended C-terminally by one to five aa were processed and presented, with presentation completely blocked in the presence of the proteasome inhibitor lactacystin. Furthermore, when a synthetic peptide corresponding to the SIINFEKL-precursor peptide QLESIINFEKL was incubated with cytosolic extracts, virtually no carboxypeptidase products were found (BENINGA et al. 1998). Moreover, as already mentioned, there is an excellent correlation of aa frequencies at the C terminus of proteasomal products and MHC class I ligands on the one hand, and the N terminus of proteasomal products and the immediate ligand flanking position C + 1, on the other (see Sect. 3.3; NIEDERMANN et al. 1996, 1999; ALTUVIA and MARGALIT 2000).

In contrast to the apparent lack of carboxypeptidase activity in the cytosol and ER, aminopeptidase activity is found in both. If N-terminally extended versions of epitopes are expressed as minigenes, the epitopes are almost always presented (SNYDER et al. 1994; ELLIOTT et al. 1995; CRAIU et al. 1997a; STOLTZE et al. 1998; Mo et al. 1999). For some epitopes, e.g., the 1D OVA epitope SIINFEKL, it was found that presentation was not inhibited by the proteasome inhibitor lactacystin (CRAIU et al. 1997a; Mo et al. 1999). In contrast, presentation was markedly reduced when N-acetylated versions of extended peptides such as QLESIINFEKL were introduced into the cytosol (Mo et al. 1999). N-acetylation prevents trimming by aminopeptidases. While such experiments have often been interpreted as indicating that the proteasome is not involved in liberation of the ligand N termini, we believe that in some cases the experiments have not been conclusive. For example, we have found that the Glu–Ser-peptide-bond at the SIINFEKL N terminus is a major site used by purified 20S proteasomes, 20S–PA28 complexes or 26S proteasomes, and cleavage at this site, which is likely to be performed by the

Table 1. Epitope generation by purified proteasomes in vitro

aa Sequence of CTL epitope/ MHC class I ligand with natural flanking regions*	MHC molecule	Source protein	Length of digested fragment (aa)	Type of proteasome	References
Major product (no N-terminally elongated precursors)					
WNNF **SYFPEITHI** VIK	H-2K^d	JAK1	19/21	Mouse 20S***	Dick et al. 1996, 1998b
WNNF **SYFPEITHI** VIK	H-2K^d	JAK1	21	*Drosophila* 20S	Niedermann et al. 1997
PSEL **TLWVDPYEV** SYR	HLA-A2	BTG1	24	Mouse/yeast 20S	Niedermann et al. 1997, 1999
ACYE **FLWGPRALV** ETS	HLA-A2	MAGE-3	15–24	Mouse 20S + lacta	Valmori et al. 1999
APIL **STLPETTVVRR** RGR	HLA-Aw68	HbcAg	32	Mouse 20S	Sijts et al. 2000a
ISKL **VNIRNCCYI** SGN	H-2D^b	Ad5	40	Mouse 20S	Sijts et al. 2000b
GVRY **PLTFGWCYKL** VPV	HLA-A2	HIV-Nef	30/protein**	Human 20S	Lucchiari-Hartz et al. 2000
GVRY **PLTFGWCYKLV** PVE	HLA-A2	HIV-Nef	30/protein**	Human 20S	Lucchiari-Hartz et al. 2000
WQNY **TPGPGVRYPL** TFG	HLA-B7	HIV-Nef	30/protein**	Human 20S	Lucchiari-Hartz et al. 2000
LELR **SRYWAITRTR** SGG	HLA-B27	IV-NP	27	Enriched human 20S	Svensson et al. 1996
Major product with precursors					
EQLE **SIINFEKL** TEW	H-2K^b	OVA	Protein 22/44	Bovine 20S Mouse/yeast 20S***	Dick et al. 1994 Niedermann et al. 1995, 1996, 1997

Immunological Functions of the Proteasome

Epitope	MHC	Source	Copies	Proteasome	Reference
WQNY TPGPGVRY PLT	HLA-B7	HIV-Nef	30/protein**	Human 20S	Lucchiari-Hartz et al. 2000
PGVR **YPLTFGWCY** KLV	HLA-B7	HIV-Nef	30/protein**	Human 20S	Lucchiari-Hartz et al. 2000
Minor product with precursors					
MYDM **YPHFMPTNL** GPS	H-2L^d	MCMV-pp89	25	Mouse 20S***	Groettrupp et al. 1995; Dick et al. 1996 Niedermann et al. 1995, 1996
TQIN **KVVRFDKL** PGF	H-2K^b	OVA	22/44	Mouse 20S	
LSDS **SGVENPGGVYCL** TKW	H-2D^b	LVMV-GP	25	Mouse 20S	Schwarz et al. 2000a
MRTE **RPQASGVYM** GNV	H-2L^d	LCMV-NP	25	Mouse 20S	Schwarz et al. 2000b
SSGN **LLGRNSFEV** RVC	HLA-A2	p53	27	Human 20S	Theobald et al. 1998
N-terminus not liberated					
GLFN **KSPWFTTL** IST	H-2K^b	MuLV-p15E	26	Mouse 20S	Ossendorp et al. 1996
TLSI **IPGLPLSL** GAT	H-2L^d	c-Akt	29	Rat 20S***	Shimbara et al. 1997
LSDL **RGYVYQGL** KSG	H-2K^b	VSV-NP	18/27	Mouse 20S	Stolze et al. 1998
GSTA **VPYGSFKHV** DTR	HLA-B51	RU1	19	Human 20S	Morel et al. 2000
WKPS **SSWDFITV** SNN	H-2K^b	MuLV-Env	26	Mouse 20S	Beekman et al. 2000
WKPS **SSWDFITV** NNN					

*, Epitope/MHC-ligand in bold; **, G. Niedermann et al., unpublished data; ***, direct epitope/epitope precursor production accelerated by PA28.

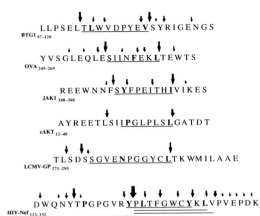

Fig. 2. 20S proteasomal cleavage sites in CTL epitope-containing protein fragments. Cleavage sites are indicated by *arrows*; the size of the arrows corresponds to the strength of cleavage. CTL epitopes/MHC-class I ligands are *underlined*. Anchor residues for MHC-binding are **bold** (see also Sect. 2.1). For references see Table 1. Only selected cleavage sites are shown for the LCMV GP fragment, according to quantitative data from SCHWARZ et al. 2000a

postglutamyl activity, is not inhibitable by lactacystin (E. Geier and G. Niedermann, unpublished data). The postglutamyl activity is generally not efficiently inhibited by the available proteasome inhibitors, with lactacystin being an exceptionally weak inhibitor of this activity (BOGYO et al. 1997). The presentation of an epitope can even be improved in the presence of lactacystin. A well documented example is the epitope FLWGPRALV derived from the melanoma antigen MAGE3 (VALMORI et al. 1999). This epitope is normally cryptic because proteasomes do not cut at its C terminus, but rather within the epitope sequence. Presentation from the MAGE3 protein or from C-terminally extended constructs was only possible in the presence of lactacystin. 20S proteasomal digests in vitro revealed that lactacystin treatment reduced intra-epitopic cleavages by lactacystin-sensitive sites and enabled cleavage at the C terminus by a lactacystin-insensitive proteasomal activity. Similar to SIINFEKL, the MAGE-epitope FLWGPRALV is preceded by a glutamate (E) (Table 1) and cleavage by purified proteasomes at its N terminus is not inhibited by lactacystin (VALMORI et al. 1999). In addition, and perhaps generally more important, slightly extended substrates such as the 11-mer QLESIINFEKL have a size that is typical for proteasomal products rather than for substrates. They are therefore likely to be ignored by proteasomes in the cytosol and to be attacked rapidly by amino- and oligopeptidases (see Sect. 3.5). This may generally be the main reason why epitope production is not inhibited by proteasome inhibitors when constructs with only short N-terminal extensions are expressed. We have indeed found that synthetic QLESIINFEKL peptide is virtually not attacked when incubated with purified 20S proteasomes, but is readily degraded by purified tripeptidyl peptidase II or leucine aminopeptidase. The same is true for the subdominant OVA epitope KVVRFDKL (E. Geier and G. Niedermann, unpublished data). In our opinion, the minigene studies do not allow conclusions as to the role of proteasomes in the liberation of the ligand N termini from natural substrates. Nevertheless, these results imply that epitopes can be liberated by nonproteasomal trimming from N-terminally extended precursor peptides.

A candidate for a trimming enzyme is the cytosolic leucine aminopeptidase. The enzyme is upregulated by IFN-γ (HARRIS et al. 1992), and some SIINFEKL has been shown to be produced from synthetic QLESIINFEKL in cytosol preparations from IFN-γ-treated, but not untreated HeLa cells (BENINGA et al. 1998). Specific trimming enzymes in the ER have not yet been identified, although N-terminal trimming of TAP-transported peptides in the ER has been observed in model experiments (LAUVAU et al. 1999; PAZ et al. 1999). Also in one of these two studies, the OVA epitope SIINFEKL was used, though not in its natural sequence context (PAZ et al. 1999). The epitope variant SIINFEHL was appended to the carboxyl end of the invariant chain and, to facilitate epitope detection, a trypsin cleavage site was introduced at the epitope N terminus by replacing the naturally occurring, preceding glutamate (E) with a lysine (K). Upon expression of this construct, only small amounts of SIINFEHL were detected in the absence of the restricting class I molecule (H-2K^b) and the precursor KSIINFEHL was the predominant product arriving in the ER. When K^b-molecules were co-expressed, much more SIINFEHL and less KSIINFEHL was recovered from the cells. While this experiment suggests that SIINFEHL was produced predominantly by N-terminal trimming in the ER, it is not pertinent to the generation of SIINFEKL upon processing of OVA. Firstly, in contrast to glutamate, lysine is not preferred at P1 by proteasomes (OSSENDORP et al. 1996; NUSSBAUM et al. 1998; GEIER et al. 1999). Secondly, when natural full-length OVA was expressed, SIINFEKL was found bound to cytosolic HSPs (BRELOER et al. 1998). Thirdly, in a reconstituted cytosolic proteolytic system, SIINFEKL was also recovered upon digestion of full-length OVA (BEN-SHAHAR et al. 1997). We feel that some MHC ligands may be produced by trimming in the ER, but this is not essential for others. For example, the K^b-binding vesicular stomatitis virus (VSV) NP-derived ID epitope RGYVYQGL has been isolated from gp96 of VSV-infected H-2^d-cells (NIELAND et al. 1996).

As discussed above, compelling correlations exist between the frequencies of aa found at the C termini of proteasomal fragments and MHC class I ligands. Moreover, the N termini as well the C + 1 immediate flanking position of class I ligands show aa distributions similar to the N termini of proteasomal fragments (see Sect. 3.3). In contrast, almost any aa can be found in the immediate flanking position preceding class I ligands (N − 1). The absence of a striking similarity of aa frequencies at N − 1 of class I ligands and at the C terminus of proteasomal fragments has been used as an argument against the generation of epitope N termini by proteasomes. We believe that this argument does not stand up to critical examination. Firstly, previous discussions have neglected that proline and threonine are rare in the N − 1 position of acid-eluted MHC class I ligands (NIEDERMANN et al. 1996, 1999; ALTUVIA and MARGALIT 2000). These residues are also rare at P1 of proteasomal fragments. Secondly, aa at N − 1 of class I ligands are rather expected to exhibit a lower degree of selectivity. Leucine, the most frequent aa in the C-terminal position of class I ligands, is enriched compared to the natural frequencies occurring in proteins by less than a factor of three. Basic C-terminal aa, which are preferred by certain class I alleles, are not significantly enriched at all when all known acid-eluted ligands for various alleles are analyzed (NIEDERMANN et al. 1996, 1999; ALTUVIA and MARGALIT 2000). These relatively unimpressive enrichment factors represent peptides not only made in large numbers and large amounts by proteasomes, but also peptides selected by both TAP and the MHC class I molecules (see Sects. 2.1 and 2.2). In the N–1 position, the aa frequencies are not influenced by selection by peptide-binding molecules, or, if they are (e.g., in the case of the binding of a precursor to TAP), by interactions different from those acting on the C terminus. Furthermore, as outlined above, proteasomes have a rather broad cleavage site specificity, permitting cleavages after aa other than hydrophobic and charged residues, although less frequently and usually less efficiently. Nevertheless, it is conceivable that in many cases, both the direct production by proteasomes and trimming of precursor peptides made by proteasomes provide presented copies of a given epitope. The proportions in which the pathways are used may vary considerably depending on the individual epitope and depending on the class I allele.

6.2 Factors Influencing the Magnitude of Epitope Generation by Proteasomes

The amounts of epitopes and potential precursor peptides generated by proteasomes are determined largely by cleavage site distribution and efficiency of proteasomal cleavage within epitopes and at the epitope boundaries/immediate flanking regions (NIEDERMANN et al. 1995). Epitopes and epitope precursors which are produced in large amounts often have preferred P1/P1′ combinations (see Sect. 3.3) at one or both ends. Typical examples are the epitopes F-*SYFPEITHI*-V, E-*SIINFEKL*-T, L-*TLWVDPYEV*-S, Y-*TPGPGVRY*-P, Y-*TPGPGVRYPL*-T, and the epitope precursors L-SI*IPGLPLSL*-G and D-S*SGVENPGGYCL*-T (Table 1, Fig. 2, the epitopes are underlined). Other epitopes and epitope precursors are produced by proteasomes with lesser efficiencies. For almost all epitopes studied so far, proteasome cleavage sites have been found not only at the epitope boundaries and in the flanking regions, but also within the epitopes. Such intra-epitopic sites can be minor sites, but epitopes can harbor several major sites and are nevertheless produced by proteasomes. For example, the dominant epitopes SYFPEITHI (JAK1-derived) and SIINFEKL (OVA-derived) as well as the high-copy HIV Nef epitope TPGPGVRY contain only minor sites (Fig. 2). In contrast, the subdominant OVA epitope KVVRFDKL, the ID MCMV-pp89 epitope YPHFMPTNL and the overlapping Nef epitopes TPGPGVRYPL, YPLTFGWCY, PLTFGWCYKL and PLTFGWCYKLV harbor major sites (Table 1, Fig. 2). In fact, many proteasomal products overlap with others. This may be in part due to the partial differences in cleavages between housekeeping and immunoproteasomes (see Sects. 4.3 and 4.4). However, homogeneous 20S proteasome preparations, e.g., those purified from yeast or erythrocytes, also produce overlapping peptides (NUSSBAUM et al. 1998; EMMERICH et al. 2000). One must therefore conclude that proteasomes do not attack every individual substrate molecule at the same potential sites. Proteasomal proteolysis therefore allows presentation of overlapping epitopes (LUCCHIARI-HARTZ et al. 2000), apparently even by the same cell, as has been observed for the overlapping HIV Gag epitopes SLYNTVATL and TLYCVHQR (HARRER et al. 1998). This means that epitopes are not absolutely preserved. In many cases, a proportion of substrate molecules are always degraded such that the epitope is destroyed. In so doing, the C terminus of an overlapping epitope can be generated, e.g., proteasomes cut the HLA-A2-binding Nef ligand(s) PLTFGWCYKL/V at several sites within the epitope. The cleavage between the leucine (L), which is the N-terminal anchor of the A2 ligands, and the threonine (T) creates the C terminus of the B7-binding ligand TPGPGVRYPL, and cleavage between the tyrosine (Y) and the lysine (K) creates that of the B7-binding epitope YPLTFGWCY (Fig. 2) (LUCCHIARI-HARTZ et al. 2000).

It has been proposed that proline protects epitopes from destruction (SHIMBARA et al. 1998), e.g., the c-Akt epitope precursor SI*IPGLPLSL* is made by proteasomes in large amounts and contains virtually no internal cleavage sites (Fig. 2). Because proline at P4, P3 or P1' can promote cleavage by proteasomes (see Sect. 3.3) it is conceivable that protection requires several prolines within the epitope sequence. Interestingly, the N-terminal epitope anchor, which is often the second aa from the N terminus, is in the case of many MHC class I alleles a hydrophobic (aliphatic/aromatic) or charged aa (RAMMENSEE et al. 1997), i.e., a typical preferred P1 residue of proteasomes. Surprisingly, there is usually only a minor or no cleavage site after such N-terminal anchors (Fig. 2). In our opinion, this may be due to the fact that there is often a large hydrophobic aa (either F, W, L, I, V, or Y) in the third position of such MHC class I ligands (the P1'-position of the N-terminal anchor; Table 1, Fig. 2). In contrast to a small aa, a large hydrophobic aa at P1' is usually counterproductive for efficient cleavage (see Sect. 3.3). In addition, a distance of 8 to 11aa, the typical distance between the C and N termini (or N-terminal flanking region) of an epitope, is probably more optimal for efficient cleavage than the distance of 6 or 7aa, typical between the C- and N-terminal anchor positions. Peptides with a large hydrophobic aa in P1' of a hydrophobic or charged N-terminal anchor may hence often be efficiently presented, and TAP and MHC class I molecules may have evolved to present such peptides efficiently (see Sect. 2).

Certain potential epitopes or epitope precursors are not or only poorly generated by proteasomes. Such epitopes may be subdominant or even cryptic. We and others have shown that this may occur by dominant proteasomal attack within the epitope sequence (NIEDERMANN et al. 1995, 1996; OSSENDORP et al. 1996; YELLEN-SHAW et al. 1997). In all these cases, proteasomes have been shown to attack the epitopes when embedded in the sequence context of both their natural flanking sequences. The existence of this mechanism has been questioned on the basis of experiments in which the SD epitope was expressed as the minimal epitope or with only one short flank (Mo et al. 2000). Short substrates of this type are, however, inadequate to assess proteasomal processing of natural substrates. For example, it has been shown that only presentation to CTLs of epitopes in translation products of 21aa or longer, but not of 17aa or shorter, was sensitive to proteasome inhibitors (YANG et al. 1996; see also Sect. 3.4).

In accordance with the importance of the P1'-position for the determination of the sites and rate of cleavage, it has also been observed that point mutations in the immediate flanking position C + 1 can strongly impair epitope presentation (NIEDERMANN et al. 1995; YELLEN-SHAW et al. 1997; THEOBALD et al. 1998; BEEKMAN et al. 2000). In addition, it has been reported for epitopes from influenza virus NP that single proline substitutions in the N − 1 position can impair their presentation from the full-length protein (YELLEN-SHAW et al. 1997). Such effects of mutations in the flanking position N − 1 have, however, not been observed as often.

Taken together, it appears that proteasomal specificity, although broad, determines the repertoire of peptides from which the MHC class I molecules can select peptides for presentation, and that quantitative differences in epitope or precursor production contribute to the immunohierarchy among potential epitopes (NIEDERMANN et al. 1995, 1999; VAN HALL et al. 2000).

7 Summary

Proteasomes are highly abundant cytosolic and nuclear protease complexes that degrade most intracellular proteins in higher eukaryotes and appear to play a major role in the cytosolic steps of MHC class I antigen processing. This review summarizes the knowledge of the role of proteasomes in antigen processing and the impact of proteasomal proteolysis on T cell-mediated immunity.

Acknowledgements. I am particularly grateful to Dr. Klaus Eichmann for the many helpful discussions and for critical reading of the manuscript. I also thank Peter van Endert, Ian Haidl and Trixi Mentz for their comments on the manuscript. Work from our laboratory was supported by grant NI 368/2-2 of the Deutsche Forschungsgemeinschaft and by grants from the European Union.

References

Ahn K, Erlander M, Leturcq D, Peterson PA, Fruh K, Yang Y (1996) In vivo characterization of the proteasome regulator PA28. J Biol Chem 271:18237–18242

Aki M, Shimbara N, Takashina M, Akiyama K, Kagawa S, Tamura T, Tanahashi N, Yoshimura T, Tanaka K, Ichihara A (1994) Interferon-gamma induces different subunit organizations and functional diversity of proteasomes. J Biochem 115:257–269

Akopian TN, Kisselev AF, Goldberg AL (1997) Processive degradation of proteins and other catalytic properties of the proteasome from Thermoplasma acidophilum. J Biol Chem 272:1791–1798

Altuvia Y, Margalit H (2000) Sequence signals for generation of antigenic peptides by the proteasome: implications for proteasomal cleavage mechanism. J Mol Biol 295:879–890

Arendt CS, Hochstrasser M (1997) Identification of the yeast 20S proteasome catalytic centers and subunit interactions required for active-site formation. Proc Natl Acad Sci USA 94:7156–7161

Arnold D, Driscoll J, Androlewicz M, Hughes E, Cresswell P, Spies T (1992) Proteasome subunits encoded in the MHC are not generally required for the processing of peptides bound by MHC class I molecules. Nature 360:171–174

Bai A, Forman J (1997) The effect of the proteasome inhibitor lactacystin on the presentation of transporter associated with antigen processing (TAP)-dependent and TAP-independent peptide epitopes by class I molecules. J Immunol 159:2139–2146

Banchereau J, Steinman RM (1998) Dendritic cells and the control of immunity. Nature 392:245–252

Baumeister W, Walz J, Zuhl F, Seemüller E (1998) The proteasome: paradigm of a self-compartmentalizing protease. Cell 92:367–380

Beekman NJ, van Veelen PA, van Hall T, Neisig A, Sijts A, Camps M, Kloetzel PM, Neefjes JJ, Melief CJ, Ossendorp F (2000) Abrogation of CTL epitope processing by single amino acid substitution flanking the C-terminal proteasome cleavage site. J Immunol 164:1898–1905

Ben-Shahar S, Cassouto B, Novak L, Porgador A, Reiss Y (1997) Production of a specific major histocompatibility complex class I-restricted epitope by ubiquitin-dependent degradation of modified ovalbumin in lymphocyte lysate. J Biol Chem 272:21060–21066

Benham AM, Gromme M, Neefjes J (1998) Allelic differences in the relationship between proteasome activity and MHC class I peptide loading. J Immunol 161:83–89

Beninga J, Rock KL, Goldberg AL (1998) Interferon-gamma can stimulate post-proteasomal trimming of the N terminus of an antigenic peptide by inducing leucine aminopeptidase. J Biol Chem 273:18734–18742

Bochtler M, Ditzel L, Groll M, Hartmann C, Huber R (1999) The proteasome. Ann Rev Biophys Biomol Struct 28:295–317

Bodmer JG, Marsh SG, Albert ED, Bodmer WF, Bontrop RE, Charron D, Dupont B, Erlich HA, Fauchet R, Mach B, Mayr WR, Parham P, Sasazuki T, Schreuder GM, Strominger JL, Svejgaard A, Terasaki PI (1997) Nomenclature for factors of the HLA system, 1996. Tissue Antigens 49:297–321

Boehm U, Klamp T, Groot M, Howard JC (1997) Cellular responses to interferon-gamma. Annu Rev Immunol 15:749–795

Boes B, Hengel H, Ruppert T, Multhaup G, Koszinowski UH, Kloetzel PM (1994) Interferon gamma stimulation modulates the proteolytic activity and cleavage site preference of 20S mouse proteasomes. J Exp Med 179:901–909

Bogyo M, McMaster JS, Gaczynska M, Tortorella D, Goldberg AL, Ploegh H (1997) Covalent modification of the active site threonine of proteasomal beta subunits and the *Escherichia coli* homolog HslV by a new class of inhibitors. Proc Natl Acad Sci USA 94:6629–6634

Bogyo M, Shin S, McMaster JS, Ploegh HL (1998) Substrate binding and sequence preference of the proteasome revealed by active-site-directed affinity probes. Chem Biol 5:307–320

Boon T, Van Pel A (1989) T cell-recognized antigenic peptides derived from the cellular genome are not protein degradation products but can be generated directly by transcription and translation of short subgenic regions. A hypothesis [see comments]. Immunogenetics 29:75–79

Breloer M, Marti T, Fleischer B, von Bonin A (1998) Isolation of processed, H-2Kb-binding ovalbumin-derived peptides associated with the stress proteins HSP70 and gp96. Eur J Immunol 28:1016–1021

Cardozo C, Vinitsky A, Michaud C, Orlowski M (1994) Evidence that the nature of amino acid residues in the P3 position directs substrates to distinct catalytic sites of the pituitary multicatalytic proteinase complex (proteasome). Biochemistry 33:6483–6489

Cardozo C, Chen WE, Wilk S (1996) Cleavage of Pro-X and Glu-X bonds catalyzed by the branched chain amino acid preferring activity of the bovine pituitary multicatalytic proteinase complex (20S proteasome). Arch Biochem Biophys 334:113–120

Cardozo C, Kohanski RA (1998) Altered properties of the branched chain amino acid-preferring activity contribute to increased cleavages after branched chain residues by the 'immunoproteasome'. J Biol Chem 273:16764–16770

Cerundolo V, Kelly A, Elliott T, Trowsdale J, Townsend A (1995) Genes encoded in the major histocompatibility complex affecting the generation of peptides for TAP transport [published erratum appears in Eur J Immunol 1995 25:1485]. Eur J Immunol 25:554–562

Cerundolo V, Benham A, Braud V, Mukherjee S, Gould K, Macino B, Neefjes J, Townsend A (1997) The proteasome-specific inhibitor lactacystin blocks presentation of cytotoxic T lymphocyte epitopes in human and murine cells. Eur J Immunol 27:336–341

Chisari FV, Ferrari C (1995) Hepatitis B virus immunopathogenesis. Annu Rev Immunol 13:29–60

Ciechanover A, Orian A, Schwartz AL (2000) Ubiquitin-mediated proteolysis: biological regulation via destruction. Bioessays 22:442–451

Coux O, Tanaka K, Goldberg AL (1996) Structure and functions of the 20S and 26S proteasomes. Annu Rev Biochem 65:801–847

Craiu A, Akopian T, Goldberg A, Rock KL (1997a) Two distinct proteolytic processes in the generation of a major histocompatibility complex class I-presented peptide. Proc Natl Acad Sci USA 94:10850–10855

Craiu A, Gaczynska M, Akopian T, Gramm CF, Fenteany G, Goldberg AL, Rock KL (1997b) Lactacystin and clasto-lactacystin beta-lactone modify multiple proteasome beta-subunits and inhibit intracellular protein degradation and major histocompatibility complex class I antigen presentation. J Biol Chem 272:13437–13445

Daniel S, Brusic V, Caillat-Zucman S, Petrovsky N, Harrison L, Riganelli D, Sinigaglia F, Gallazzi F, Hammer J, Van Endert PM (1998) Relationship between peptide selectivities of human transporters associated with antigen processing and HLA class I molecules. J Immunol 161:617–624

Dick LR, Moomaw CR, DeMartino GN, Slaughter CA (1991) Degradation of oxidized insulin B chain by the multiproteinase complex macropain (proteasome). Biochemistry 30:2725–2734

Dick LR, Aldrich C, Jameson SC, Moomaw CR, Pramanik BC, Doyle CK, DeMartino GN, Bevan MJ, Forman JM, Slaughter CA (1994) Proteolytic processing of ovalbumin and beta-galactosidase by the proteasome to a yield antigenic peptides. J Immunol 152:3884–3894

Dick TP, Ruppert T, Groettrup M, Kloetzel PM, Kuehn L, Koszinowski UH, Stevanovic S, Schild H, Rammensee HG (1996) Coordinated dual cleavages induced by the proteasome regulator PA28 lead to dominant MHC ligands. Cell 86:253–262

Dick TP, Nussbaum AK, Deeg M, Heinemeyer W, Groll M, Schirle M, Keilholz W, Stevanovic S, Wolf DH, Huber R, Rammensee HG, Schild H (1998a) Contribution of proteasomal beta-subunits to the cleavage of peptide substrates analyzed with yeast mutants. J Biol Chem 273:25637–25646

Dick TP, Stevanovic S, Keilholz W, Ruppert T, Koszinowski U, Schild H, Rammensee HG (1998b) The making of the dominant MHC class I ligand SYFPEITHI. Eur J Immunol 28:2478–2486

Ditzel L, Huber R, Mann K, Heinemeyer W, Wolf DH, Groll M (1998) Conformational constraints for protein self-cleavage in the proteasome. J Mol Biol 279:1187–1191

Ditzel L, Stock D, Lowe J (1997) Structural investigation of proteasome inhibition. Biol Chem 378:239–247

Dolenc I, Seemüller E, Baumeister W (1998) Decelerated degradation of short peptides by the 20S proteasome. FEBS Lett 434:357–361

Driscoll J, Brown MG, Finley D, Monaco JJ (1993) MHC-linked LMP gene products specifically alter peptidase activities of the proteasome [see comments]. Nature 365:262–264

Dubiel W, Pratt G, Ferrell K, Rechsteiner M (1992) Purification of an 11S regulator of the multicatalytic protease. J Biol Chem 267:22369–22377

Dubiel W and Kloetzel PM (2000) The 20S proteasome activator PA28 or 11S regulator. In: Hilt W, Wolf DH (eds) The world of regulatory proteolysis. Eureka.com/Landes Bioscience, Georgetown, pp 129–136

Eggers M, Boes-Fabian B, Ruppert T, Kloetzel PM, Koszinowski UH (1995) The cleavage preference of the proteasome governs the yield of antigenic peptides. J Exp Med 182:1865–1870

Ehring B, Meyer TH, Eckerskorn C, Lottspeich F, Tampe R (1996) Effects of major-histocompatibility-complex-encoded subunits on the peptidase and proteolytic activities of human 20S proteasomes. Cleavage of proteins and antigenic peptides. Eur J Biochem 235:404–415

Eisenlohr LC, Bacik I, Bennink JR, Bernstein K, Yewdell JW (1992) Expression of a membrane protease enhances presentation of endogenous antigens to MHC class I-restricted T lymphocytes. Cell 71:963–972

Eleuteri AM, Kohanski RA, Cardozo C, Orlowski M (1997) Bovine spleen multicatalytic proteinase complex (proteasome). Replacement of X, Y, and Z subunits by LMP7, LMP2, and MECL1 and changes in properties and specificity. J Biol Chem 272:11824–11831

Elliott T, Willis A, Cerundolo V, Townsend A (1995) Processing of major histocompatibility class I-restricted antigens in the endoplasmic reticulum. J Exp Med 181:1481–1491

Emmerich NPN, Nussbaum AK, Stevanovic S, Priemer M, Toes REM, Rammensee HG, Schild HJ (2000) The human 26S and 20S proteasomes generate overlapping but different sets of peptide fragments from a model protein substrate. J Biol Chem 275:21140–21148

Enenkel C, Lehmann H, Kipper J, Guckel R, Hilt W, Wolf DH (1994) PRE3, highly homologous to the human major histocompatibility complex-linked LMP2 (RING12) gene, codes for a yeast proteasome subunit necessary for the peptidylglutamyl-peptide hydrolyzing activity. FEBS Lett 341:193–196

Enenkel C, Lehmann A, Kloetzel PM (1998) Subcellular distribution of proteasomes implicates a major location of protein degradation in the nuclear envelope-ER network in yeast. EMBO J 17:6144–6154

Fehling HJ, Swat W, Laplace C, Kuhn R, Rajewsky K, Muller U, von Boehmer H (1994) MHC class I expression in mice lacking the proteasome subunit LMP-7. Science 265:1234–1237

Frisan T, Levitsky V, Polack A, Masucci MG (1998) Phenotype-dependent differences in proteasome subunit composition and cleavage specificity in B cell lines. J Immunol 160:3281–3289

Gaczynska M, Rock KL, Goldberg AL (1993) Gamma-interferon and expression of MHC genes regulate peptide hydrolysis by proteasomes [see comments] [published erratum appears in Nature 1995 374(6519):290]. Nature 365:264–267

Gaczynska M, Goldberg AL, Tanaka K, Hendil KB, Rock KL (1996) Proteasome subunits X and Y alter peptidase activities in opposite ways to the interferon-gamma-induced subunits LMP2 and LMP7. J Biol Chem 271:17275–17280

Geier E, Pfeifer G, Wilm M, Lucchiari-Hartz M, Baumeister W, Eichmann K, Niedermann G (1999) A giant protease with potential to substitute for some functions of the proteasome. Science 283:978–981

Gileadi U, Moins-Teisserenc HT, Correa I, Booth BL Jr., Dunbar PR, Sewell AK, Trowsdale J, Phillips RE, Cerundolo V (1999) Generation of an immunodominant CTL epitope is affected by proteasome subunit composition and stability of the antigenic protein. J Immunol 163:6045–6052

Gomard E, Sitbon M, Toubert A, Begue B, Levy JP (1984) HLA-B27, a dominant restricting element in antiviral responses? Immunogenetics 20:197–204

Goulder PJ, Phillips RE, Colbert RA, McAdam S, Ogg G, Nowak MA, Giangrande P, Luzzi G, Morgan B, Edwards A, McMichael AJ, Rowland-Jones S (1997) Late escape from an immunodominant cytotoxic T-lymphocyte response associated with progression to AIDS. Nature Med 3:212–217

Gray CW, Slaughter CA, DeMartino GN (1994) PA28 activator protein forms regulatory caps on proteasome stacked rings. J Mol Biol 236:7–15

Griffin TA, Nandi D, Cruz M, Fehling HJ, Kaer LV, Monaco JJ, Colbert RA (1998) Immunoproteasome assembly: cooperative incorporation of interferon gamma (IFN-gamma)-inducible subunits. J Exp Med 187:97–104

Groettrup M, Ruppert T, Kuehn L, Seeger M, Standera S, Koszinowski U, Kloetzel PM (1995) The interferon-gamma-inducible 11S regulator (PA28) and the LMP2/LMP7 subunits govern the peptide production by the 20S proteasome in vitro. J Biol Chem 270:23808–23815

Groettrup M, Soza A, Eggers M, Kuehn L, Dick TP, Schild H, Rammensee HG, Koszinowski UH, Kloetzel PM (1996a) A role for the proteasome regulator PA28alpha in antigen presentation. Nature 381:166–168

Groettrup M, Soza A, Kuckelkorn U, Kloetzel PM (1996b) Peptide antigen production by the proteasome: complexity provides efficiency. Immunol Today 17:429–435

Groettrup M, Standera S, Stohwasser R, Kloetzel PM (1997) The subunits MECL-1 and LMP2 are mutually required for incorporation into the 20S proteasome. Proc Natl Acad Sci USA 94:8970–8975

Groll M, Ditzel L, Lowe J, Stock D, Bochtler M, Bartunik HD, Huber R (1997) Structure of 20S proteasome from yeast at 2.4A resolution [see comments]. Nature 386:463–471

Grune T, Reinheckel T, Davies KJ (1997) Degradation of oxidized proteins in mammalian cells. FASEB Journal 11:526–534

Harrer T, Harrer E, Barbosa P, Kaufmann F, Wagner R, Bruggemann S, Kalden JR, Feinberg M, Johnson RP, Buchbinder S, Walker BD (1998) Recognition of two overlapping CTL epitopes in HIV-1 p17 by CTL from a long-term nonprogressing HIV-1-infected individual. J Immunol 161:4875–4881

Harris C, Hunte B, Krauss MR, Taylor A, Epstein LB (1992) Induction of leucine aminopeptidase by interferon-gamma. Identification by protein microsequencing after purification by preparative two-dimensional gel electrophoresis. J Biol Chem 267:6865–6869

Heinemeyer W, Gruhler A, Mohrle V, Mahe Y, Wolf DH (1993) PRE2, highly homologous to the human major histocompatibility complex-linked RING10 gene, codes for a yeast proteasome subunit necessary for chrymotryptic activity and degradation of ubiquitinated proteins. J Biol Chem 268:5115–5120

Heinemeyer W, Fischer M, Krimmer T, Stachon U, Wolf DH (1997) The active sites of the eukaryotic 20S proteasome and their involvement in subunit precursor processing. J Biol Chem 272:25200–25209

Hendil KB, Khan S, Tanaka K (1998) Simultaneous binding of PA28 and PA700 activators to 20S proteasomes. Biochem J 332:749–754

Hirsch C, Ploegh HL (2000) Intracellular targeting of the proteasome. Trends Cell Biol 10:268–271

Howard JC (1995) Supply and transport of peptides presented by class I MHC molecules. Curr Opin Immunol 7:69–76

Huczko EL, Bodnar WM, Benjamin D, Sakaguchi K, Zhu NZ, Shabanowitz J, Henderson RA, Appella E, Hunt DF, Engelhard VH (1993) Characteristics of endogenous peptides eluted from the class I MHC molecule HLA-B7 determined by mass spectrometry and computer modeling. J Immunol 151:2572–2587

Ishii T, Udono H, Yamano T, Ohta H, Uenaka A, Ono T, Hizuta A, Tanaka N, Srivastava PK, Nakayama E (1999) Isolation of MHC class I-restricted tumor antigen peptide and its precursors associated with heat shock proteins hsp70, hsp90, and gp96. J Immunol 162:1303–1309

Kärre K (1997) NK cells, MHC class I antigens and missing self. Immunol Rev 155:5–221

Kasahara M (1998) What do the paralogous regions in the genome tell us about the origin of the adaptive immune system? Immunol Rev 166:159–175

Kaslow RA, Carrington M, Apple R, Park L, Munoz A, Saah AJ, Goedert JJ, Winkler C, O'Brien SJ, Rinaldo C, Detels R, Blattner W, Phair J, Erlich H, Mann DL (1996) Influence of combinations of human major histocompatibility complex genes on the course of HIV-1 infection [see comments]. Nature Med 2:405–411

Kaufman J, Wallny HJ (1996) Chicken MHC molecules, disease resistance and the evolutionary origin of birds. Curr Top Microbiol Immunol 212:129–141

Kisselev AF, Akopian TN, Woo KM, Goldberg AL (1999) The sizes of peptides generated from protein by mammalian 26 and 20S proteasomes. Implications for understanding the degradative mechanism and antigen presentation. J Biol Chem 274:3363–3371

Klein L, Kyewski B (2000) Self-antigen presentation by thymic stromal cells: a subtle division of labor. Curr Opin Immunol 12:179–186

Knowlton JR, Johnston SC, Whitby FG, Realini C, Zhang Z, Rechsteiner M, Hill CP (1997) Structure of the proteasome activator REGalpha (PA28alpha). Nature 390:639–643

Koopmann JO, Post M, Neefjes JJ, Hämmerling GJ, Momburg F (1996) Translocation of long peptides by transporters associated with antigen processing (TAP). Eur J Immunol 26:1720–1728

Kopito RR (1997) ER quality control: the cytoplasmic connection. Cell 88:427–430

Korber BTM, Brander C, Moore JP, Walker BD, Koup R, Haynes BF (1997) HIV Molecular Immunology Database (Los Alamos National Laboratory, Theoretical Biology and Biophysics, Los Alamos, New Mexico)

Kuckelkorn U, Frentzel S, Kraft R, Kostka S, Groettrup M, Kloetzel PM (1995) Incorporation of major histocompatibility complex-encoded subunits LMP2 and LMP7 changes the quality of the 20S proteasome polypeptide processing products independent of interferon-gamma. Eur J Immunol 25:2605–2611

Langman R (2000) Self-nonself-discrimination revisited. Semin Immunol 12:159–344

Lauvau GKK, Niedermann G, Ostankovitch M, Yotnda P, Firat H, Chisari FV, and van Endert PM (1999) Human transporters associated with antigen processing (TAPs) select epitope precursor peptides for processing in the endoplasmic reticulum and presentation to T cells. J Exp Med 190:1227–1240

Levitskaya J, Sharipo A, Leonchiks A, Ciechanover A, Masucci MG (1997) Inhibition of ubiquitin/proteasome-dependent protein degradation by the Gly-Ala repeat domain of the Epstein–Barr virus nuclear antigen 1. Proc Natl Acad Sci USA 94:12616–12621

Lin L, Ghosh S (1996) A glycine-rich region in NF-κB p105 functions as a processing signal for the generation of the p50 subunit. Mol Cell Biol 16:2248–2254

Lobigs M, Mullbacher A (1993) Recognition of vaccinia virus-encoded major histocompatibility complex class I antigens by virus immune cytotoxic T cells is independent of the polymorphism of the peptide transporters. Proc Natl Acad Sci USA 90:2676–2680

Lord JM, Davey J, Frigerio L, Roberts LM (2000) Endoplasmic reticulum-associated protein degradation. Semin Cell Devel Biol 11:159–164

Lucchiari-Hartz M, van Endert P, Lauvau G, Mayer R, Meyerhans A, Eichmann K, Niedermann G (2000) CTL-epitopes of HIV-1-Nef: generation of multiple definitive epitopes by proteasomes. J Exp Med 191:239–252

Ma CP, Slaughter CA, DeMartino GN (1992) Identification, purification, and characterization of a protein activator (PA28) of the 20S proteasome (macropain). J Biol Chem 267:10515–10523

Macagno A, Gilliet M, Sallusto F, Lanzavecchia A, Nestle FO, Groettrup M (1999) Dendritic cells up-regulate immunoproteasomes and the proteasome regulator PA28 during maturation. Eur J Immunol 29:4037–4042

Madden DR (1995) The three-dimensional structure of peptide–MHC complexes. Annu Rev Immunol 13:587–622

Malarkannan S, Horng T, Shih PP, Schwab S, Shastri N (1999) Presentation of out-of-frame peptide/MHC class I complexes by a novel translation initiation mechanism. Immunity 10:681–690

Mayrand SM, Green WR (1998) Non-traditionally derived CTL epitopes: exceptions that prove the rules? Immunol Today 19:551–556

McCormack TA, Cruikshank AA, Grenier L, Melandri FD, Nunes SL, Plamondon L, Stein RL, Dick LR (1998) Kinetic studies of the branched chain amino acid preferring peptidase activity of the 20S

proteasome: development of a continuous assay and inhibition by tripeptide aldehydes and clasto-lactacystin beta-lactone. Biochemistry 37:7792–7800
McCusker D, Wilson M, Trowsdale J (1999) Organization of the genes encoding the human proteasome activators PA28 alpha and beta. Immunogenetics 49:438–445
Minami Y, Kawasaki H, Minami M, Tanahashi N, Tanaka K, Yahara I (2000) A critical role for the proteasome activator PA28 in the Hsp90-dependent protein refolding. J Biol Chem 275:9055–9061
Mo AX, van Lelyveld SF, Craiu A, Rock KL (2000) Sequences that flank subdominant and cryptic epitopes influence the proteolytic generation of MHC class I-presented peptides. J Immunol 164:4003–4010
Mo XY, Cascio P, Lemerise K, Goldberg AL, Rock K (1999) Distinct proteolytic processes generate the C and N termini of MHC class I-binding peptides. J Immunol 163:5851–5859
Momburg F, Ortiz-Navarrete V, Neefjes J, Goulmy E, van de Wal Y, Spits H, Powis SJ, Butcher GW, Howard JC, Walden P, et al. (1992) Proteasome subunits encoded by the major histocompatibility complex are not essential for antigen presentation. Nature 360:174–177
Momburg F, Roelse J, Howard JC, Butcher GW, Hämmerling GJ, Neefjes JJ (1994) Selectivity of MHC-encoded peptide transporters from human, mouse and rat. Nature 367:648–651
Momburg F, Hämmerling GJ (1998) Generation and TAP-mediated transport of peptides for major histocompatibility complex class I molecules. Adv Immunol 68:191–256
Monaco JJ (1992) A molecular model of MHC class-I-restricted antigen processing. Immunol Today 13:173–179
Monaco JJ, McDevitt HO (1986) The LMP antigens: a stable MHC-controlled multisubunit protein complex. Hum Immunol 15:416–426
Morel S, Levy F, Burlet-Schiltz O, Brasseur F, Probst-Kepper M, Peitrequin AL, Monsarrat B, Van Velthoven R, Cerottini JC, Boon T, Gairin JE, Van den Eynde BJ (2000) Processing of some antigens by the standard proteasome but not by the immunoproteasome results in poor presentation by dendritic cells. Immunity 12:107–117
Murata S, Kawahara H, Tohma S, Yamamoto K, Kasahara M, Nabeshima Y, Tanaka K, Chiba T (1999) Growth retardation in mice lacking the proteasome activator PA28gamma. J Biol Chem 274:38211–38215
Nandi D, Jiang H, Monaco JJ (1996) Identification of MECL-1 (LMP-10) as the third IFN-gamma-inducible proteasome subunit. J Immunol 156:2361–2364
Nandi D, Marusina K, Monaco JJ (1998) How do endogenous proteins become peptides and reach the endoplasmic reticulum. Curr Top Microbiol Immunol 232:15–47
Neefjes J, Gottfried E, Roelse J, Grommé M, Obst R, Hämmerling GJ, Momburg F (1995) Analysis of the fine specificity of rat, mouse and human TAP peptide transporters. Eur J Immunol 25:1133–1136
Neisig A, Roelse J, Sijts AJ, Ossendorp F, Feltkamp MC, Kast WM, Melief CJ, Neefjes JJ (1995) Major differences in transporter associated with antigen presentation (TAP)-dependent translocation of MHC class I-presentable peptides and the effect of flanking sequences. J Immunol 154:1273–1279
Niedermann G, Butz S, Ihlenfeldt HG, Grimm R, Lucchiari M, Hoschutzky H, Jung G, Maier B, Eichmann K (1995) Contribution of proteasome-mediated proteolysis to the hierarchy of epitopes presented by major histocompatibility complex class I molecules. Immunity 2:289–299
Niedermann G, King G, Butz S, Birsner U, Grimm R, Shabanowitz J, Hunt DF, Eichmann K (1996) The proteolytic fragments generated by vertebrate proteasomes: structural relationships to major histocompatibility complex class I binding peptides. Proc Natl Acad Sci USA 93:8572–8577
Niedermann G, Grimm R, Geier E, Maurer M, Realini C, Gartmann C, Soll J, Omura S, Rechsteiner MC, Baumeister W, Eichmann K (1997) Potential immunocompetence of proteolytic fragments produced by proteasomes before evolution of the vertebrate immune system. J Exp Med 186:209–220
Niedermann G, Geier E, Lucchiari-Hartz M, Hitziger N, Rampsperger A, Eichmann K (1999) The specificity of proteasomes: impact on processing and presentation of antigens. Immunol Rev 172:29–48
Nieland TJ, Tan MC, Monne-van Muijen M, Koning F, Kruisbeek AM, van Bleek GM (1996) Isolation of an immunodominant viral peptide that is endogenously bound to the stress protein GP96/GRP94. Proc Natl Acad Sci USA 93:6135–6139
Nussbaum AK, Dick TP, Keilholz W, Schirle M, Stevanovic S, Dietz K, Heinemeyer W, Groll M, Wolf DH, Huber R, Rammensee HG, Schild H (1998) Cleavage motifs of the yeast 20S proteasome beta subunits deduced from digests of enolase 1. Proc Natl Acad Sci USA 95:12504–12509
Orlowski M, Michaud C (1989) Pituitary multicatalytic proteinase complex. Specificity of components and aspects of proteolytic activity. Biochemistry 28:9270–9278

Orlowski M, Cardozo C, Michaud C (1993) Evidence for the presence of five distinct proteolytic components in the pituitary multicatalytic proteinase complex. Properties of two components cleaving bonds on the carboxyl side of branched chain and small neutral amino acids. Biochemistry 32:1563–1572

Ossendorp F, Eggers M, Neisig A, Ruppert T, Groettrup M, Sijts A, Mengede E, Kloetzel PM, Neefjes J, Koszinowski U, Melief C (1996) A single residue exchange within a viral CTL epitope alters proteasome-mediated degradation resulting in lack of antigen presentation. Immunity 5:115–124

Palmer A, Rivett AJ, Thomson S, Hendil KB, Butcher GW, Fuertes G, Knecht E (1996) Subpopulations of proteasomes in rat liver nuclei, microsomes and cytosol. Biochem J 316:401–407

Pamer E, Cresswell P (1998) Mechanisms of MHC class I-restricted antigen processing. Annu Rev Immunol 16:323–358

Paz P, Brouwenstijn N, Perry R, Shastri N (1999) Discrete proteolytic intermediates in the MHC class I antigen processing pathway and MHC I-dependent peptide trimming in the ER. Immunity 11:241–251

Pieters J (2000) MHC class II-restricted antigen processing and presentation. Adv Immunol 75:159–192

Preckel T, Fung-Leung WP, Cai Z, Vitiello A, Salter-Cid L, Winqvist O, Wolfe TG, Von Herrath M, Angulo A, Ghazal P, Lee JD, Fourie AM, Wu Y, Pang J, Ngo K, Peterson PA, Fruh K, Yang Y (1999) Impaired immunoproteasome assembly and immune responses in PA28−/− mice. Science 286:2162–2165

Rammensee H-G, Bachmann J, Stevanovic S (1997) The function. In: Rammensee H-G (ed) MHC Ligands and Peptide Motifs. Landes Biosciences, Austin, Texas, pp 217–369

Realini C, Rogers SW, Rechsteiner M (1994) KEKE motifs. Proposed roles in protein-protein association and presentation of peptides by MHC class I receptors. FEBS Lett 348:109–113

Rechsteiner M, Realini C, Ustrell V (2000) The proteasome activator 11S REG (PA28) and class I antigen presentation. Biochem J 345:1–15

Reits EAJ, Benham AM, Plougastel B, Neefjes J, Trowsdale J (1997) Dynamics of proteasome distribution in living cells. EMBO J 16:6087–6094

Reits EA, Vos JC, Gromme M, Neefjes J (2000) The major substrates for TAP in vivo are derived from newly synthesized proteins [see comments]. Nature 404:774–778

Rivett AJ (1985) Purification of a liver alkaline protease which degrades oxidatively modified glutamine synthetase. Characterization as a high molecular weight cysteine proteinase. J Biol Chem 260:12600–12606

Rivett AJ, Palmer A, Knecht E (1992) Electron microscopic localization of the multicatalytic proteinase complex in rat liver and in cultured cells. J Histochem Cytochem 40:1165–1172

Rock KL, Gramm C, Rothstein L, Clark K, Stein R, Dick L, Hwang D, Goldberg AL (1994) Inhibitors of the proteasome block the degradation of most cell proteins and the generation of peptides presented on MHC class I molecules. Cell 78:761–771

Rock KL, Goldberg AL (1999) Degradation of cell proteins and the generation of MHC class I-presented peptides. Annu Rev Immunol 17:739–779

Rowland-Jones S, Tan R, McMichael A (1997) Role of cellular immunity in protection against HIV infection. Adv Immunol 65:277–346

Schubert U, Anton LC, Gibbs J, Norbury CC, Yewdell JW, Bennink JR (2000) Rapid degradation of a large fraction of newly synthesized proteins by proteasomes [see comments]. Nature 404:770–774

Schwarz K, de Giuli R, Schmidtke G, Kostka S, van den Broek M, Kim KB, Crews CM, Kraft R, Groettrup M (2000a) The selective proteasome inhibitors lactacystin and epoxomicin can be used to either up- or down-regulate antigen presentation at nontoxic doses. J Immunol 164:6147–6157

Schwarz K, van Den Broek M, Kostka S, Kraft R, Soza A, Schmidtke G, Kloetzel PM, Groettrup M (2000b) Overexpression of the proteasome subunits LMP2, LMP7, and MECL-1, but not PA28 alpha/beta, enhances the presentation of an immunodominant lymphocytic choriomeningitis virus T cell epitope. J Immunol 165:768–778

Seemüller E, Lupas A, Stock D, Lowe J, Huber R, Baumeister W (1995) Proteasome from Thermoplasma acidophilum: a threonine protease [see comments]. Science 268:579–582

Sewell AK, Price DA, Teisserenc H, Booth BL Jr., Gileadi U, Flavin FM, Trowsdale J, Phillips RE, Cerundolo V (1999) IFN-gamma exposes a cryptic cytotoxic T lymphocyte epitope in HIV-1 reverse transcriptase. J Immunol 162:7075–7079

Shimbara N, Nakajima H, Tanahashi N, Ogawa K, Niwa S, Uenaka A, Nakayama E, Tanaka K (1997) Double-cleavage production of the CTL epitope by proteasomes and PA28: role of the flanking region. Genes to Cells 2:785–800

Shimbara N, Ogawa K, Hidaka Y, Nakajima H, Yamasaki N, Niwa S, Tanahashi N, Tanaka K (1998) Contribution of proline residue for efficient production of MHC class I ligands by proteasomes. J Biol Chem 273:23062–23071

Sibille C, Gould KG, Willard-Gallo K, Thomson S, Rivett AJ, Powis S, Butcher GW, De Baetselier P (1995) LMP2$^+$ proteasomes are required for the presentation of specific antigens to cytotoxic T lymphocytes. Curr Biol 5:923–930

Sijts AJ, Ruppert T, Rehermann B, Schmidt M, Koszinowski U, Kloetzel PM (2000a) Efficient generation of a hepatitis B virus cytotoxic T lymphocyte epitope requires the structural features of immunoproteasomes. J Exp Med 191:503–514

Sijts AJ, Standera S, Toes RE, Ruppert T, Beekman NJ, van Veelen PA, Ossendorp FA, Melief CJ, Kloetzel PM (2000b) MHC class I antigen processing of an adenovirus CTL epitope is linked to the levels of immunoproteasomes in infected cells. J Immunol 164:4500–4506

Snyder HL, JW Yewdell, JR Bennink (1994) Trimming of antigenic peptides in an early secretory compartment. J Exp Med 180:2389–2394

Song X, Mott JD, von Kampen J, Pramanik B, Tanaka K, Slaughter CA, DeMartino GN (1996) A model for the quaternary structure of the proteasome activator PA28. J Biol Chem 271: 26410–26417

Stein RL, Melandri F, Dick L (1996) Kinetic characterization of the chymotryptic activity of the 20S proteasome. Biochemistry 35:3899–3908

Stevanovic S, Schild H (1999) Quantitative aspects of T cell activation–peptide generation and editing by MHC class I molecules. Semin Immunol 11:375–384

Stoltze L, Dick TP, Deeg M, Pommerl B, Rammensee HG, Schild H (1998) Generation of the vesicular stomatitis virus nucleoprotein cytotoxic T lymphocyte epitope requires proteasome-dependent and -independent proteolytic activities. Eur J Immunol 28:4029–4036

Svensson K, Levy F, Sundberg U, Boman HG, Hendil KB, Kvist S (1996) Proteasomes generate in vitro a natural peptide of influenza-A nucleoprotein functional in HLA-B27 antigen assembly. Int Immunol 8:467–478

Tanaka K, Kasahara M (1998) The MHC class I ligand-generating system: roles of immunoproteasomes and the interferon-gamma-inducible proteasome activator PA28. Immunol Rev 163:161–176

Theobald M, Ruppert T, Kuckelkorn U, Hernandez J, Haussler A, Ferreira EA, Liewer U, Biggs J, Levine AJ, Huber C, Koszinowski UH, Kloetzel PM, Sherman LA (1998) The sequence alteration associated with a mutational hotspot in p53 protects cells from lysis by cytotoxic T lymphocytes specific for a flanking peptide epitope. J Exp Med 188:1017–1028

Thompson MW, Singh SK, Maurizi MR (1994) Processive degradation of proteins by the ATP-dependent Clp protease from Escherichia coli. Requirement for the multiple array of active sites in ClpP but not ATP hydrolysis. J Biol Chem 269:18209–18215

Tokunaga F, Goto T, Koide T, Murakami Y, Hayashi S, Tamura T, Tanaka K, Ichihara A (1994) ATP- and antizyme-dependent endoproteolysis of ornithine decarboxylase to oligopeptides by the 26S proteasome. J Biol Chem 269:17382–17385

Tsubuki S, Saito Y, Kawashima S (1994) Purification and characterization of an endogenous inhibitor specific to the Z-Leu-Leu-Leu-MCA degrading activity in proteasome and its identification as heat-shock protein 90. FEBS Lett 344:229–233

Udaka K, Tsomides TJ, Walden P, Fukusen N, Eisen HN (1993) A ubiquitous protein is the source of naturally occurring peptides that are recognized by a CD8$^+$ T-cell clone. Proc Natl Acad Sci USA 90:11272–11276

Uebel S, Kraas W, Kienle S, Wiesmuller KH, Jung G, Tampe R (1997) Recognition principle of the TAP transporter disclosed by combinatorial peptide libraries. Proc Natl Acad Sci USA 94:8976–8981

Uenaka A, Ono T, Akisawa T, Wada H, Yasuda T, Nakayama E (1994) Identification of a unique antigen peptide pRL1 on BALB/c RL male 1 leukemia recognized by cytotoxic T lymphocytes and its relation to the Akt oncogene. J Exp Med 180:1599–1607

Ustrell V, Pratt G, Rechsteiner M (1995) Effects of interferon gamma and major histocompatibility complex-encoded subunits on peptidase activities of human multicatalytic proteases [published errata appear in Proc Natl Acad Sci USA 1995, 92:3632 and 1995, 92:7605]. Proc Natl Acad Sci USA 92:584–588

Valmori D, Gileadi U, Servis C, Dunbar PR, Cerottini JC, Romero P, Cerundolo V, Levy F (1999) Modulation of proteasomal activity required for the generation of a cytotoxic T lymphocyte-defined peptide derived from the tumor antigen MAGE-3. J Exp Med 189:895–906

Van Endert PM (1996) Peptide selection for presentation by HLA class I: a role for the human transporter associated with antigen processing? Immunol Res 15:265–279

Van Endert PM, Riganelli D, Greco G, Fleischhauer K, Sidney J, Sette A, Bach JF (1995) The peptide-binding motif for the human transporter associated with antigen processing. J Exp Med 182:1883–1895

Van Hall T, Sijts A, Camps M, Offringa R, Melief C, Kloetzel PM, Ossendorp F (2000) Differential influence on cytotoxic T lymphocyte epitope presentation by controlled expression of either proteasome immunosubunits or PA28. J Exp Med 192:483–494

Van Kaer L, Ashton-Rickardt PG, Eichelberger M, Gaczynska M, Nagashima K, Rock KL, Goldberg AL, Doherty PC, Tonegawa S (1994) Altered peptidase and viral-specific T cell response in LMP2 mutant mice. Immunity 1:533–541

Villadangos JA, Ploegh HL (2000) Proteolysis in MHC class II antigen presentation: who's in charge? Immunity 12:233–239

Vinitsky A, Cardozo C, Sepp-Lorenzino L, Michaud C, Orlowski M (1994) Inhibition of the proteolytic activity of the multicatalytic proteinase complex (proteasome) by substrate-related peptidyl aldehydes. J Biol Chem 269:29860–29866

Vinitsky A, Anton LC, Snyder HL, Orlowski M, Bennink JR, Yewdell JW (1997) The generation of MHC class I-associated peptides is only partially inhibited by proteasome inhibitors: involvement of nonproteasomal cytosolic proteases in antigen processing? J Immunol 159:554–564

Wenzel T, Eckerskorn C, Lottspeich F, Baumeister W (1994) Existence of a molecular ruler in proteasomes suggested by analysis of degradation products. FEBS Lett 349:205–209

Wilk S, Orlowski M (1980) Cation-sensitive neutral endopeptidase: isolation and specificity of the bovine pituitary enzyme. J Neurochem 35:1172–1182

Yague J, Alvarez I, Rognan D, Ramos M, Vazquez J, de Castro JA (2000) An N-acetylated natural ligand of human histocompatibility leukocyte antigen (HLA)-B39. Classical major histocompatibility complex class I proteins bind peptides with a blocked NH(2) terminus in vivo. J Exp Med 191:2083–2092

Yang B, Hahn YS, Hahn CS, Braciale TJ (1996) The requirement for proteasome activity class I major histocompatibility complex antigen presentation is dictated by the length of preprocessed antigen. J Exp Med 183:1545–1552

Yang Y, Waters JB, Fruh K, Peterson PA (1992) Proteasomes are regulated by interferon gamma: implications for antigen processing. Proc Natl Acad Sci USA 89:4928–4932

Yellen-Shaw AJ, Wherry EJ, Dubois GC, Eisenlohr LC (1997) Point mutation flanking a CTL epitope ablates in vitro and in vivo recognition of a full-length viral protein. J Immunol 158:3227–3234

Yewdell JW, Esquivel F, Arnold D, Spies T, Eisenlohr LC, Bennink JR (1993) Presentation of numerous viral peptides to mouse major histocompatibility complex (MHC) class I-restricted T lymphocytes is mediated by the human MHC-encoded transporter or by a hybrid mouse-human transporter. J Exp Med 177:1785–1790

Yewdell J, Lapham C, Bacik I, Spies T, Bennink J (1994) MHC-encoded proteasome subunits LMP2 and LMP7 are not required for efficient antigen presentation. J Immunol 152:1163–1170

Yewdell JW, Anton LC, Bennink JR (1996) Defective ribosomal products (DRiPs): a major source of antigenic peptides for MHC class I molecules? J Immunol 157:1823–1826

Yewdell JW, Bennink JR (1999a) Immunodominance in major histocompatibility complex class I-restricted T lymphocyte responses. Annu Rev Immunol 17:51–88

Yewdell J, Anton LC, Bacik I, Schubert U, Snyder HL, Bennink JR (1999b) Generating MHC class I ligands from viral gene products. Immunol Rev 172:97–108

Yukawa M, Sakon M, Kambayashi J, Shiba E, Kawasaki T, Ariyoshi H, Mori T (1991) Proteasome and its novel endogeneous activator in human platelets. Biochem Biophys Res Commun 178:256–262

Zhou X, Momburg F, Liu T, Abdel Motal UM, Jondal M, Hämmerling GJ, Ljunggren HG (1994) Presentation of viral antigens restricted by H-2Kb, Db or Kd in proteasome subunit LMP2- and LMP7-deficient cells. Eur J Immunol 24:1863–1868

Zinkernagel RM, Ehl S, Aichele P, Oehen S, Kundig T, Hengartner H (1997) Antigen localisation regulates immune responses in a dose- and time-dependent fashion: a geographical view of immune reactivity. Immunol Rev 156:199–209

Zwickl P, Baumeister W (1999) AAA-ATPases at the crossroads of protein life and death [news]. Nature Cell Biol 1:E97–E98

Natural Substrates of the Proteasome and Their Recognition by the Ubiquitin System

H.D. ULRICH

1	Introduction	137
2	Natural Substrates of the Proteasome	139
3	The Ubiquitin Conjugation System	141
3.1	Ubiquitin-Activating Enzyme	141
3.2	Ubiquitin-Conjugating Enzymes and E2-Like Proteins	141
3.3	Ubiquitin Protein Ligases	143
4	Substrate Recognition by the Ubiquitin System	146
4.1	Recognition by HECT E3s	147
4.2	Recognition by RING E3s	149
4.2.1	The SCF and Related Complexes	149
4.2.2	The APC/Cyclosome	152
4.2.3	UBR1	153
4.2.4	Other RING Finger Proteins	154
4.3	E3 Proteins as Substrates for Ubiquitylation	156
4.4	Recognition of Hydrophobic Surfaces	157
4.4.1	The MATα2 Repressor	158
4.4.2	ER Degradation and the Unfolded Protein Response	158
5	Targeting of Ubiquitylated Proteins to the Proteasome	160
5.1	Recognition of Multiubiquitin Chains by the Proteasome	160
5.2	Potential Targeting Factors	162
6	Recognition of Proteins without Ubiquitylation	164
7	Summary	165
References		166

1 Introduction

The 26S proteasome is recognized as the principal mediator of intracellular proteolysis in eukaryotes. As a consequence, its influence on cellular metabolism is as complex and manifold as are the proteins degraded by this protease, and new natural substrates are being discovered in ever increasing numbers. It has long been

---

Max Planck Institute for Terrestrial Microbiology, Department of Organismic Interactions, Karl-von-Frisch-Strasse, 35043 Marburg/Lahn, Germany

realized that the modulation of the steady-state levels of proteins can occur at the level of their synthesis as well as their degradation (SCHIMKE 1973). Thus, one major area of proteasome function is the control of basic cellular processes such as cell cycle progression, signal transduction, and transcription via the degradation of short-lived regulatory factors. In addition, the proteasome plays a central role in the removal of misfolded, aberrant, or damaged proteins, which is a critical aspect of the cellular stress response. Last but not least, the mammalian proteasome is responsible for the generation of antigenic peptides presented on the cell surface by major histocompatibility complex (MHC) class I molecules as an integral part of the immune system (see the chapter by Niedermann, this volume). Tight control of proteasome activity is essential to guarantee the correctly timed removal of short-lived regulatory proteins but at the same time prevent the untimely destruction of other important cellular components not targeted for degradation. The system that distinguishes between stable proteins and those destined for breakdown and thus ensures the fidelity of selective proteolysis is the ubiquitin system. Accordingly, malfunctions or absence of components of this intricate enzymatic machinery lead to a variety of inherited or acquired diseases (SCHWARTZ and CIECHANOVER 1999).

With very few exceptions (MURAKAMI et al. 1992; JARIEL-ENCONTRE et al. 1995), potential substrates are marked for proteasomal degradation by the covalent attachment of ubiquitin, an abundant, highly conserved protein of 76 amino acids. The conjugation reaction is mediated by a battery of enzymes collectively known as the ubiquitin conjugation system (Fig. 1). In an initial ATP-dependent step, the C terminus of ubiquitin is activated by a high-energy thiolester linkage to an enzyme known as E1, or ubiquitin-activating enzyme. Ubiquitin is then transferred to one of several E2s or ubiquitin-conjugating enzymes (UBCs), which mediate the attachment of ubiquitin's C terminus to the ε-amino group of an internal lysine

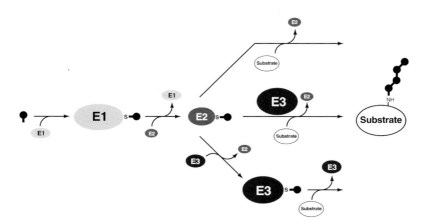

Fig. 1. The ubiquitin conjugation system. Ubiquitin (*black lollipop symbol*) is activated by E1 and transferred to an E2 as a covalent thiolester adduct. Conjugation to an internal lysine of the substrate protein by an isopeptide bond is often mediated by an E3 enzyme, which may take part in the thiolester relay. Successive rounds of ubiquitin conjugation result in the formation of multiubiquitin chains

residue of the substrate via an isopeptide linkage, in many cases assisted by additional factors called E3s or ubiquitin protein ligases. Successive rounds of ubiquitin conjugation to a lysine residue of the preceding ubiquitin moiety usually result in the formation of long multiubiquitin chains, which serve as a signal for degradation by the 26S proteasome. Thus, the targeting of a protein to the proteasome is a two-step process: first the identification of the substrate by the ubiquitylation machinery; and subsequently the recognition of the multiubiquitylated protein by the 26S proteasome. This chapter is dedicated to the proteasome's natural substrates. A number of comprehensive reviews have appeared about the regulatory roles of protein breakdown within the cell (HOCHSTRASSER 1995; HERSHKO and CIECHANOVER 1998; KORNITZER and CIECHANOVER 2000), and rather than repeating the examples given there or attempting to present an exhaustive compilation of substrates, which would rapidly become outdated, I will focus primarily on the principles and mechanisms of how the ubiquitin system directs its substrates to the 26S proteasome. Following an overview about classes of substrate proteins and the ubiquitin conjugation system I will describe in detail recent findings about targeting to the proteasome, considering both the substrate recognition factors of the ubiquitylation machinery and the delivery of multiubiquitylated proteins to the 26S proteasome.

2 Natural Substrates of the Proteasome

It is generally argued that the removal of a regulatory factor by proteolysis as opposed to the modulation of its activity by post-translational modification or sequestration is the most effective way of ensuring the irreversible nature of the respective biological process. Possibly due to this notion the most prominent function of the ubiquitin/proteasome system may be the control of the eukaryotic cell cycle (reviewed in NASMYTH 1996; TYERS and JORGENSEN 2000). Progression through the cell cycle is mediated by cyclin-dependent kinases (Cdk) in association with their regulatory cyclin subunits. The cyclins activate the kinases and control their specificity in the different stages of the cell cycle. For example, in yeast the G1-specific cyclins CLN1–CLN3 initiate budding, the S-phase-specific cyclins CLB5 and CLB6 control the initiation of DNA replication, while the mitotic cyclins CLB1–CLB4 are required for mitosis. Thus, the cyclins need to be synthesized, but also inactivated in a timely manner. The resulting periodic oscillation of cyclin abundance is controlled on one hand at the level of transcription, on the other hand by ubiquitin-mediated proteolysis (GLOTZER et al. 1991). In addition, Cdk activity is regulated by specific inhibitors, such as *Saccharomyces cerevisiae* SIC1, FAR1, or mammalian $p27^{Kip1}$, all of which are also subject to regulated proteasomal degradation. Apart from cyclins and Cdk-inhibitors, cell cycle-relevant substrates of the proteasome include the anaphase inhibitor PDS1 from *S. cerevisiae*, the phosphatase Cdc25, which acts as an upstream regulator of Cdk activity, the replication initiator protein CDC6, and the polo-like kinase CDC5, which activates the mitotic cyclin destruction machinery (see Sect. 4). Thus, ubiquitin-dependent proteolysis plays a role in virtually all aspects of cell cycle progression. Most of these regulators remain stable during part of the cell cycle, but are targeted for destruction at another stage; in many cases their degradation is initiated by phosphorylation.

A second area in which the ubiquitin/proteasome system plays a major regulatory role is the control of cell growth by selective proteolysis of transcription factors and repressors, tumor suppressor proteins and other members of signal transduction pathways. Common to many transcription factors of lower and higher eukaryotes is their short half-life. Examples are GCN4 and MET4 from yeast as well as mammalian c-jun, c-fos, c-myc, E2F-1, β-catenin, and p53 (reviewed in HERSHKO and CIECHANOVER 1998). While the ubiquitin-dependent degradation of some of these factors is modulated by phosphorylation or other signals, in other cases constitutive turnover can serve as an alternative to regulated destruction as it provides the cell with the option to quickly adjust the abundance of important regulatory factors at the level of their synthesis.

Activation of the nuclear factor (NF)-κB pathway provides another paradigm for the role of the ubiquitin/proteasome system in controlling cellular responses to external stimuli. NF-κB is an inducible transcription factor that regulates the inflammatory response in mammals. It is normally sequestered in the cytoplasm in an inactive form by means of association with an inhibitory protein of the IκB family, e.g., IκBα, which masks the nuclear localization signal of NF-κB (reviewed in KARIN and BEN-NERIAH 2000). Upon induction by an inflammatory signal IκBα is rapidly phosphorylated, ubiquitylated and subsequently degraded by the proteasome, thus allowing translocation of active NF-κB into the nucleus. In addition to IκBα, NF-κB itself is subject to ubiquitin-dependent proteolysis; however, in this case ubiquitylation does not result in complete degradation of the protein, but rather in the controlled processing of the NF-κB precursor protein, p105, to yield the mature transcription factor, p50 (PALOMBELLA et al. 1994). The mechanism by which complete proteolysis is prevented in this system is not fully understood. It is believed, however, that ubiquitylation of p105 induces the degradation of only the C-terminal portion, whereas the N-terminal domain, yielding p50, is protected from destruction by a glycine-rich region shortly upstream of the cleavage sites (LIN and GHOSH 1996).

Proteasomal processing of precursor proteins has long been regarded as an exceptional event limited to the NF-κB protein. However, recently another study has provided evidence for a very similar system in yeast: HOPPE et al. (2000) found that the transcription factors SPT23 and MGA2, responsible for the control of unsaturated fatty acid levels by induction of a fatty acid desaturase, are synthesized as inactive endoplasmic reticulum (ER) membrane-anchored precursor proteins. Activation of SPT23 and MGA2 occurs by ubiquitylation and proteasomal processing, which causes the release of soluble, active transcription factor. Interestingly, SPT23 and MGA2 are even structurally related to the NF-κB protein family.

Complementary to its regulatory function is the role of the ubiquitin/proteasome system in the removal of misfolded or damaged proteins (PARAG et al. 1987; SEUFERT and JENTSCH 1990, 1992). Recognition of aberrant proteins by the ubiquitylation machinery as part of the stress response implies that in principle any nonnative protein can become a target for proteasomal degradation. Particularly well studied is the removal of misfolded integral membrane as well as lumenal proteins of the ER, which involves the retrograde transport of target proteins to the cytoplasmic ubiquitylation and degradation machinery (reviewed in SOMMER and WOLF 1997; KOPITO 1997). Activity of the ER degradation system is tightly linked to the unfolded protein response, suggesting that this pathway plays a critical role in the survival under environmental stress (see Sect. 4.4.2).

3 The Ubiquitin Conjugation System

3.1 Ubiquitin-Activating Enzyme

Ubiquitin-activating enzyme, or E1, catalyzes the first step of the conjugation pathway – the activation of ubiquitin as a high-energy thiolester. E1 is a highly conserved protein of roughly 100kDa, which in most organisms, such as yeast and man, is encoded by a single-copy, essential gene (reviewed in HAAS and SIEPMANN 1997). An exception is wheat, which possesses three related genes. All E1 enzymes have in common a nucleotide binding motif, $G-x-G-x_2-G$ (where x is any amino acid), as well as a conserved cysteine residue defining the active site for ubiquitin thiolester formation. Ubiquitin activation by E1 is a multistep reaction (HAAS and ROSE 1982). In the first step, E1 catalyzes the formation of a ubiquitin adenylate intermediate at the expense of ATP. This intermediate is converted to the covalent thiolester adduct by transfer to E1's active-site cysteine and release of AMP. E1 is active as a homodimer (CIECHANOVER et al. 1982). Each subunit can simultaneously accommodate both the adenylate intermediate and the covalent thiolester adduct in a stable ternary complex, which necessitates the existence of two distinct domains mediating the two steps of the activation reaction (HAAS and ROSE 1982). Temperature-sensitive mutants of E1 have been instrumental in the analysis of ubiquitin-dependent proteolysis in mammalian cell culture (FINLEY et al. 1984; KULKA et al. 1988).

3.2 Ubiquitin-Conjugating Enzymes and E2-Like Proteins

Activated ubiquitin is passed as a thiolester from E1 to the active site cysteine of a ubiquitin-conjugating enzyme, E2, which mediates the successive transfer to the ε-amino group of an internal lysine residue of the substrate protein. The E2 enzymes, or UBCs, constitute a conserved protein family, with identities of 35%–40% within the catalytic domain bearing the cysteine engaged in thiolester formation. The high degree of sequence homology allows the identification of novel E2s on the basis of sequence alignments alone. By this criterion *S. cerevisiae* possesses 13 genes encoding UBCs, two of which mediate the conjugation not of ubiquitin itself, but of the ubiquitin-like modifiers SMT3 and RUB1, respectively (TANAKA et al. 1998). Several crystal structures show a common fold of the approximately 160 amino acid catalytic UBC domain, consisting of four α-helices and a four-stranded anti-parallel β-sheet. Although it has been observed that UBCs can form homo- as well as heterodimers (CHEN et al. 1993; GWOZD et al. 1995), most reported structures show a monomeric arrangement in the crystals. The most prominent differences among the E2s are observed in a variety of N- and C-terminal extensions. While class I E2s, such as the UBC4/5 subfamily, are small proteins consisting of the UBC domain only, class II enzymes possess C-terminal tails of varying lengths. Five of the 13 yeast E2s belong to this class, and their tails are believed to influence substrate specificity or subcellular localization. Class III E2s, which are absent in *S. cerevisiae*, are characterized by N-terminal extensions (MATUSCHEWSKI et al. 1996), and finally, class IV enzymes possess additional N- as well as C-terminal sequences. Striking examples of this type of E2 are the murine BRUCE and its human homolog, 530kDa proteins that possess an N-terminal BIR domain in addition to the UBC domain, suggesting a role in the regulation of apoptosis (HAUSER et al. 1998; CHEN et al. 1999). Another large E2 isolated from reticulocyte lysate, E2-230K, combines its ubiquitin-conjugating activity, mediated by a conserved UBC domain, with E3-like properties (BERLETH and PICKART 1996). These examples illustrate that E2s can in some cases supplement their conjugating activity with additional features in the same protein that may normally be achieved by the recruitment of other factors.

In contrast to E1, most UBCs are nonessential proteins. In fact, among the yeast E2s responsible for the conjugation of ubiquitin, only UBC3 (CDC34), is essential for survival (GOEBL et al. 1988). This observation may reflect the notion that there is some redundancy among the UBCs with respect to their substrate specificities and interactions with E3s. As a consequence, some combinations of different yeast *ubc* mutants result in synthetic effects in their sensitivity to environmental stress or even inviability, suggesting their involvement in a common, essential process (SEUFERT et al. 1990; JUNGMANN et al. 1993). Despite this partial overlap in function, different UBCs usually have distinct properties, and correspondingly mutants exhibit a variety of different phenotypes, indicating a high degree of specialization (CHEN et al. 1993).

A class of proteins whose function is not yet fully understood is the family of E2-like proteins, or ubiquitin-conjugating enzyme variants (UEVs). UEVs are highly related to the UBCs in sequence and predicted secondary and tertiary structure, but lack the active-site cysteine essential for catalytic activity. Members of this protein family are conserved from yeast to mammals, found in a variety of different contexts, and were originally postulated to represent dominant negative regulators of UBC activity (KOONIN and ABAGYAN 1997; PONTING et al. 1997). Human *CROC-1*, expressed in at least four different splice variants, was first identified as an activator of the c-*fos* promoter (ROTHOFSKY and LIN 1997), but was independently cloned several times by differential display methods (FRITSCHE et al. 1997; SANCHO et al. 1998). A closely related gene, *hMMS2*, was subsequently identified by expressed sequence tag (EST) database searches. Its yeast homolog, the DNA repair gene *MMS2*, has recently provided a first clue about a possible mechanism of action for the UEV proteins. HOFMANN and PICKART (1999) isolated MMS2 in association with the ubiquitin-conjugating enzyme UBC13 as a heterodimeric complex capable of assembling multiubiquitin chains that are linked in a nonstandard way, via Lys63. This type of chain had previously been implicated in DNA repair (SPENCE et al. 1995). Interestingly, whereas UBC13 is a canonical ubiquitin-conjugating enzyme bearing the conserved active-site cysteine, it remains catalytically inactive unless paired with MMS2. Thus, it may well turn out that UEVs generally function as positive rather than negative regulators of UBC activity. However, the biochemical functions of the other members of the UEV family remain to be elucidated.

3.3 Ubiquitin Protein Ligases

Ubiquitin protein ligases (E3s) mediate the transfer of ubiquitin from the E2 to an internal lysine residue of the substrate protein. In contrast to the UBCs, however, no common mechanism of action has been assigned to them. Thus, even the exact definition of an E3 is still a matter of debate. While some researchers postulate that a true E3 should participate directly in catalysis by means of an essential enzyme intermediate, others prefer a broader definition which only requires the E3 to bring the substrate into contact with the E2, thereby facilitating the transfer reaction (HERSHKO and CIECHANOVER 1998). Considering the diversity of proteins marked for degradation by the ubiquitin conjugation system, E3s as the main determinants of substrate specificity should play a prominent role in ubiquitin-dependent proteolysis. Yet, as most E3s are not easily recognizable as such by their sequence, only a few proteins had been identified as ubiquitin ligases until recently. Today, substrate recognition by the ubiquitin system has become one of the most important aspects of ubiquitin-related research, and with the discovery of more and more E3 enzymes a general principle is now emerging that allows a classification of ubiquitin ligases into at least two distinct categories (Fig. 2).

One class of E3 is represented by the human E6-associated protein (E6-AP), which in combination with the papillomavirus E6 oncoprotein mediates the ubiquitylation and degradation of p53 (SCHEFFNER et al. 1993). SCHEFFNER et al. (1995) showed that ubiquitylation by the E6-AP ligase occurs by a mechanism that involves an essential thiolester intermediate between ubiquitin and the E3 itself. In this case ubiquitin is first transferred from the E2 to the E3 and subsequently to the substrate, and E6-AP thus fits the narrow definition of an E3 involved in the catalytic mechanism as discussed above. The cysteine responsible for thiolester formation resides in a C-terminal domain of approximately 350 amino acids, now called the HECT domain (for homologous to the E6-AP C-terminus), which is

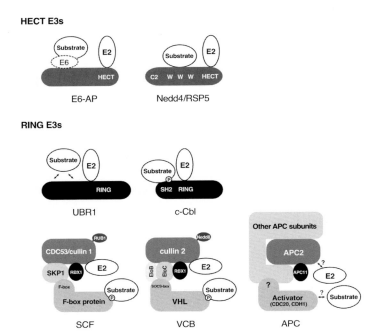

Fig. 2. Classes of ubiquitin ligases. Representative members of HECT and RING E3 families are shown schematically to illustrate their substrate recognition motifs and E2 binding sites. Phosphorylation of substrate proteins is indicated where it is a prerequisite for recognition by the E3

conserved in a large family of related proteins, including five members in *S. cerevisiae* and at least 20 in humans (SCHWARZ et al. 1998). Wherever tested, HECT proteins have been capable of thiolester formation like E6-AP, strongly suggesting that they all function as ubiquitin protein ligases (HUIBREGTSE et al. 1995; NUBER et al. 1996; SCHWARZ et al. 1998).

Several additional types of E3s have been identified, which will be described in more detail in Section 4 (Fig. 2). Among the best characterized are the yeast UBR1 protein and its mammalian homolog E3α (HERSHKO et al. 1983), which promote the ubiquitylation of several short-lived proteins (see Sect. 4.2.3). While UBR1 and E3α exist as monomers or homodimers (cited in KORNITZER and CIECHANOVER 2000), other ubiquitin ligases are large multiprotein complexes that possess specialized subunits for interaction with E2, substrate recognition, and catalytic function. A high molecular weight, multisubunit E3 involved in the ubiquitylation of G1 cyclins and Cdk inhibitors in yeast and higher eukaryotes is the SCF complex, named after its subunits Skp1, Cdc53 (or cullin), and F-box protein (reviewed in PATTON et al. 1998b; DESHAIES 1999; see Sect. 4.2.1). Closely related to the SCF complex is the mammalian VCB complex, based on the von Hippel–Lindau tumor suppressor protein VHL and the Cdc53 homolog cullin-2. Another multisubunit E3, the anaphase-promoting complex (APC), or cyclosome, plays an essential role in cell cycle progression by ubiquitylation of the mitotic cyclins and anaphase inhibitors (reviewed in PETERS 1999; PAGE and HIETER 1999; TYERS and JORGENSEN 2000). The APC consists of at least 10 constitutive subunits and associates with a number of additional regulatory factors that determine its substrate specificity (PAGE and HIETER 1999; see Sect. 4.2.2).

Based on a number of observations, it has recently been suggested that all these non-HECT E3s should be grouped together in a single category (KORNITZER and CIECHANOVER 2000). This notion seems attractive, as in addition to parallels in the architecture among the multimeric complexes (see below), they all contain an essential subunit that carries a RING finger, and this domain is also present in UBR1. The RING domain is a zinc binding motif that was first identified in the protein encoded by the human gene *RING-1* (*really interesting new gene*), but is also present in a multitude of proteins from fungi, animals, and plants (reviewed in SAURIN et al. 1996). The consensus motif can be described as a series of conserved cysteines and histidines, according to the pattern: $C-x_2-C-x_{(9-39)}-C-x_{(1-3)}-H-x_{(2-3)}-C/H-x_2-C-x_{(4-48)}-C-x_2-C$. Structural studies indicate that these residues coordinate two zinc ions in a unique 'cross-brace' arrangement that differs from regular zinc finger motifs (BORDEN and FREEMONT 1996). Despite its abundance the biological function of the RING domain remained enigmatic. A participation in the assembly of multiprotein complexes was suggested (SAURIN et al. 1996), but more recently an involvement in the ubiquitin system has become apparent (see Sect. 4.2.4). A role in the ubiquitin conjugation step was first suggested by Bachmair and coworkers (POTUSCHAK et al. 1998), and direct evidence for an E3 function came from studies of the RING finger protein c-Cbl, which mediates the ubiquitin-dependent downregulation of receptor tyrosine kinases (JOAZEIRO et al. 1999; YOKOUCHI et al. 1999; LEVKOWITZ et al. 1999), as well as in vitro studies by Weissman and co-workers (LORICK et al. 1999). Several other RING finger proteins have subsequently been identified as ubiquitin protein ligases (FREEMONT 2000). In many cases the RING finger has been demonstrated to mediate interaction with the E2 (LORICK et al. 1999; MARTINEZ-NOEL et al. 1999; MOYNIHAN et al. 1999). However, a stable complex with the conjugating enzyme is apparently not required for ligase activity (LORICK et al. 1999), and in at least one case the E2 contacts the ligase, UBR1, at a site adjacent to the RING domain, even though the intact RING finger is necessary for E3 activity (XIE and VARSHAVSKY 1999). The question of whether the RING domain is directly involved in the catalytic mechanism has not been answered definitively. A thiolester intermediate involving the RING domain, however, is unlikely, as RING E3s have in most cases proven insensitive to thiol-alkylating agents when compared to HECT E3s (HERSHKO et al. 1983, PATTON et al. 1998a; SEOL et al. 1999; FANG et al. 2000).

Thus, the presence of a RING domain seems to be a unifying theme for the non-HECT E3s. Naturally, this notion does not answer the question whether additional, yet unidentified families of ubiquitin protein ligases exist that obey different rules. Similarly, the reverse question remains unanswered: are all RING finger proteins ubiquitin ligases? While it is difficult to disprove an involvement in the ubiquitin conjugation pathway in the absence of biochemical information,

RING finger proteins have been implicated in a variety of cellular contexts not known to be associated with the ubiquitin system, such as immunoglobulin gene recombination by RAG1 (RODGERS et al. 1996) or transcriptional repression by TIF1β (PENG et al. 2000). In these cases, the RING domain was found to mediate di- or multimerization, as was also shown for the oncoprotein BRCA1 (BRZOVIC et al. 1998). Interestingly, the RING finger of BRCA1 also binds to the ubiquitin hydrolase BAP1, suggesting a functional involvement of the ubiquitin system (JENSEN et al. 1998). LORICK et al. (1999) have demonstrated that several RING finger proteins, including BRCA1, can be subject to RING-dependent ubiquitylation in vitro. In the case of the IAP apoptosis inhibitors the RING domain was found to be required in vivo for their own downregulation via proteasomal degradation (YANG et al. 2000). In these instances, it remains unclear whether the RING finger proteins are simply substrates of the ubiquitin system or whether they are true ubiquitin ligases in the sense that they can also mediate the ubiquitylation of other proteins. In summary, an involvement of many RING finger proteins in ubiquitin-dependent proteolysis is indisputable; however, further studies are clearly needed to prove E3 activity in some cases or disprove a participation in the ubiquitin pathway in others.

4 Substrate Recognition by the Ubiquitin System

As outlined above, the ubiquitin conjugation system follows a hierarchical organization. A single E1 is responsible for the transfer of activated ubiquitin to several E2s with differing substrate preferences and subcellular localizations. Conjugation to the substrate proteins by the E2 in turn is aided by an even larger number of E3s, which fine-tune the specificity set forth by the E2s. Interaction of a set of E2s with a variety of different E3s (NUBER et al. 1996; KUMAR et al. 1997; SCHWARZ et al. 1998; see also HOCHSTRASSER 1995) thus endows the ubiquitin system with a combinatorial diversity that allows an exquisite specificity in the recognition of potential substrates among thousands of cellular proteins and the flexibility to respond to environmental cues for degradation. The diversity in the components of the ubiquitin system is complemented by a variety of constitutive and regulatable degradation signals on the substrate proteins, and accurate specificity results from the combination of these two aspects. As an added level of complexity, recognition of substrates by the ubiquitin system is often regulated in response to internal or environmental cues. This modulation can be achieved by a variety of different strategies, for example by modification of the substrates via phosphorylation, by regulating the activity or abundance of the ubiquitin ligases themselves, by preventing or allowing the accessibility of degradation signals to the conjugation machinery, or finally at the level of deubiquitination (WILKINSON 1997). As substrate recognition is a task that lies largely in the hands of the ubiquitin ligases, it will be discussed here from the perspective of the different types of E3s, as it relates to their specific architectures. A number of representative substrates are summarized in Table 1 according to specific E3s and their recognition motifs. Special emphasis is given to the budding yeast, as in many cases it represents the system for which the most detailed information is available.

4.1 Recognition by HECT E3s

HECT E3s seem to follow a number of different strategies in substrate recognition. While their C-terminal HECT domain is responsible for catalytic activity as well as recruitment of the E2 (Kumar et al. 1997; Huang et al. 1999), substrate binding depends on their N-terminal parts. E6-AP itself was found to mediate the ubiquitylation of its most prominent substrate, the tumor suppressor protein p53, exclusively in a complex with the papillomavirus oncoprotein E6 (Scheffner et al. 1993). In fact, E6-AP was first identified due to its association with E6 (Huibregtse et al. 1991). Thus, p53 recognition by E6-AP is not direct, but involves the viral protein as an obligatory cofactor. In a similar E6-dependent fashion, the replication licensing factor Mcm7 is ubiquitylated by E6-AP (Kühne and Banks 1998). In this case, recognition of Mcm7 depends on a homotypic interaction motif, termed the L2G box, that is present in both E6 and Mcm7. On the other hand, E6-AP is active as a ubiquitin ligase even in the absence of the viral E6 protein, as HHR23A, a human homolog of the yeast DNA repair protein RAD23, was identified as an E6-independent substrate of E6-AP (Kumar et al. 1999). By what motif the E6-AP N-terminus mediates recognition of HHR23A, however, is not fully understood.

A distinct subgroup within the family of HECT E3s is formed by the mammalian Nedd4 and its homologs (Harvey and Kumar 1999). Common to the members of this group are an N-terminal Ca^{2+}/lipid-binding C2 domain and two to four copies of a protein–protein interaction module known as the WW domain. Named after two conserved tryptophan residues, this approximately 35 amino acid domain consists of a hydrophobic core surrounded by β-sheets (Sudol 1996) and is known to bind preferentially to proline-rich (PY) sequences with the consensus P-P-x-Y. While the C2 domain probably functions in plasma membrane localization of Nedd4 (Plant et al. 1997), several studies have provided compelling evidence that substrate recognition is mediated by the WW motifs. The first Nedd4 substrate to be identified was the epithelial sodium channel ENaC, consisting of three subunits, all of which bear PY motifs recognized by Nedd4 (Staub et al. 1996). Mutations or deletions of the PY motifs stabilize the channel and cause Liddle's syndrome, an inherited form of hypertension (Abriel et al. 1999). Similarly, the yeast Nedd4 homolog RSP5 is involved in the ubiquitylation of a number of substrates by means of its WW domains, such as the large subunit of RNA polymerase II (Huibregtse et al. 1997; Chang et al. 2000) and the membrane-anchored transcription factors SPT23 and MGA2 (Hoppe et al. 2000). Interestingly, two proteins involved in stress resistance, BUL1 and BUL2, were found to interact with yeast RSP5 by means of PY motifs, even though they do not seem to be ubiquitylated, raising the possibility that they might cooperate with RSP5 as

Table 1. Ubiquitin ligases and their substrate recognition motifs

Ubiquitin ligase	Substrate recognition	Substrate	References
HECT E3s			
E6-AP (H.s.)	E6	p53	Scheffner et al. 1993
		Mcm7	Kühne and Banks 1998
	Direct	HHR23A	Kumar et al. 1999
		E6-AP	Nuber et al. 1998
Nedd4 (H.s.)	WW domains	ENaC	Staub et al. 1996
RSP5 (S.c.)	WW domains	Pol II	Huibregtse et al. 1997; Chang et al. 2000
		SPT23, MGA2	Hoppe et al. 2000
Pub1 (S.p.)	WW domains	Cdc25	Nefsky and Beach 1996
RING E3s			
SCF^{CDC4} (S.c.)	WD40	SIC1	Skowyra et al. 1997; Feldman et al. 1997
		FAR1	Henchoz et al. 1997
		CDC6	Drury et al. 1997
		GCN4	Meimoun et al. 2000
		CDC4	Zhou and Howley 1998; Galan and Peter 1999; Mathias et al. 1999
SCF^{GRR1} (S.c.)	LRR	CLN1, 2	Skowyra et al. 1997
		GIC2	Cited in Patton et al. 1998b
		GRR1	Galan and Peter 1999
SCF^{MET30} (S.c.)	WD40	MET4	Rouillon et al. 2000
		SWE1	Kaiser et al. 1998
		MET30	Galan and Peter 1999
SCF$^{Pop1/2}$ (S.p.)	WD40	Rum1	Kominami and Toda 1997
		Cdc18	Kominami and Toda 1997
SCF^{βTrCP} (H.s.)	WD40	IκBα	Yaron et al. 1998; Spencer et al. 1999; Winston et al. 1999
		β-catenin	Winston et al. 1999; Liu et al. 1999
	Vpu	CD4	Margottin et al. 1998
SCF^{SKP2} (H.s.)	LRR	E2F-1	Marti et al. 1999
		p27^{Kip1}	Tsvetkov et al. 1999; Carrano et al. 1999
VCB^{VHL} (H.s.)	b-Domain	HIF1-α	Maxwell et al. 1999
VCB^{SOCS1} (H.s.)	SH2	Vav	De Sepulveda et al. 2000
APC^{CDC20} (S.c.)	WD40	PDS1	Cohen-Fix et al. 1996; Shirayama et al. 1999
		CLB3, 5	Shirayama et al. 1999
APC^{CDH1} (S.c.)	WD40	CLB1, 2, 3	Schwab et al. 1997; Schwab et al. 1997
		CDC5	Shirayama et al. 1998
		ASE1	Juang et al. 1997
		DBF4	Cheng et al. 1999
		CDC20	Fang et al. 1998
APC^{CDH1} (X.l.)	WD40	Nek2, B99	Pfleger and Kirschner 2000
UBR1	Type I, II sites	Synthetic N-end rule substrates	Bartel et al. 1990
		SCC1	Rao et al. 2001
		REC8	Buonomo et al. 2000
	Body site	CUP9	Byrd et al. 1998
		GPA1	Madura and Varshavsky 1994

Ubiquitin ligase	Substrate recognition	Substrate	References
c-Cbl (H.s.)	SH2	EGF-R	LEVKOWITZ et al. 1999; YOKOUCHI et al. 1999
		PDGF-R	JOAZEIRO et al. 1999
Sina (D.m.)	Phyllopod	Tramtrack	LI et al. 1997; TANG et al. 1997
Siah-1 (H.s.)	C-terminal domain	DCC	HU et al. 1997; HU and FEARON 1999
		Siah-1	HU and FEARON 1999
Mdm2 (H.s.)	?	p53	HONDA et al. 1997; HONDA and YASUDA 2000; FANG et al. 2000
		Mdm2	FANG et al. 2000
HRD1 (S.c.)	Hydrophobic sites?	HMG2	GARDNER et al. 2001
DOA10 (S.c.)	Deg1	MATa2	SWANSON et al. 2001

Listed are representative E3s, including their substrate recognition motifs or subunits (where known) and relevant substrates. D.m., *Drosophila melanogaster*; H.s., *Homo sapiens*; S.c., *Saccharomyces cerevisiae*; S.p., *Schizosaccharomyces pombe*; X.l., *Xenopus laevis*.

adaptors in the ubiquitylation of other proteins (YASHIRODA et al. 1996, 1998). WW domains display some degree of substrate selectivity both in vitro and in vivo (SUDOL 1996; HARVEY et al. 1999; WANG et al. 1999; CHANG et al. 2000; HOPPE et al. 2000), suggesting that the existence of multiple copies in the Nedd4-like ligases provides the enzymes with a means to recognize a number of different substrates. Pub1, the fission yeast homolog of Nedd4, is involved in the degradation of the cell cycle phosphatase Cdc25; however, no details of substrate recognition have been reported (NEFSKY and BEACH 1996).

Yeast RSP5 not only targets proteins to proteasomal degradation – it is also involved in the downregulation of several plasma membrane proteins by endocytosis (reviewed in HICKE 1999). While it is evident that ubiquitylation of the respective proteins is a prerequisite for endocytosis and the ubiquitin ligase activity of RSP5 is essential for this process, substrate recognition in this case remains poorly understood.

4.2 Recognition by RING E3s

4.2.1 The SCF and Related Complexes

Today, the SCF complex best exemplifies the paradigm of a multimeric E3 whose different tasks – ligase activity, substrate recognition, and E2 binding – are delegated to specialized subunits. A number of excellent reviews have appeared on the structure and the substrates of this type of ligase (PATTON et al. 1998b; DESHAIES 1999). The SCF complex is conserved from yeast to mammals and was first recognized as an E3 in *S. cerevisiae* by its ability to ubiquitylate the Cdk inhibitor SIC1 (SKOWYRA et al. 1997; FELDMAN et al. 1997). Its constant core consists of a cullin subunit, in yeast known as CDC53, which binds the E2 CDC34, and an adaptor protein called SKP1, which recruits the substrate recognition subunit (SKOWYRA et al. 1997; PATTON et al. 1998a). Although CDC34 seems to be the E2 most commonly utilized in cooperation with the SCF, members of the UBC4/5 family are also able to support ubiquitin-conjugation (OHTA et al. 1999). A RING finger protein known as RBX1, ROC1, or HRT1, which contacts the cullin, was later shown to be another essential component (OHTA et al. 1999; SEOL et al. 1999; SKOWYRA et al. 1999). Interestingly, all cullins are found to be modified in vivo by the ubquitin-like protein Nedd8 or its budding yeast homolog RUB1 (LAMMER et al. 1998; LIAKOPOULOS et al. 1998; HORI et al. 1999), and this modification has been reported to stimulate or even be required for the ligase activity of the SCF complex in mammalian systems (MORIMOTO et al. 2000; PODUST et al. 2000; READ et al. 2000). The SCF's most distinctive feature is its flexible substrate recognition by means of exchangeable subunits, the F-box proteins. These proteins contain a conserved domain called the F-box, which contacts the SCF core via SKP1, and a variety of additional protein–protein interaction modules for substrate binding. For example, the yeast F-box proteins CDC4 and MET30 carry multiple copies of a motif known as the WD40 repeat (NEER et al. 1994), whereas yeast GRR1 bears a series of leucine-rich repeats (LRR) (KOBE and DEISENHOFER 1994). A number of other motifs, such as tetratricopeptide (TPR) repeats, are also found in F-box proteins from higher eukaryotes. The nematode *Caenorhabditis elegans* possesses more than 60 different F-box proteins, indicating the enormous versatility of SCF-type ubiquitin ligases (PATTON et al. 1998b).

The SCF complex is instrumental for the cell cycle transition from G1 to S phase. SCF^{GRR1} of budding yeast targets the G1 cyclins CLN1 and CLN2 (SKOWYRA et al. 1997) and the GTPase-associated proteins GIC1 and GIC2, which control bud emergence (cited in PATTON et al. 1998b). Substrates of SCF^{CDC4} include the Cdk inhibitors SIC1 and FAR1 (SKOWYRA et al. 1997; FELDMAN et al. 1997; HENCHOZ et al. 1997), the replication factor CDC6 (DRURY et al. 1997) as well as the transcription factor GCN4 (MEIMOUN et al. 2000). SCF^{MET30} ubiquitylates the transcriptional activator MET4 (ROUILLON et al. 2000) and the cell cycle kinase SWE1 (KAISER et al. 1998). Similarly, the fission yeast homologs of CDC4, Pop1 and Pop2, mediate degradation of the *S. pombe* Cdk inhibitor Rum1 and the CDC6 homolog Cdc18 (KOMINAMI and TODA 1997). In humans, the first F-box protein for which E3 function was demonstrated was β-TrCP, which associates with cullin-1 and human SKP1 via WD40 repeats for ubiquitylation of the signaling factor β-catenin and the NF-κB inhibitor IκBα (YARON et al. 1998; LIU et al. 1999; SPENCER et al. 1999; WINSTON et al. 1999). The human F-box protein SKP2 is responsible for ubiquitylation of the transcription factor E2F-1 and the Cdk inhibitor $p27^{Kip1}$ by means of leucine-rich repeats (MARTI et al. 1999; TSVETKOV et al. 1999; CARRANO et al. 1999). Other F-box proteins are believed to mediate proteolysis in a variety of cellular contexts ranging from floral development in plants to amino acid biosynthesis and a number of signaling pathways (reviewed in PATTON et al. 1998b). Like HECT E3s, the SCF complex can be subverted by viral proteins to ubiquitylate cellular proteins that are not normally subject to degradation: the human CD4 receptor is targeted for proteolysis by the HIV-1 Vpu protein, which associates with the F-box protein β-TrCP and in a ternary complex recruits CD4 to the SCF ligase for ubiquitylation (MARGOTTIN et al. 1998).

Despite the variations in substrate recognition motifs among F-box proteins, common to all of them is their specificity for phosphorylated substrates. The strict phosphorylation-dependence of SCF-mediated ubiquitylation in fact provides an important means to regulate the activity of many cell cycle factors, as it links the stability of cyclins and Cdk inhibitors to the catalytic activity of the cyclin-dependent kinases themselves. G1 cyclins were first shown to be degraded in a manner dependent on a so-called 'PEST motif', a region rich in proline, glutamate, serine, and threonine, which had previously been predicted to represent a degradation signal (ROGERS et al. 1986). As it turned out, phosphorylation of serine or threonine residues within this region by cyclin-dependent kinase is the signal actually recognized by the ubiquitylation machinery (SALAMA et al. 1994; YAGLOM et al. 1995; LANKER et al. 1996). While most PEST regions do not confirm to a strict consensus and their function as a degradation signal relies rather on amino acid composition than on sequence, interestingly a conserved motif was found in three different proteins recognized by the human F-box protein β-TrCP, namely IκBα, β-catenin, and human immunodeficiency virus (HIV) Vpu (reviewed in LANEY and HOCHSTRASSER 1999). In each case, phosphorylation of two conserved serines within this motif was required for β-TrCP association. As outlined above, recognition of the viral protein by the F-box component does not lead to ubiquitylation of Vpu itself, but rather of the associated CD4. In the case of IκBα this conserved sequence may not be the only feature that confers specificity to the degradation signal. In fact, the phosphorylation pattern of IκBα is more complex and involves constitutive modification by other kinases within a C-terminal PEST region of the protein, which is also required for turnover (reviewed in MAY and GHOSH 1998; KARIN and BEN-NERIAH 2000). Thus, several distinct ubiquitylation signals may, in combination, provide the flexibility necessary for the regulation of IκBα stability and thereby NF-κB activation in response to environmental conditions.

Mammalian cells contain at least six different cullins, which are all capable of association with the RING finger protein ROC1/RBX1 (OHTA et al. 1999; KAMURA et al. 1999), suggesting that the cullin–RBX1 complex may be the defining feature of a series of ubiquitin ligases. In fact, a cullin-2-based complex, involving the VHL tumor suppressor protein and named VCB after its subunits VHL, elongin C, and elongin B, was recently identified as an E3 enzyme based on its ability to ubiquitylate the hypoxia-inducible factor HIF1-α (COCKMAN et al. 2000). Mutations in the VHL protein are responsible for a number of spontaneous as well as hereditary types of carcinomas (WHALEY et al. 1994). The architecture of the VHL complex is reminiscent of the SCF ligase itself in that it also contains a SKP1 homolog, elongin C. Unlike SKP1, however, elongin C is present in a complex with elongin B, a protein containing a ubiquitin-like domain. Substrate recognition is mediated in a similar modular manner by the VHL protein, which binds to elongin BC via a conserved motif present in a number of other factors, the SOCS box (KAMURA et al. 1998; STEBBINS et al. 1999). Substrate recognition by VHL is achieved through its β-domain and is abrogated by a number of tumor-associated mutations involving this domain (COCKMAN et al. 2000). It is likely that other SOCS box proteins, for example SOCS-1 itself, may target other substrates for ubiquitylation by means of different protein–protein interaction motifs such as the SH2 domain (KAMURA et al. 1998; DE SEPULVEDA et al. 2000). Recent reports suggest that analogous complexes involving other cullins can also act as ubiquitin ligases (SINGER et al. 1999; MAEDA et al. 2001; KAMURA et al. 2001). However, cullin-3 has already been implicated in some aspects of cyclin E degradation in mouse (SINGER et al. 1999).

4.2.2 The APC/Cyclosome

The APC, or cyclosome, was first discovered as a high molecular weight complex in fractions of clam oocyte extracts based on its ability to reconstitute the ubiquitination of cyclin B in combination with E1 and E2 (HERSHKO et al. 1994; SUDAKIN et al. 1995). Even today little is known about its architecture on the molecular level (for reviews see PETERS 1999; PAGE and HIETER 1999). Neither its catalytic center nor the domains responsible for interaction with the E2 have yet been localized to any particular subunit. However, based on a comparison with the SCF complex, specific roles can be tentatively assigned to a subset of APC subunits. With respect to the subunit composition, the budding yeast complex has been studied most extensively (IRNIGER et al. 1995; ZACHARIAE et al. 1996, 1998a; HWANG and MURRAY 1997). It consists of 11 core subunits present in stoichiometric amounts, 8 of which are essential for viability. This number varies somewhat in other species like vertebrates or *Xenopus laevis*, but it is unknown at this point whether smaller or loosely associated proteins have all been identified in these cases. Remarkably, one of the essential components, APC2, contains a region homologous to the cullins. A function of this subunit in E2 recruitment is therefore likely. Consistent with the classification of E3s into RING and HECT types, the APC11 subunit belongs to the family of RING finger proteins. Several other subunits contain TPR repeats, which are protein–protein interaction domains found in a variety of contexts (GOEBL and YANAGIDA 1991). These domains are likely to fulfill a scaffold function, but a role in substrate interaction cannot be excluded. The E2s involved in APC-dependent ubiquitylation have not been unambiguously defined in *S. cerevisiae*. It is likely that members of the UBC4/5 family cooperate with the APC, as they can mediate ubiquitin-conjugation in vitro, and a mutant allele of *CDC23*, encoding one of the APC subunits, is synthetically lethal with the *ubc4* null mutation (IRNIGER et al. 1995). However, due to the redundancy within this E2 family even the *ubc4/ubc5* double mutant is viable and does not exhibit any cell cycle defects. In vertebrates, an additional family of UBC enzymes, represented by the clam E2-C or human UbcH10, was observed to cooperate with the APC (ARISTARKHOV et al. 1996).

The most prominent substrates of the APC are the B-type mitotic cyclins, whose destruction is essential for the transition from metaphase to anaphase and also the exit from mitosis (reviewed in PETERS 1999; TYERS and JORGENSEN 2000).

Separation of sister chromatids is initiated by the APC-dependent destruction of the anaphase inhibitor PDS1 (COHEN-FIX et al. 1996). Similarly, ASE1, a microtubule-associated protein required for spindle elongation, is ubiquitylated by the APC and degraded at the end of mitosis (JUANG et al. 1997). APC activity persists into G1 phase to mediate the degradation of the S phase-promoting factor DBF4 prior to START (CHENG et al. 1999). Finally, the kinase CDC5 is an activator, but at the same time a substrate of the APC (SHIRAYAMA et al. 1998).

A number of accessory factors, the yeast CDC20 and CDH1 (or HCT1), in *Drosophila* known as fizzy and fizzy-related, respectively, were found to be peripherally associated with the APC. Although an interaction with substrate proteins has not been directly demonstrated, these factors are promising candidates for the substrate recognition units, as they activate the APC and determine its specificity (SCHWAB et al. 1997; VISINTIN et al. 1997; FANG et al. 1998). Moreover, similar to some of the F-box components of the SCF ligases both CDC20 and CDH1 bear WD40 repeats for protein–protein interaction. Thus, it is likely that the APC relies on exchangeable substrate recognition modules in the same manner as the SCF. Consistent with this idea, the two forms of the APC, APC^{CDC20} and APC^{CDH1}, are believed to function in distinct steps during mitosis by ubiquitylation of different sets of substrates: APC^{CDC20} is active at the metaphase to anaphase transition and is responsible for the ubiquitylation of PDS1 and CLB5 (VISINTIN et al. 1997; SHIRAYAMA et al. 1999). Subsequently, APC^{CDH1} acts during exit from mitosis and in G1 and is specific for CLB1, CLB2, ASE1, and DBF4 (VISINTIN et al. 1997; SCHWAB et al. 1997; ZACHARIAE et al. 1998b; JUANG et al. 1997; CHENG et al. 1999).

CDC20 and CDH1 each confer a distinct substrate specificity to the APC. All CDC20-dependent substrates bear a conserved motif known as destruction box, which was first discovered in the mitotic cyclins and obeys the consensus sequence $R\text{-}x_2\text{-}L\text{-}x_4\text{-}N$ (GLOTZER et al. 1991), while specificity of CDH1-recognition is somewhat more relaxed (FANG et al. 1998). In fact, PFLEGER and KIRSCHNER (2000) have recently identified a second motif, which they coined KEN box, according its conserved amino acid sequence K-E-N, that serves as a targeting signal for APC^{CDH1}. Based on this motif, two additional cell cycle proteins, Nek2 and B99, were newly identified as CDH1-dependent APC substrates in *Xenopus*. In contrast to the SCF-dependent ligase activity, ubiquitination by the APC is not dependent on the phosphorylation of the substrate proteins. Correctly timed degradation is in this case achieved by the cell cycle-dependent regulation of APC activity itself, via phosphorylation of many of its subunits as well as control of the abundance or affinity of its activating subunits CDC20 and CDH1 (see Sect. 4.3).

4.2.3 UBR1

UBR1 was identified as the recognition component of a degradation pathway known as the N-end rule pathway (BARTEL et al. 1990). The N-end rule relates the in vivo half-life of a protein to the identity of its N-terminal residue (reviewed in VARSHAVSKY 1996, 1997). It was found that artificial fusion proteins bearing destabilizing N-termini liberated by in vivo cleavage undergo rapid UBR1-mediated ubiquitylation and degradation (BARTEL et al. 1990). The E2 responsible for cooperation with UBR1 was shown to be RAD6 (DOHMEN et al. 1991), and to date RAD6 remains the only UBC known to physically associate with UBR1. In contrast to the HECT proteins, the E2 binding site of UBR1 is distinct from the catalytically relevant domain, the RING finger (XIE and VARSHAVSKY 1999). Recognition of N-end rule substrates by UBR1 and its mammalian counterpart E3α (KWON et al. 1998) occurs through at least two distinct, noncompeting sites, which bind substrates directly by their N-termini: basic residues are recognized by a type I site and bulky hydrophobic amino acids are recognized by a type II site (REISS et al. 1988; VARSHAVSKY 1997). A related protein, E3β, exhibits specificity for small uncharged residues (HELLER and HERSHKO 1990). The affinity of N-end rule substrates for the binding sites is moderate, but is considerably enhanced if a lysine residue is present in sufficient proximity to become a target for ubiquitylation by RAD6 (VARSHAVSKY 1996). Substrates with a ubiquitylation site can even be targeted to UBR1 in trans as part of an oligomeric complex in which a different subunit bears the destabilizing N-terminus (JOHNSON et al. 1990). However, a number of proteins lacking destabilizing N-termini are also subject to UBR1-dependent ubiquitylation, such as the Gα subunit of a heterotrimeric G protein involved in cell differentiation, GPA1 (MADURA and VARSHAVSKY 1994), which is targeted not through its N-terminus, but through a different feature, also called 'body site' that is apparently recognized soon after the protein is synthesized (SCHAUBER et al. 1998a). Another example is the transcriptional repressor CUP9, which controls the activity of the peptide transporter PTR2 (BYRD et al. 1998). Given the observation that di- and tripeptides act as inhibitors of the N-end rule pathway, UBR1's peptide binding sites may actually function in a regulatory way to control the rate of peptide import through PTR2 transcription. On the other hand, the search for natural substrates whose degradation in vivo is indeed dependent on a destabilizing N terminus is still ongoing. One potential candidate substrate, the GTPase-activating protein RGS4, was shown to be degraded by an N-end rule pathway in reticulocyte lysate, but mutation of the N terminus did not inhibit proteolysis in vivo, indicating that another degradation signal was operative (DAVYDOV and VARSHAVSKY 2000). However, true physiological N-end rule substrates were finally discovered in yeast: cleavage of the mitotic and meiotic cohesin subunits SCC1 and REC8, respectively, during anaphase results in proteolytic fragments that bear destabilizing N-termini and are subject to UBR1-dependent degradation (RAO et al. 2001; BUONOMO et al. 2000).

4.2.4 Other RING Finger Proteins

Several other RING finger proteins have recently been identified as ubiquitin ligases. BACHMAIR and coworkers first suggested that the conserved RING domain might be a hallmark of many E3s based on their identification of the *Arabidopsis* N-end rule-specific ligase, PTR1, and sequence comparison with several other components of the ubiquitin system (POTUSCHAK et al. 1998). Since then, the number of RING finger proteins for which ubiquitin ligase activity could be demonstrated has grown steadily. Like UBR1 most of these RING E3s seem to combine substrate recognition and E2 binding in a single polypeptide chain, although in many cases it has not been fully established whether they cooperate with additional cofactors in vivo.

One system in which substrate recognition is particularly well understood is the ligand-induced ubiquitylation of receptor protein tyrosine kinases (RPTKs) by the proto-oncogene product c-Cbl. In vivo, over-expression of c-Cbl enhances the ubiquitylation and downregulation of platelet-derived growth factor receptor (PDGF-R) and epidermal growth factor receptor (EGF-R), and both ubiquitylation and proteasomal degradation have been reconstituted in vitro using recombinant c-Cbl and immunopurified receptor (JOAZEIRO et al. 1999; LEVKOWITZ et al. 1999; YOKOUCHI et al. 1999). Unlike UBR1, c-Cbl recruits its cognate E2s via the RING domain. E2 binding and in vitro ubiquitylation were demonstrated for Ubc4 (JOAZEIRO et al. 1999), UbcH5 (LEVKOWITZ et al. 1999), and UbcH7 (YOKOUCHI et al. 1999), although conflicting results were reported with respect to c-Cbl's selectivity between UbcH5 and UbcH7. Both PDGF-R and EGF-R are recognized by the N-terminal SH2-domain of c-Cbl, suggesting that as in the case of SCF substrates phosphorylation of the RPTK is a prerequisite for E3 binding and ubiquitylation. Indeed, LEVKOWITZ et al. (1999) identified residue Y1045, a minor phosphorylation site within EGF-R as the phosphorylation-dependent docking site for the ligase. Interestingly, c-Cbl itself is subject to phosphorylation by the RPTK on residue Y371, adjacent to the RING domain, and this modification is required for the activation of c-Cbl's ubiquitin ligase activity (LEVKOWITZ et al.1999). Thus, ligand-induced phosphorylation of both the receptor and – upon docking – the E3 seems to provide a two-step safeguard mechanism for desensitization towards the growth hormone signal. Consequently, mutations that abolish ubiquitin ligase activity or its ability to bind the receptor confer oncogenic potential to c-Cbl.

Another RING finger protein that was identified as a ubiquitin ligase is the proto-oncogene product Mdm2, which is responsible for E6-independent downregulation of the tumor suppressor protein p53. Although initially a HECT-like activity was suspected (HONDA et al. 1997), it was later shown that indeed the C-terminal RING finger is responsible for E3 activity of Mdm2 (HONDA and YASUDA 2000). Over-expression of Mdm2 in vivo enhances the degradation of p53, and in vitro ubiquitylation was demonstrated with cellular-derived or in vitro translated p53, purified E1 and E2 (UbcH5), and recombinant Mdm2 of bacterial origin, indicating that the ligase activity is inherent in the Mdm2 protein (FANG et al. 2000). Exchange of the Mdm2 RING finger (which does not exactly match the consensus sequence) for the corresponding Praja-1-derived domain allowed p53 binding and self-ubiquitylation of the Mdm2 protein (see Sect. 4.3), but abolished ligase activity towards p53, suggesting that the identity of the RING finger may have some influence on substrate specificity (FANG et al. 2000).

Other RING finger proteins implicated in ubiquitin ligation are the *Drosophila* seven in absentia (Sina), which is involved in eye development and in association with phyllopod targets the transcriptional repressor tramtrack for ubiquitin-dependent proteolysis (LI et al. 1997; TANG et al. 1997), and its closely related human homolog Siah-1, which binds to and downregulates the DCC (deleted in colorectal cancer) protein (Hu et al. 1997; Hu and FEARON 1999). In yeast, the HRD1/DER3 protein is a transmembrane RING finger protein involved in ER degradation. Its RING domain is required for function (BORDALLO et al. 1999), and E3 activity has indeed been demonstrated (BAYS et al. 2001). Similarly, the herpes simplex virus protein Vmw110 mediates the ubiquitin/proteasome-dependent degradation of the centromere-associated CENP-C protein by means of its RING finger, thereby inhibiting cell cycle progression in virus-infected cells (LOMONTE and EVERETT 1999). In other cases, RING finger proteins are likely to function as E3s, although ubiquitin ligase activity has not been directly demonstrated. For example, in postreplicative DNA repair in yeast, the two DNA-binding RING finger proteins RAD18 and RAD5 cooperate to recruit the UBCs RAD6 and the UBC13-MMS2 dimer to chromatin (ULRICH and JENTSCH 2000). While RAD18 – like UBR1 – binds RAD6 at a site distinct from the RING finger (BAILLY et al. 1997), RAD5 contacts UBC13 via its RING finger (ULRICH and JENTSCH 2000). As the proteins ubiquitylated in the context of DNA repair have not been identified, however, it is unknown whether RAD18 and RAD5 have ubiquitin ligase function. This scenario is reminiscent of a number of other RING finger proteins that were identified by their ability to interact with ubiquitin-conjugating enzymes, but whose function remains to be elucidated (LORICK et al. 1999; MARTINEZ-NOEL et al. 1999; MOYNIHAN et al. 1999).

4.3 E3 Proteins as Substrates for Ubiquitylation

Recognition of ubiquitin ligases themselves or components thereof as substrates for auto-ubiquitylation is emerging as an important means for the regulation of E3 activity. HUIBREGTSE et al. (1995) first noted that the yeast HECT E3 RSP5 is capable of auto-ubiquitylation in an intramolecular reaction, and similar observations were made for E6-AP, even though in this case an intermolecular mechanism was postulated (NUBER et al. 1998). Ubiquitylation of E6-AP indeed leads to proteasomal degradation, and an intact HECT domain is required for this function. Interestingly, ubiquitylated forms of E6-AP fail to bind to p53, but association with p53 in turn inhibits the modification of E6-AP, suggesting that auto-ubiquitylation may be a means to inactivate any pool of free E6-AP (NUBER et al. 1998).

Similar to E6-AP, LORICK et al. (1999) found that in vitro many of the monomeric RING finger proteins are capable of RING-dependent auto-ubiquitylation when supplied with E1, E2, and ubiquitin. While it remains to be established whether this process indeed reflects a physiological reaction in all cases, there is evidence that both Siah-1 and Mdm2 utilize auto-ubiquitylation for their own downregulation (Hu and FEARON 1999; FANG et al. 2000). While both proteins are normally extremely short-lived in vivo, the respective RING finger mutants are stable, such that mutant Siah-1 can act in a dominant negative fashion to inhibit the degradation of the DCC protein (Hu and FEARON 1999). With respect to Mdm2, ubiquitylation probably contributes to an autoregulatory feedback loop by which the levels and activities of p53 and Mdm2 are tightly controlled (FANG et al. 2000).

In the case of multimeric RING E3s, selective proteolysis of the subunits responsible for substrate recognition can be used as a means to regulate substrate specificity. Several studies have independently demonstrated that the F-box components of the SCF ligase, CDC4, GRR1, and MET30, are all intrinsically short-lived and subject to proteasomal degradation (Zhou and Howley 1998; Galan and Peter 1999; Mathias et al. 1999). As the SCF complex itself is responsible for the ubiquitylation of its substrate recognition modules, F-box-mediated binding to SKP1 is a prerequisite for their downregulation. On the other hand, a mutant F-box protein that retains the ability to bind the SCF complex but cannot be ubiquitylated acts as a dominant negative allele and inhibits the degradation of physiological SCF substrates (Zhou and Howley 1998). Mathias et al. (1999) identified a short sequence adjacent to the F-box within CDC4, which they termed the 'R-motif', as a transferable signal for ubiquitylation and downregulation. Thus, the inherent instability of F-box proteins is believed to enhance the flexibility of substrate binding by rapid exchange of the recognition modules.

Similar to the SCF complex, the APC also takes part in the regulation of its own substrate specificity, however in a slightly different way: during the late stages of anaphase, the regulatory subunit CDC20 is exchanged for its relative CDH1 (reviewed in Page and Hieter 1999; Tyers and Jorgensen 2000); correspondingly, CDC20 abundance fluctuates during the cell cycle (Prinz et al. 1998). Fang et al. (1998) showed that in fact APC^{CDH1} is responsible for the rapid loss of CDC20 at the end of mitosis. Thus, not auto-ubiquitylation but cross-recognition by CDH1 provides the basis for downregulation of CDC20. In contrast to CDC20, CDH1 levels remain constant throughout the cell cycle. Regulation of APC^{CDH1} activity is thus believed to occur not on the level of protein stability, but rather by modulation of CDH1 affinity for the APC core via phosphorylation (Zachariae et al. 1998b).

4.4 Recognition of Hydrophobic Surfaces

The removal of misfolded and damaged proteins is one of the principal tasks of the ubiquitin/proteasome system. It was noted quite early that the accumulation of heat-denatured proteins causes a rapid burst in ubiquitin-dependent protein degradation and a decrease in the free ubiquitin pool (Parag et al. 1987). Consequently, mutations in numerous components of the ubiquitin/proteasome system, such as UBC4, UBC5, UBC7, or proteasomal subunits, cause an increased sensitivity towards heat stress or protein-damaging agents in yeast (Seufert and Jentsch 1990, 1992; Jungmann et al. 1993). Yet, in contrast to the degradation signals that elicit the ubiquitylation of short-lived regulatory factors, little is known about the recognition of abnormal proteins by the ubiquitin conjugation machinery. Based on the notion that nonnative proteins expose hydrophobic amino acids on the surface that would normally be buried within their core, it seems most likely that these hydrophobic stretches themselves serve as recognition signals for the ubiquitin system. Consistent with this idea, genetic screens for synthetic degradation signals in *S. cerevisiae* have yielded mostly hydrophobic sequences as well as motifs predicted to form amphipathic helices (SADIS et al. 1995; GILON et al. 1998). Moreover, the UBCs implicated in the ubiquitylation of fusion proteins bearing these signals coincide with those involved in stress resistance, suggesting that they might participate directly in substrate recognition. In the following section, I will briefly discuss two systems in which proteolysis elicited by hydrophobic surfaces has been studied in greater detail.

4.4.1 The MATα2 Repressor

MATα2 is a homeodomain protein involved in the determination of mating type in *S. cerevisiae*. In diploid cells, it forms a heterodimer with the MATa-encoded a1 protein and is responsible for the repression of haploid-specific genes, while in α-cells it is assisted by MCM1 in the repression of a-specific genes (reviewed in HERSKOWITZ et al. 1992). Both the a1 and the α2 protein are extremely short-lived in haploid cells, allowing rapid mating type switches in homothallic strains. Hochstrasser and coworkers have studied in detail the mechanism and regulation of α2 proteolysis (reviewed in LANEY and HOCHSTRASSER 1999). Two distinct pathways involving four UBCs mediate α2 ubiquitylation. In particular, a sequence residing in the N-terminal 67 amino acids, denoted *Deg1*, was identified as a UBC6- and UBC7-dependent degradation signal (CHEN et al. 1993). Consistent with the notion that UBC6 and UBC7 are implicated in the recognition of hydrophobic surfaces (SADIS et al. 1995; GILON et al. 1998), *Deg1* is part of a segment predicted to form an amphipathic helix, and mutations in the hydrophobic face abolish its function as a degradation signal (JOHNSON et al. 1998). Interestingly, the *Deg1* signal partially overlaps with the site involved in heterodimerization with the a1 protein, implying that the ubiquitylation machinery may compete with a1 binding. Indeed, α2 was found to be significantly stabilized in diploid cells or upon coexpression of a1, indicating that heterodimerization masks the degradation signal. Thus, the exposure of a hydrophobic surface seems to be used deliberately for the mating type-specific regulation of α2 stability. In fact, an E3 that cooperates with UBC6 and UBC7 in the Deg1-dependent ubiquitination of Matα2, but also functions in ER degradation was recently identified by Hochstrasser and coworkers (SWANSON et al. 2001).

4.4.2 ER Degradation and the Unfolded Protein Response

Integral membrane proteins and components of the endocytic, lysosomal, and secretory compartments make up a substantial fraction of total cellular protein. Upon synthesis these proteins are translocated into the ER, where folding, core glycosylation, and the formation of disulfide bonds take place. Thus, the removal of abnormal and misfolded proteins must be ensured in this compartment as well as in the cytoplasm. The system that is responsible for ER quality control has been termed ER-associated degradation (for reviews, see KOPITO 1997; SOMMER and WOLF 1997). Misfolded ER proteins do not reach the vacuole, but are transported back across the ER membrane into the cytosol, where they become substrates of the ubiquitin/proteasome system. Model substrates that have served to study mechanistic aspects of this process are proteins that fail to fold correctly, such as unassembled T-cell receptor subunits, a variant of the cystic fibrosis transmembrane conductance regulator CFTR, a mutant form of carboxypeptidase Y, or an unglycosylated version of pro-α-factor. One protein whose controlled removal by ER degradation is not due to misfolding, but serves a regulatory purpose is the integral membrane protein 3-hydroxy-3-methylglutaryl-CoA reductase, a key enzyme of sterol biosynthesis (HAMPTON et al. 1996).

ER degradation is believed to involve the same principle operative in the ubiquitylation of misfolded cytoplasmic proteins, i.e., the recognition of hydrophobic surfaces. The pathway shares some components of the ubiquitin system with the cytosolic degradation machinery, such as the E2 enzymes UBC6 and UBC7 and the proteasome, but also includes a number of ER membrane proteins specific for subsets of ER degradation substrates, like yeast DER1, DER3/HRD1, and HRD3, which may be involved in substrate recognition (WILHOVSKY et al. 2000). In particular DER3/HRD1 was shown to be an ER-specific RING type ubiquitin ligase (BAYS et al. 2001). Retrograde transport involves the translocon for protein import (PLEMPER et al. 1997), and in some cases the proteasome itself may take part in the extraction from the ER membrane (MAYER et al. 1998). Importantly, the ER-resident chaperones calnexin and BiP/KAR2 have both been implicated in the recognition of proteins for ER degradation (cited in SOMMER and WOLF 1997).

A second system that participates in the recognition of misfolded ER proteins is the unfolded protein response, or UPR (reviewed in CHAPMAN et al. 1998). Substances that interfere with folding in the ER, such as the reducing agent dithiothreitol or tunicamycin, an inhibitor of N-linked glycosylation, cause ER stress through the accumulation of unfolded proteins. This condition activates an unconventional signaling pathway that eventually leads to the production of a transcription factor, HAC1, which is responsible for the upregulation of a number of target genes involved in ER stress tolerance. Recently, a functional connection between the systems of ER degradation and the UPR has been established that may explain how the UPR signal leads to the correction of misfolding. Walter and coworkers have used a genome-wide transcription analysis under ER stress conditions for the identification of UPR target genes (TRAVERS et al. 2000). In addition to ER-resident chaperones and enzymes involved in phospholipid biosynthesis they identified several components of the ER degradation pathway, including DER1, DER3/HRD1, HRD3, and UBC7. A functional UPR was required for upregulation of these factors, and degradation of a model substrate was inhibited under conditions of ER stress, indicating that the ER degradation system is indeed responsible for UPR-induced removal of aberrant ER proteins. Conversely, mutations in ER degradation factors resulted in a constitutive activation of the UPR. Very similar results were obtained by Sommer and coworkers, who have studied the cooperation between the E2 enzymes UBC1 and UBC7 in the removal of unfolded ER proteins (FRIEDLÄNDER et al. 2000). Coordinated regulation of ER degradation and the UPR seems to be essential for survival of stress, as mutations in both pathways do not support growth at elevated temperatures.

5 Targeting of Ubiquitylated Proteins to the Proteasome

Once a substrate protein has been recognized by the ubiquitin conjugation system and is modified by the covalent attachment of a multiubiquitin chain it becomes a target for proteasomal degradation. Ubiquitylation thus serves to translate the rather diverse degradation signals present on the individual substrates into a common 'tag' recognizable by the 26S proteasome. How exactly multiubiquitin chains function as degradation tags is a topic of active research. Little is yet known about the intermediate steps between ubiquitin chain assembly and the actual proteolytic event or the accessory factors that mediate the delivery of multiubiquitinated substrates to the proteasome. Equally important is the nature of the signal conveyed by the ubiquitin chain itself, as outlined below.

5.1 Recognition of Multiubiquitin Chains by the Proteasome

There is now ample evidence that long multiubiquitin chains linked via isopeptide bonds as opposed to conjugates of multiple monoubiquitin units represent the major recognition signal for the 26S proteasome (reviewed in PICKART 1997). Several possible reasons have been suggested to explain this preference for multiubiquitin chains. One explanation would be the high local concentration of ubiquitin in a long chain that may facilitate targeting; another reason may be an increase in the fidelity of the ubiquitylation system that would require the assembly of an entire chain rather than a single conjugation event and would thus prevent the premature destruction of erroneously ubiquitylated proteins. On the other hand it has been suggested that long chains can provide some protection from disassembly by isopeptidases, thus preventing the escape of substrates from degradation. Moreover, it has become clear that monoubiquitylation often serves a purpose different from proteasomal degradation, such as endocytosis and delivery to the vacuole (reviewed in HICKE 1999). Finally, Pickart and coworkers have recently presented compelling evidence that a multiubiquitin chain represents a structurally unique recognition element for the proteasome. The crystal structure of tetraubiquitin linked via isopeptide bonds to Lys48, the principal site of chain elongation (see below), exposes a regular array of hydrophobic patches within each ubiquitin molecule on the surface (COOK et al. 1994). Made up of amino acids Lys9, Ile44, and Val70, these patches are essential for efficient recognition by the proteasome (BEAL et al. 1996). Using synthetic ubiquitin chains with defined positions of mutant ubiquitin moieties in which this signal is inactivated, Thrower et al. were able to demonstrate that the spatial arrangement of hydrophobic patches within a tetraubiquitin unit is in fact the principal determinant for recognition by the proteasome (THROWER et al. 2000). These results nicely explain the finding that the affinity of multiubiquitin chains for the proteasome rises dramatically as the number of ubiquitin moieties increases to at least four (PIOTROWSKI et al. 1997; THROWER et al. 2000).

The identity of the multiubiquitin chain receptor(s) on the 26S proteasome is still unresolved. Subunit S5a was identified as a multiubiquitin chain binding protein by probing immobilized proteasomal subunits with labeled chains, and its binding properties closely resemble that of the entire 26S particle in that it exhibits a strong preference for chains longer than four ubiquitin moieties (DEVERAUX et al. 1994; BEAL et al. 1996). However, it turned out that deletion of its yeast homolog, MCB1/RPN10, had no apparent detrimental effect on cell survival and led to the stabilization of only a subset of short-lived test substrates, suggesting that S5a/MCB1/RPN10 is probably not the only ubiquitin chain receptor of the proteasome (VAN NOCKER et al. 1996). Identification of a second recognition site on the 19S cap may prove difficult if this site is composed of more than one subunit and requires a fully assembled cap for activity.

An interesting issue, the significance of which is not fully understood, pertains to the structure of the multiubiquitin chains themselves. Ubiquitin has seven lysine residues, and five of them have been shown to be capable of serving as possible acceptor sites for chain elongation (reviewed in PICKART 1997). As these residues are distributed over the surface of ubiquitin, the resulting chains are believed to adopt very different geometries. Chains linked throughout via Lys48 represent the principal signal for proteasomal degradation (CHAU et al. 1989), and this residue is the only essential lysine in ubiquitin (FINLEY et al. 1994). However, chains polymerized in vitro that are linked via Lys6 or Lys11 can also bind the proteasomal subunit S5a (BABOSHINA and HAAS 1996). The assembly of alternatively-linked ubiquitin chains is clearly mediated by specific E2/E3 combinations. For example, yeast RAD6 in combination with the ligase UBR1 synthesizes Lys48-linked chains (CHAU et al. 1989), but in vitro isolated RAD6 can produce Lys6-linkages (BABOSHINA and HAAS 1996). In vivo, Lys29 was shown to play a significant role in the ubiquitylation of short-lived ubiquitin fusion proteins by UBC4/5 and the E3 UFD4 (JOHNSON et al. 1995); however, in this case only an initiation of the chain via this residue was shown, and it is unclear whether this linkage persists throughout the chain. An interplay between Lys29 and Lys48 in the same degradation pathway was suggested by KOEGL et al. who observed a dependence of a chain elongation factor, UFD2, on Lys48 in a system where chain initiation takes place via Lys29 (KOEGL et al. 1999). An E3 activity that assembles homogeneous Lys29-linkages was identified in rabbit reticulocyte, although the same enzyme was also capable of catalyzing linkage via Lys48 (MASTRANDREA et al. 1999). As three-dimensional structures of ubiquitin chains linked in nonstandard ways are not yet available, it is currently unclear whether the proteasome is able to accommodate different chain geometries. A hint that alternative linkages may indeed convey different signals comes from the study of Lys63-linked chains. An E2 that mediates chain assembly via this lysine was identified as a heterodimer of UBC13 and the UEV MMS2 (HOFMANN and PICKART 1999; see Sect. 3.2). Lys63 has been implicated in the repair of DNA damage, in endocytosis, and in mitochondrial inheritance, as a yeast strain in which ubiquitin is replaced by a mutant bearing arginine at this site shows defects in these processes (SPENCE et al. 1995; GALAN and HAGUENAUER-TSAPIS 1997; FISK and YAFFE 1999). In contrast, bulk protein turnover is unaffected by this mutation (SPENCE et al. 1995). Moreover, the UV sensitivity or endocytosis defect associated with this lysine mutant is not shared by mutants in proteasomal subunits (DOR et al. 1996), suggesting that indeed Lys63-linked chains do not signal primarily to the proteasome, even though in vitro a test substrate modified by Lys63-linked chains is recognized as a substrate and degraded (HOFFMANN and PICKART 2001).

5.2 Potential Targeting Factors

In vitro studies in well-defined systems show that purified 26S proteasome can degrade multiubiquitylated proteins in the absence of any additional components, indicating that it is directly involved in the recognition of its substrates. However, in vivo degradation pathways seem to be considerably more complex and require the participation of multiple downstream factors in addition to the ubiquitin conjugation machinery. Indeed, a significant fraction of the multiubiquitin chain receptor MCB1/RPN10 exists in a pool of free protein not associated with the proteasome, suggesting that it may fulfill a shuttling function (VAN NOCKER et al. 1996). Another example is the system responsible for the degradation of uncleavable ubiquitin fusion proteins in yeast, termed the UFD pathway (JOHNSON et al. 1995). Five genes, *UFD1–5*, were isolated in a screen for mutants that stabilize an otherwise short-lived ubiquitin-β-galactosidase fusion protein. Only one of these, *UFD4*, encoding a HECT E3, turned out to be directly involved in the conjugation process. *UFD5* was identified as *SON1*, encoding a proteasomal subunit. The remaining three genes encode accessory factors with an indirect effect on the stability of UFD substrates, and their functions remain poorly understood. However, their general importance is implied by the fact that they are involved in a variety of other degradative pathways as well. *UFD1* is an essential gene required for a post-ubiquitylation step. *UFD3* has an influence on the pool of free ubiquitin, as mutants can be suppressed by over-expression of ubiquitin (JOHNSON et al. 1995). *UFD2* was recently shown to encode a chain elongation factor, termed E4, which facilitates the polymerization of long multiubiquitin chains in cooperation with E1, E2, and E3 (KOEGL et al. 1999).

Recently, a series of physical and genetic interactions has emerged that reveals an intricate network in which the UFD proteins may cooperate. A factor central in their coordination turned out to be CDC48, an AAA-type ATPase that is also involved in homotypic membrane fusion. CDC48 function is required for the degradation of a UFD substrate. Furthermore, the protein was found independently in complexes with UFD2 and UFD3 (KOEGL et al. 1999; GHISLAIN et al. 1996). Like UFD1, CDC48 acts at a post-ubiquitylation step, as UFD substrates are stabilized in *cdc48–6* mutants with long multiubiquitin chains. Intriguingly, CDC48 antagonizes the binding of UFD2 to immobilized multiubiquitylated substrate, indicating that it may function in a targeting step after completion of chain assembly (KOEGL et al. 1999). This notion is supported by the observation that the mammalian homolog of CDC48, p97 or VCP (for valosin-containing protein), apparently participates in the degradation of IκBα by providing a physical link between ubiquitylated IκBα and the 26S proteasome (DAI et al. 1998). Moreover, p97 was found in a ternary complex with the mammalian homologs of UFD1 and NPL4 (MEYER et al. 2000). As NPL4 is an essential component of the nuclear pore in yeast, this association suggests a link between the ubiquitin/proteasome system and nuclear transport. A similar connection is apparent in the UFD1-dependent processing of SPT23 and MGA2, the ER membrane-bound precursors of nuclear transcription factors involved in the regulation of membrane fluidity (HITCHCOCK et al. 2001): conditions that rescue the temperature sensitivity of a *ufd1–2* mutant also suppress the lethality of a *npl4–2* mutant (HOPPE et al. 2000). Taken together, these observations argue for a central function of CDC48 and its associated factors in the delivery of multiubiquitylated proteins to the 26S proteasome, and perhaps even an influence in determining the localization of the degradation event. Further analysis of the role of NPL4 and nuclear transport in proteolysis may help to answer the question of whether degradation of certain proteins may indeed be restricted to specific cellular compartments.

Another class of proteins that may control access to the proteasome are the molecular chaperones and heat shock factors. As they are responsible for the binding and refolding of nonnative proteins, they may play an important role at the interface between protein folding and degradation. As mentioned above, the chaperones BiP and calnexin have been implicated in ER-associated degradation (SOMMER and WOLF 1997). More recently, the chaperone cofactor BAG-1 was shown to provide a direct physical link between the chaperones Hsc70/Hsp70 and the 26S proteasome (LÜDERS et al. 2000). BAG-1, which bears a ubiquitin-like domain at its N-terminus, was found to enhance association between Hsc70/Hsp70 and the 26S proteasome in an ATP-dependent manner. As the Hsc70/Hsp70 factors specialize in the stabilization of nonnative proteins, BAG-1 may thus facilitate their transfer from the chaperones to the proteasome for degradation (LÜDERS et al. 2000). Intriguingly, BAG-1 had been previously implicated in the inhibition of the ubiquitin ligase Siah-1, further supporting a function in ubiquitin-mediated proteolysis (MATSUZAWA et al. 1998). BAG-1 is not the only ubiquitin-related protein with a connection to the chaperone system: a family of structurally related proteins were recently found to interact with the Hsp70-like Stch protein (KAYE et al. 2000). Another protein bearing an N-terminal ubiquitin-like domain, the yeast DNA repair protein RAD23, associates with the MCB1/RPN10 subunit of the proteasome in a manner dependent on its ubiquitin-like domain (SCHAUBER et al. 1998b; HIYAMA et al. 1999). Overlapping roles of RAD23 and MCB1/RPN10 are indeed suggested by the pleiotropic effects of the respective double mutants (LAMBERTSON et al. 1999). Surprisingly, although interaction with the 19S cap is essential for RAD23's activity in nucleotide excision repair, proteolytic activity of the proteasome is not required for this function (RUSSELL et al. 1999). On the other hand, the human homolog of RAD23, HHR23A, is subject to proteasomal degradation (KUMAR et al. 1999). Thus, the precise role of RAD23's ubiquitin-like domain and its interaction with the proteasome remains unclear (ORTOLAN et al. 2000).

Finally, the intriguing possibility was raised by XIE and VARSHAVSKY (2000) that ubiquitin ligases may participate directly in the delivery of substrates to the proteasome. The RING E3 UBR1 as well as the HECT E3 UFD4, components of the N-end rule and the UFD pathway, respectively, were found to associate with a number of proteasomal 19S subunits, and a functional significance was suggested by effects of overproduction of these subunits. Even UBCs have recently been found in association with the 26S proteasome (TONGAONKAR et al. 2000). It remains to be demonstrated that the associations of these enzymes with the proteasome are indeed essential for the activity of the respective degradation pathway. Another open question is whether an E3-bound multiubiquitinated substrate is required for the docking of the ligase to the proteasome, or whether chain formation and targeting occur independently. Nevertheless, an E3-mediated substrate delivery mechanism is an attractive model that deserves further investigation.

6 Recognition of Proteins without Ubiquitylation

Although multiubiquitylated proteins make up the major fraction of proteasome substrates, proteins can be degraded by the proteasome without the participation of the ubiquitin system. The prototype substrate for ubiquitin-independent degradation is mammalian ornithine decarboxylase (ODC), the rate-limiting enzyme involved in polyamine biosynthesis (reviewed in MURAKAMI et al. 2000 and references cited therein). As a tight regulation of polyamine levels is essential for cell viability and growth, the activity of ODC must be equally well balanced. The key factor responsible for the control of ODC is a protein known as antizyme (AZ), whose biosynthesis is in turn regulated by polyamine levels by means of a programmed ribosomal frame-shifting event. AZ counteracts ODC in two distinct ways: on the one hand, association with AZ causes ODC's reversible inhibition, presumably by disruption of the catalytically active ODC homodimer; on the other hand, AZ is thought to act in a manner analogous to the ubiquitin system by targeting ODC to the 26S proteasome (MURAKAMI et al. 1992). An ATP-dependent, irreversible inactivation of the enzyme that is believed to be due to a sequestration of ODC inside the 26S particle precedes the actual degradation step (MURAKAMI et al. 1999). Despite many studies on the regulation of AZ-mediated ODC degradation, the mechanism of AZ-dependent delivery to the proteasome remains poorly understood. An essential element within ODC for recognition by the proteasome is a short C-terminal sequence, which becomes exposed upon association with AZ and presumably acts as a recognition element for the 26S proteasome. A sequence within the N-terminus of AZ not involved in ODC binding acts as a transferable signal that renders the degradation of a short-lived protein ubiquitin independent (LI et al. 1996). However, it is presently unclear whether the targeting event depends solely on the ODC C-terminal element exposed by AZ binding, or whether AZ itself contributes to the recognition. If indeed ODC alone is recognized by the proteasome, it is hard to imagine how the N terminus of AZ can act as a transferable degradation signal; on the other hand, association of the ODC–AZ complex with the 26S proteasome does not depend on this N-terminal region (LI et al. 1996).

A second case in which ubiquitin-independent proteasomal degradation was reported is the downregulation of the transcription factor c-jun (JARIEL-ENCONTRE et al. 1995). It was shown in a cell-free degradation system that ubiquitylation was not a prerequisite for ATP-dependent proteolysis, even though the proteasome was responsible for the degradation. On the other hand, in vivo ubiquitylation of c-jun has clearly been demonstrated (TREIER et al. 1994). Destruction of c-jun is apparently very complex, possibly involving several different degradation systems such as calpains or the lysosome (SALVAT et al. 1999). Thus, the importance of the ubiquitin system in this process remains an open question.

7 Summary

The multitude of natural substrates of the 26S proteasome demonstrates convincingly the diversity and flexibility of the ubiquitin/proteasome system: at the same time, the number of pathways in which ubiquitin-dependent degradation is involved highlights the importance of regulated proteolysis for cellular metabolism. This review has addressed recent advances in our understanding of the principles that govern the recognition and targeting of potential substrates. While the mechanism of ubiquitin activation and conjugation is largely understood, the determination of substrate specificity by ubiquitin protein ligases remains a field of active research. Several conserved degradation signals within substrate proteins have been identified, and it is becoming increasingly clear that these serve as docking sites for specific sets of E3s, which in turn adhere to a number of well-defined strategies for the recognition of these motifs. In particular, RING finger proteins are now emerging as a new and apparently widespread class of ubiquitin ligases. The discovery of more and more E3s will undoubtedly reveal even better the common principles in architecture and mechanisms of this class of enzymes. In contrast to substrate recognition by the ubiquitin conjugation system, the way in which a ubiquitylated protein is delivered to the 26S proteasome is poorly understood. There is no doubt that multiubiquitin chains serve as the principal determinant for recognition by the proteasome, and a number of receptors and candidate targeting factors are known, some of which are associated with the proteasome itself; however, unresolved issues are the significance of the different geometries that alternatively linked multiubiquitin chains can adopt, the role of transport between subcellular compartments, as well as the participation of chaperones in the delivery step. Finally, the analysis of ubiquitin-independent, substrate-specific targeting mechanisms, such as the AZ-dependent degradation of ODC, may provide unexpected answers to questions about protein recognition by the 26S proteasome.

Acknowledgements. I would like to thank Regine Kahmann and Stefan Jentsch for generous support and Jörg Höhfeld and George Pyrowolakis for helpful comments and discussions. I wish to apologize to all those researchers whose work on relevant issues could not be cited due to space constraints or the time taken to publish this manuscript.

References

Abriel H, Loffing J, Rebhun JF, Pratt JH, Schild L, Horisberger JD, Rotin D, Staub O (1999) Defective regulation of the epithelial Na^+ channel by Nedd4 in Liddle's syndrome. J Clin Invest 103:667–673

Aristarkhov A, Eytan E, Moghe A, Admon A, Hershko A, Ruderman JV (1996) E2-C, a cyclin-selective ubiquitin carrier protein required for the destruction of mitotic cyclins. Proc Natl Acad Sci USA 93:4294–4299

Baboshina OV, Haas AL (1996) Novel multiubiquitin chain linkages catalyzed by the conjugating enzymes E2-EPF and RAD6 are recognized by the 26S proteasome subunit 5. J Biol Chem 271: 2823–2831

Bailly V, Prakash S, Prakash L (1997) Domains required for dimerization of yeast RAD6 ubiquitin-conjugating enzyme and RAD18 DNA binding protein. Mol Cell Biol 17:4536–4543

Bartel B, Wünning I, Varshavsky A (1990) The recognition component of the N-end rule pathway. EMBO J 9:3179–3189

Bays NW, Gardner RG, Seelig LP, Joazeiro CA, Hampton RY (2001) Hrd1p/Der3p is a membrane-anchored ubiquitin ligase required for ER-associated degradation. Nature Cell Biol 3:24–29

Beal R, Deveraux Q, Xia G, Rechsteiner M, Pickart CM (1996) Surface hydrophobic residues of multiubiquitin chains essential for proteolytic targeting. Proc Natl Acad Sci USA 93:861–866

Berleth ES, Pickart CM (1996) Mechanism of ubiquitin conjugating enzyme E2–230K: catalysis involving a thiol relay? Biochemistry 35:1664–1671

Bordallo J, Wolf DH (1999) A RING-H2 finger motif is essential for the function of Der3/Hrd1 in endoplasmic reticulum associated protein degradation in the yeast *Saccharomyces cerevisiae*. FEBS Lett 448:244–248

Borden KL, Freemont PS (1996) The RING finger domain: a recent example of a sequence-structure family. Curr Opin Struct Biol 6:395–401

Brzovic PS, Meza J, King MC, Klevit RE (1998) The cancer-predisposing mutation C61G disrupts homodimer formation in the NH_2-terminal BRCA1 RING finger domain. J Biol Chem 273:7795–7799

Buonomo SB, Clyne RK, Fuchs J, Loidl J, Uhlmann F, Nasmyth K (2000) Disjunction of homologous chromosomes in meiosis I depends on proteolytic cleavage of the meiotic cohesin Rec8 by separin. Cell 103:387–398

Byrd C, Turner GC, Varshavsky A (1998) The N-end rule pathway controls the import of peptides through degradation of a transcriptional repressor. EMBO J 17:269–277

Carrano AC, Eytan E, Hershko A, Pagano M (1999) SKP2 is required for ubiquitin-mediated degradation of the CDK inhibitor p27. Nature Cell Biol 1:193–199

Chang A, Cheang S, Espanel X, Sudol M (2000) Rsp5 WW domains interact directly with the carboxy-terminal domain of RNA polymerase II. J Biol Chem 275:20562–20571

Chapman R, Sidrauski C, Walter P (1998) Intracellular signaling from the endoplasmic reticulum to the nucleus. Annu. Rev. Cell Dev Biol 14:459–485

Chau V, Tobias JW, Bachmair A, Marriott D, Ecker DJ, Gonda DK, Varshavsky A (1989) A multi-ubiquitin chain is confined to specific lysine in a targeted short-lived protein. Science 243:1576–1583

Chen P, Johnson P, Sommer T, Jentsch S, Hochstrasser M (1993) Multiple ubiquitin-conjugating enzymes participate in the in vivo degradation of the yeast MATα2 repressor. Cell 74:357–369

Chen Z, Naito M, Hori S, Mashima T, Yamori T, Tsuruo T (1999) A human IAP-family gene, apollon, expressed in human brain cancer cells. Biochem Biophys Res Commun 264:847–854

Cheng L, Collyer T, Hardy CF (1999) Cell cycle regulation of DNA replication initiator factor Dbf4p. Mol Cell Biol 19:4270–4278

Ciechanover A, Elias S, Heller H, Hershko A (1982) 'Covalent affinity' purification of ubiquitin activating enzyme. J Biol Chem 257:2537–2542

Cockman ME, Masson N, Mole DR, Jaakkola P, Chang GW, Clifford SC, Maher ER, Pugh CW, Ratcliffe PJ, Maxwell PH (2000) Hypoxia inducible factor-alpha binding and ubiquitylation by the von Hippel-Lindau tumor suppressor protein. J Biol Chem 275:25733–25741

Cohen-Fix O, Peters JM, Kirschner MW, Koshland D (1996) Anaphase initiation in *Saccharomyces cerevisiae* is controlled by the APC-dependent degradation of the anaphase inhibitor Pds1p. Genes Dev 10:3081–3093

Cook WJ, Jeffrey LC, Kasperek E, Pickart CM (1994) Structure of tetraubiquitin shows how multiubiquitin chains can be formed. J Mol Biol 236:601–609

Dai R-M, Chen E, Longo DL, Gorbea CM, Li C-C (1998) Involvement of valosin-containing protein, an ATPase co-purified with IkBa and 26S proteasome, in ubiquitin-proteasome-mediated degradation of IkBa. J Biol Chem 273:3562–3573

Davydov IV, Varshavsky A (2000) RGS4 is arginylated and degraded by the N-end rule pathway in vitro. J Biol Chem 275:22931–22941

De Sepulveda P, Ilangumaran S, Rottapel R (2000) Suppressor of cytokine signaling-1 inhibits VAV function through protein degradation. J Biol Chem 275:14005–14008

Deshaies RJ (1999) SCF and Cullin/RingH2-based ubiquitin ligases. Annu Rev Cell Dev Biol 15:435–467

Deveraux Q, Ustrell V, Pickart CM, Rechsteiner M (1994) A 26S protease subunit that binds ubiquitin conjugates. J Biol Chem 269:7059–7961

Dohmen RJ, Madura K, Bartel B, Varshavsky A (1991) The N-end rule is mediated by the UBC2(RAD6) ubiquitin-conjugating enzyme. Proc Natl Acad Sci USA 88:7351–7355

Dor Y, Raboy B, Kulka RG (1996) Role of the conserved carboxy-terminal alpha-helix of Rad6p in ubiquitination and DNA repair. Mol Microbiol 21:1197–1206

Drury LS, Perkins G, Diffley JF (1997) The Cdc4/34/53 pathway targets Cdc6p for proteolysis in budding yeast. EMBO J 16:5966–5976

Fang G, Yu H, Kirschner MW (1998) Direct binding of CDC20 protein family members activates the anaphase-promoting complex in mitosis and G1. Mol Cell 2:163–171

Fang S, Jensen JP, Ludwig RL, Vousden KH, Weissman AM (2000) Mdm2 is a RING finger-dependent ubiquitin protein ligase for itself and p53. J Biol Chem 275:8945–8951

Feldman RM, Correll CC, Kaplan KB, Deshaies RJ (1997) A complex of Cdc4p, Skp1p, and Cdc53p/cullin catalyzes ubiquitination of the phosphorylated CDK inhibitor Sic1p. Cell 91:221–30

Finley D, Ciechanover A, Varshavsky A (1984) Thermolability of ubiquitin-activating enzyme from the mammalian cell cycle mutant ts85. Cell 37:43–55

Finley D, Sadis S, Monia BP, Boucher P, Ecker DJ, Crooke ST, Chau V (1994) Inhibition of proteolysis and cell cycle progression in a multiubiquitin-deficient yeast mutant. Mol Cell Biol 14:5501–5509

Fisk HA, Yaffe MP (1999) A role for ubiquitination in mitochondrial inheritance in *Saccharomyces cerevisiae*. J Cell Biol 145:1199–1208

Freemont PS (2000) Ubiquitination: RING for destruction? Curr Biol 10:R84–R87

Friedländer R, Jarosch E, Urban J, Volkwein C, Sommer T (2000) A regulatory link between ER-associated protein degradation and the unfolded-protein response. Nature Cell Biol 2:379–384

Fritsche J, Rehli M, Krause SW, Andreesen R, Kreutz M (1997) Molecular cloning of a 1a,25-dihydroxyvitamin D_3-inducible transcript (DDVit1) in human blood monocytes. Biochem. Biophys Res Commun 235:407–412

Galan J, Haguenauer-Tsapis R (1997) Ubiquitin lys63 is involved in ubiquitination of a yeast plasma membrane protein. EMBO J 16:5847–5854

Galan J-M, Peter M (1999) Ubiquitin-dependent degradation of multiple F-box proteins by an autocatalytic mechanism. Proc Natl Acad Sci USA 96:9124–9129

Ghislain M, Dohmen RJ, Levy F, Varshavsky A (1996) Cdc48p interacts with Ufd3p, a WD repeat protein required for ubiquitin-mediated proteolysis in *Saccharomyces cerevisiae*. EMBO J 15:4884–4899

Gilon T, Chomsky O, Kulka RG (1998) Degradation signals for ubiquitin system proteolysis in *Saccharomyces cerevisiae*. EMBO J 17:2759–2766

Glotzer M, Murray AW, Kirschner MW (1991) Cyclin is degraded by the ubiquitin pathway. Nature 349:132–138

Goebl M, Yanagida M (1991) The TPR snap helix: a novel protein repeat motif from mitosis to transcription. Trends Biochem Sci 16:173–177

Goebl MG, Yochem J, Jentsch S, McGrath JP, Varshavsky A, Byers B (1988) The yeast cell cycle gene CDC34 encodes a ubiquitin-conjugating enzyme. Science 241:1331–1335

Gwozd CS, Arnason TG, Cook WJ, Chau V, Ellison MJ (1995) The yeast UBC4 ubiquitin conjugating enzyme monoubiquitinates itself in vivo. Evidence for an E2-E2 homointeraction. Biochemistry 34:6296–6302

Haas AL, Rose IA (1982) The mechanism of ubiquitin activating enzyme. J Biol Chem 257:10329–10337

Haas AL, Siepmann TJ (1997) Pathways of ubiquitin conjugation. FASEB J 11:1257–1268

Hampton RY, Gardner RG, Rine J (1996) Role of the 26S proteasome and HRD genes in the degradation of 3-hydroxy-3-methylglutaryl-CoA reductase, an integral endoplasmic reticulum membrane protein. Mol Biol Cell 7:2029–2044

Harvey KF, Dinudom A, Komwatana P, Jolliffe CN, Day ML, Parasivam G, Cook DI, Kumar S (1999) All three WW domains of murine Nedd4 are involved in the regulation of epithelial sodium channels by intracellular Na^+. J Biol Chem 274:12525–12530

Harvey KF, Kumar S (1999) Nedd4-like proteins: an emerging family of ubiquitin-protein ligases implicated in diverse cellular functions. Trends Cell Biol 9:166–169

Hauser HP, Bardroff M, Pyrowolakis G, Jentsch S (1998) A giant ubiquitin-conjugating enzyme related to IAP apoptosis inhibitors. J Cell Biol 141:1415–1422

Heller H, Hershko A (1990) A ubiquitin-protein ligase specific for type III protein substrates. J Biol Chem 265:6532–6535

Henchoz S, Chi Y, Catarin B, Herskowitz I, Deshaies RJ, Peter M (1997) Phosphorylation- and ubiquitin-dependent degradation of the cyclin-dependent kinase inhibitor Far1p in budding yeast. Genes Dev 11:3046–3060

Hershko A, Ciechanover A (1998) The ubiquitin system. Annu Rev Biochem 67:425–479

Hershko A, Ganoth D, Sudakin V, Dahan A, Cohen LH, Luca FC, Ruderman JV, Eytan E (1994) Components of a system that ligates cyclin to ubiquitin and their regulation by the protein kinase cdc2. J Biol Chem 269:4940–4946

Hershko A, Heller H, Elias S, Ciechanover A (1983) Components of ubiquitin-protein ligase system. Resolution, affinity purification, and role in protein breakdown. J Biol Chem 258:8206–8214

Herskowitz I, Rine J, Strathern J (1992) Mating-type determination and mating-type interconversion in *Saccharomyces cerevisiae*. In: Jones EW, Pringle JR, Broach JR (eds) The Molecular and Cellular Biology of the Yeast *Saccharomyces*: gene expression. Cold Spring Harbor Laboratory Press, Cold Spring Harbor, New York, pp 583–656

Hicke L (1999) Gettin' down with ubiquitin: turning off cell-surface receptors, transporters and channels. Trends Cell Biol 9:107–112

Hitchcock AL, Krebber H, Frietze S, Lin A, Latterich M, Silver PA (2001) The conserved npl4 protein complex mediates proteasome-dependent membrane-bound transcription factor activation. Mol Biol Cell 12:3226–3241

Hiyama H, Yokoi M, Masutani C, Sugasawa K, Maekawa T, Tanaka K, Hoeijmakers JH, Hanaoka F (1999) Interaction of hHR23 with S5a. The ubiquitin-like domain of hHR23 mediates interaction with S5a subunit of 26S proteasome. J Biol Chem 274:28019–28025

Hochstrasser M (1995) Ubiquitin, proteasomes, and the regulation of intracellular protein degradation. Curr Opin Cell Biol 7:215–223

Hofmann RM, Pickart CM (1999) Noncanonical *MMS2*-encoded ubiquitin-conjugating enzyme functions in assembly of novel polyubiquitin chains for DNA repair. Cell 96:645–653

Hofmann RM, Pickart CM (2001) In vitro assembly and recognition of Lys-63 polyubiquitin chains. J Biol Chem 276:27936–27943

Honda R, Tanaka H, Yasuda H (1997) Oncoprotein MDM 2 is a ubiquitin ligase E3 for tumor suppressor p53. FEBS Lett 420:25–27

Honda R, Yasuda H (2000) Activity of MDM 2, a ubiquitin ligase, toward p53 or itself is dependent on the RING finger domain of the ligase. Oncogene 19:1473–1476

Hoppe T, Matuschewski K, Rape M, Schlenker S, Ulrich HD, Jentsch S (2000) Activation of a membrane-bound transcription factor by regulated ubiquitin/proteasome-dependent processing. Cell 102:577–586

Hori T, Osaka F, Chiba T, Miyamoto C, Okabayashi K, Shimbara N, Kato S, Tanaka K (1999) Covalent modification of all members of human cullin family proteins by NEDD8. Oncogene 18:6829–6834

Hu G, Fearon ER (1999) Siah-1 N-terminal RING domain is required for proteolysis function, and C-terminal sequences regulate oligomerization and binding to target proteins. Mol Cell Biol 19:724–732

Hu G, Zhang S, Vidal M, Baer JL, Xu T, Fearon ER (1997) Mammalian homologs of seven in absentia regulate DCC via the ubiquitin-proteasome pathway. Genes Dev 11:2701–2714

Huang L, Kinnucan E, Wang G, Beaudenon S, Howley PM, Huibregtse JM, Pavletich NP (1999) Structure of an E6AP-UbcH7 complex: insights into ubiquitination by the E2-E3 enzyme cascade. Science 286:1321–1326

Huibregtse JM, Scheffner M, Beaudenon S, Howley PM (1995) A family of proteins structurally and functionally related to the E6-AP ubiquitin-protein ligase. Proc Natl Acad Sci USA 92:2563–2567

Huibregtse JM, Scheffner M, Howley PM (1991) A cellular protein mediates association of p53 with the E6 oncoprotein of human papillomavirus type 16 or 18. EMBO J 10:4129–4135

Huibregtse JM, Yang JC, Beaudenon SL (1997) The large subunit of RNA polymerase II is a substrate of the Rsp5 ubiquitin-protein ligase. Proc Natl Acad Sci USA 94:3656–3661

Hwang LH, Murray AM (1997) A novel yeast screen for mitotic arrest mutants identifies *DOC1*, a new gene involved in cyclin proteolysis. Mol Cell Biol 8:1877–1887

Irniger S, Piatti S, Michaelis C, Nasmyth K (1995) Genes involved in siter chromatid separation are needed for B-type cyclin proteolysis in budding yeast. Cell 81:269–278

Jariel-Encontre I, Pariat M, Martin F, Carillo S, Salvat C, Piechaczyk M (1995) Ubiquitinylation is not an absolute requirement for degradation of c-Jun protein by the 26S proteasome. J Biol Chem 270:11623–11627

Jensen DE, Proctor M, Marquis ST, Gardner HP, Ha SI, Chodosh LA, Ishov AM, Tommerup N, Vissing H, Sekido Y, Minna J, Borodovsky A, Schultz DC, Wilkinson KD, Maul GG, Barlev N, Berger SL, Prendergast GC, Rauscher FJ (1998) BAP1: a novel ubiquitin hydrolyse which binds to the BRCA1 RING finger and enhances BRCA1-mediated cell growth suppression. Oncogene 16:1097–1112

Joazeiro CA, Wing SS, Huang H, Leverson JD, Hunter T, Liu YC (1999) The tyrosine kinase negative regulator c-Cbl as a RING-type, E2-dependent ubiquitin-protein ligase. Science 286:309–312

Johnson ES, Gonda DK, Varshavsky A (1990) *Cis-trans* recognition and subunit-specific degradation of short-lived proteins. Nature 346:287–291

Johnson ES, Ma PC, Ota IM, Varshavsky A (1995) A proteolytic pathway that recognizes ubiquitin as a degradation signal. J Biol Chem 270:17442–17456

Johnson PR, Swanson R, Rakhilina L, Hochstrasser M (1998) Degradation signal masking by heterodimerization of MATa2 and MATa1 blocks their mutual destruction by the ubiquitin-proteasome pathway. Cell 94:217–227

Juang YL, Huang J, Peters JM, McLaughlin ME, Tai CY, Pellman D (1997) APC-mediated proteolysis of Ase1 and the morphogenesis of the mitotic spindle. Science 275:1311–1314

Jungmann J, Reins HA, Schobert C, Jentsch S (1993) Resistance to cadmium mediated by ubiquitin-dependent proteolysis. Nature 361:369–371

Kaiser P, Sia RA, Bardes EG, Lew DJ, Reed SI (1998) Cdc34 and the F-box protein Met30 are required for degradation of the Cdk-inhibitory kinase Swe1. Genes Dev 12:2587–2597

Kamura T, Burian D, Yan Q, Schmidt SL, Lane WS, Querido E, Branton PE, Shilatifard A, Conaway RC, Conaway JW (2001) Muf1, a novel Elongin BC-interacting leucine-rich repeat protein that can assemble with Cul5 and Rbx1 to reconstitute a ubiquitin ligase. J Biol Chem 276:29748–29753

Kamura T, Koepp DM, Conrad MN, Skowyra D, Moreland RJ, Iliopoulos O, Lane WS, Jr. WGK, Elledge SJ, Conaway RC, Harper JW, Conaway JW (1999) Rbx1, a component of the VHL tumor suppressor complex and SCF ubiquitin ligase. Science 284:657–661

Kamura T, Sato S, Haque D, Liu L, Kaelin WG, Conaway RC, Conaway JW (1998) The Elongin BC complex interacts with the conserved SOCS-box motif present in members of the SOCS, ras WD-40 repeat, and ankyrin repeat families. Genes Dev 12:3872–3881

Karin M, Ben-Neriah Y (2000) Phosphorylation meets ubiquitination: the control of NF-κB activity. Annu Rev Immunol 18:621–663

Kaye FJ, Modi S, Ivanovska I, Koonin EV, Thress K, Kubo A, Kornbluth S, Rose MD (2000) A family of ubiquitin-like proteins binds the ATPase domain of Hsp70-like Stch. FEBS Lett 467:348–352

Kobe B, Deisenhofer J (1994) The leucine-rich repeat: a versatile binding motif. Trends Biochem Sci 19:415–421

Koegl M, Hoppe T, Schlenker S, Ulrich HD, Mayer TU, Jentsch S (1999) A novel ubiquitination factor, E4, is involved in multiubiquitin chain assembly. Cell 96:635–644

Kominami K, Toda T (1997) Fission yeast WD-repeat protein pop1 regulates genome ploidy through ubiquitin-proteasome-mediated degradation of the CDK inhibitor Rum1 and the S-phase initiator Cdc18. Genes Dev 11:1548–1560

Koonin EV, Abagyan RA (1997) TSG101 may be the prototype of a class of dominant negative ubiquitin regulators. Nature Genet 16:330–331

Kopito RR (1997) ER quality control: the cytoplasmic connection. Cell 88:427–430

Kornitzer D, Ciechanover A (2000) Modes of regulation of ubiquitin-mediated protein degradation. J Cell Phys 182:1–11

Kühne C, Banks L (1998) E3-ubiquitin ligase/E6-AP links multicopy maintenance protein 7 to the ubiquitination pathway by a novel motif, the L2G box. J Biol Chem 273:34302–34309

Kulka RG, Raboy B, Schuster R, Parag HA, Diamond G, Ciechanover A, Marcus M (1988) A Chinese hamster cell cycle mutant arrested at G2 phase has a temperature-sensitive ubiquitin-activating enzyme, E1. J Biol Chem 263:15726–15731

Kumar S, Kao WH, Howley PM (1997) Physical interaction between specific E2 and Hect E3 enzymes determines functional cooperativity. J Biol Chem 272:13548–13554

Kumar S, Talis AL, Howley PM (1999) Identification of HHR23 A as a substrate for E6-associated protein-mediated ubiquitination. J Biol Chem 274:18785–18792

Kwon YT, Reiss Y, Fried VA, Hershko A, Yoon JK, Gonda DK, Sangan P, Copeland NG, Jenkins NA, Varshavsky A (1998) The mouse and human genes encoding the recognition component of the N-end rule pathway. Proc Natl Acad Sci USA 95:7898–7903

Lambertson D, Chen L, Madura K (1999) Pleiotropic defects caused by loss of the proteasome-interacting factors Rad23 and Rpn10 of *Saccharomyces cerevisiae*. Genetics 153:69–79

Lammer D, Mathias N, Laplaza JM, Jiang W, Liu Y, Callis J, Goebl M, Estelle M (1998) Modification of yeast Cdc53p by the ubiquitin-related protein Rub1p affects function of the SCF^{Cdc4} complex. Genes Dev 12:914–926

Laney JD, Hochstrasser M (1999) Substrate targeting in the ubiquitin system. Cell 97:427–430

Lanker S, Valdivieso MH, Wittenberg C (1996) Rapid degradation of the G1 cyclin Cln2 induced by CDK-dependent phosphorylation. Science 271:1597–1601

Levkowitz G, Waterman H, Ettenberg SA, Katz M, Tsygankov AY, Alroy I, Lavi S, Iwai K, Reiss Y, Ciechanover A, Lipkowitz S, Yarden Y (1999) Ubiquitin ligase activity and tyrosine phosphorylation underlie suppression of growth factor signaling by c-Cbl/Sli-1. Mol Cell 4:1029–1040

Li S, Li Y, Carthew RW, Lai Z-C (1997) Photoreceptor cell differentiation requires regulated proteolysis of the transcriptional repressor Tramtrack. Cell 90:469–478

Li X, Stebbins B, Hoffman L, Pratt G, Rechsteiner M, Coffino P (1996) The N terminus of antizyme promotes degradation of heterologous proteins. J Biol Chem 271:4441–4446

Liakopoulos D, Doenges G, Matuschewski K, Jentsch S (1998) A novel protein modification pathway related to the ubiquitin system. EMBO J 17:2208–2214

Lin L, Ghosh S (1996) A glycine-rich region in NF-κB p105 functions as a processing signal for the generation of the p50 subunit. Mol Cell Biol 16:2248–2254

Liu C, Kato Y, Zhang Z, Do VM, Yankner BA, He X (1999) b-TrCP couples b-catenin phosphorylation-degradation and regulates Xenopus axis formation. Proc Natl Acad Sci USA 96:6273–6278

Lomonte P, Everett RD (1999) Herpes simplex virus type 1 intermediate-early protein Vmw110 inhibits progression of cells through mitosis and from G into S phase of the cell cycle. J Virol 73:9456–9467

Lorick KL, Jensen JP, Fang S, Ong AM, Hatakeyama S, Weissman AM (1999) RING fingers mediate ubiquitin-conjugating enzyme (E2)-dependent ubiquitination. Proc Natl Acad Sci USA 96:11364–11369

Lüders J, Demand J, Höhfeld J (2000) The ubiquitin-related BAG-1 provides a link between the molecular chaperones Hsc70/Hsp70 and the proteasome. J Biol Chem 275:4613–4617

Madura K, Varshavsky A (1994) Degradation of Ga by the N-end rule pathway. Science 265:1454–1458

Maeda I, Ohta T, Koizumi H, Fukuda M (2001) In vitro ubiquitination of cyclin D1 by ROC1-CUL1 and ROC1-CUL3. FEBS Lett 494:181–185

Margottin F, Bour SP, Durand H, Selig L, Benichou S, Richard V, Thomas D, Strebel K, Benarous R (1998) A novel human WD protein, h-bTrCp, that interacts with HIV-1 Vpu connects CD4 to the ER degradation pathway through an F-box motif. Mol Cell 1:565–574

Marti A, Wirbelauer C, Scheffner M, Krek W (1999) Interaction between ubiquitin-protein ligase SCF^{SKP2} and E2F-1 underlies the regulation of E2F-1 degradation. Nature Cell Biol 1:14–19

Martinez-Noel G, Niedenthal R, Tamura T, Harbers K (1999) A family of structurally related RING finger proteins interacts specifically with the ubiquitin-conjugating enzyme UbcM4. FEBS Lett 454:257–261

Mastrandrea LD, You J, Niles EG, Pickart CM (1999) E2/E3-mediated assembly of lysine 29-linked polyubiquitin chains. J Biol Chem 274:27299–27306

Mathias N, Johnson S, Byers B, Goebl M (1999) The abundance of cell cycle regulatory protein Cdc4p is controlled by interactions between its F box and Skp1p. Mol Cell Biol 19:1759–1767

Matsuzawa S, Takayama S, Froesch BA, Zapata JM, Reed JC (1998) p53-inducible human homologue of Drosophila seven in absentia (Siah) inhibits cell growth: suppression by BAG-1. EMBO J 17:2736–2747

Matuschewski K, Hauser HP, Treier M, Jentsch S (1996) Identification of a novel family of ubiquitin-conjugating enzymes with distinct amino-terminal extensions. J Biol Chem 271:2789–2794

May MJ, Ghosh S (1998) Signal transduction through NF-κB. Immunol Today 19:80–88

Mayer TU, Braun T, Jentsch S (1998) Role of the proteasome in membrane extraction of a short-lived ER-transmembrane protein. EMBO J 17:3251–3257

Meimoun A, Holtzman T, Weissman Z, McBride HJ, Stillman DJ, Fink GR, Kornitzer D (2000) Degradation of the transcription factor Gcn4 requires the kinase Pho85 and the SCF(CDC4) ubiquitin-ligase complex. Mol Biol Cell 11:915–927

Meyer HH, Shorter JG, Seemann J, Pappin D, Warren GB (2000) A complex of mammalian Ufd1 and Npl4 links the AAA-ATPase, p97, to ubiquitin and nuclear transport pathways. EMBO J 19:2181–2192

Morimoto M, Nishida T, Honda R, Yasuda H (2000) Modification of cullin-1 by ubiquitin-like protein Nedd8 enhances the activity of SCF^{Skp2} toward p27^{Kip1}. Biochem Biophys Res Commun 270:1093–1096

Moynihan TP, Ardley HC, Nuber U, Rose SA, Jones PF, Markaham AF, Scheffner M, Robinson PA (1999) The ubiquitin-conjugating enzymes UbcH7 and UbcH8 interact with RING Finger/IBR motif-containing domains of HHARI and H7-AP1. J Biol Chem 274:30963–30968

Murakami Y, Matsufuji S, Hayashi S, Tanahashi N, Tanaka K (2000) Degradation of ornithine decarboxylase by the 26S proteasome. Biochem Biophys Res Commun 267:1–6

Murakami Y, Matsufuji S, Hayashi SI, Tanahashi N, Tanaka K (1999) ATP-dependent inactivation and sequestration of ornithine decarboxylase by the 26S proteasome are prerequisites for degradation. Mol Cell Biol 19:7216–7227

Murakami Y, Matsufuji S, Kameji T, Hayashi S, Igarashi K, Tamura T, Tanaka K, Ichihara A (1992) Ornithine decarboxylase is degraded by the 26S proteasome without ubiquitination. Nature 360:597–599

Nasmyth K (1996) At the heart of the budding yeast cell cycle. Trends Genet 12:405–412

Neer EJ, Schmidt CJ, Nambudripad R, Smith TF (1994) The ancient regulatory-protein family of WD-repeat proteins. Nature 371:297–300

Nefsky B, Beach D (1996) Pub1 acts as an E6-AP-like protein ubiquitin ligase in the degradation of cdc25. EMBO J 15:1301–1312

Nuber U, Schwarz S, Kaiser P, Schneider R, Scheffner M (1996) Cloning of human ubiquitin-conjugating enzymes UbcH6 and UbcH7 (E2-F1) and characterization of their interaction with E6-AP and RSP5. J Biol Chem 271:2795–2800

Nuber U, Schwarz SE, Scheffner M (1998) The ubiquitin-protein ligase E6-associated protein (E6-AP) serves as its own substrate. Eur J Biochem 254:643–649

Ohta T, Michel JJ, Schottelius AJ, Xiong Y (1999) ROC1, a homolog of APC11, represents a family of cullin partners with an associated ubiquitin ligase activity. Mol Cell 3:535–541

Ortolan TG, Tongaonkar P, Lambertson D, Chen L, Schauber C, Madura K (2000) The DNA repair protein rad23 is a negative regulator of multi-ubiquitin chain assembly. Nature Cell Biol 2:601–608

Page AM, Hieter P (1999) The anaphase-promoting complex: new subunits and regulators. Annu Rev Biochem 68:583–609

Palombella VJ, Rando OJ, Goldberg AL, Maniatis T (1994) The ubiquitin-proteasome pathway is required for processing the NF-kB1 precursor protein and the activation of NF-κB. Cell 78:773–785

Parag HA, Raboy B, Kulka RG (1987) Effect of heat shock on protein degradation in mammalian cells: involvement of the ubiquitin system. EMBO J 6:55–61

Patton EE, Willems AR, Sa D, Kuras L, Thomas D, Craig KL, Tyers M (1998a) Cdc53 is a scaffold protein for multiple Cdc34/Skp1/F-box protein complexes that regulate cell division and methionine biosynthesis in yeast. Genes Dev 12:692–705

Patton EE, Willems AR, Tyers M (1998b) Combinatorial control in ubiquitin-dependent proteolysis: don't Skp the F-box hypothesis. Trends Genet 14:236–243

Peng H, Begg GE, Schultz DC, Friedman JR, Jensen DE, Speicher DW, Rauscher FJ (2000) Reconstitution of the KRAB-KAP-1 repressor complex: a model system for defining the molecular anatomy of RING-B box-coiled-coil domain-mediated protein-protein interactions. J Mol Biol 295:1139–1162

Peters J-M (1999) Subunits and substrates of the anaphase-promoting complex. Exp Cell Res 248:339–349

Pfleger CM, Kirschner MW (2000) The KEN box: an APC recognition signal distinct from the D box targeted by Cdh1. Genes Dev 14:655–665
Pickart CM (1997) Targeting of substrates to the 26S proteasome. FASEB J 11:1055–1066
Piotrowski J, Beal R, Hoffman L, Wilkinson KD, Cohen RE, Pickart CM (1997) Inhibition of the 26S proteasome by polyubiquitin chains synthesized to have defined lengths. J Biol Chem 272:23712–23721
Plant PJ, Yeger H, Staub O, Howard P, Rotin D (1997) The C2 domain of the ubiquitin protein ligase Nedd4 mediates Ca^{2+}-dependent plasma membrane localization. J Biol Chem 272:32329–32336
Plemper RK, Böhmler S, Bordallo J, Sommer T, Wolf DH (1997) Mutant analysis links the translocon and BiP to retrograde protein transport for ER degradation. Nature 388:891–895
Podust VN, Brownell JE, Gladysheva TB, Luo RS, Wang C, Coggins MB, Pierce JW, Lightcap ES, Chau V (2000) A nedd8 conjugation pathway is essential for proteolytic targeting of $p27^{Kip1}$ by ubiquitination. Proc Natl Acad Sci USA 97:4579–4584
Ponting CP, Cai Y-D, Bork P (1997) The breast cancer gene product TSG101: a regulator of ubiquitination? J Mol Med 75:467–469
Potuschak T, Stary S, Schlogelhofer P, Becker F, Nejinskaia V, Bachmair A (1998) *PRT1* of *Arabidopsis thaliana* encodes a component of the plant N-end rule pathway. Proc Natl Acad Sci USA 95:7904–7908
Prinz S, Hwang ES, Visintin R, Amon A (1998) The regulation of Cdc20 proteolysis reveals a role for the APC components Cdc23 and Cdc27 during S phase and early mitosis. Curr Biol 8:750–760
Rao H, Uhlmann F, Nasmyth K, Varshavsky A (2001) Degradation of a cohesin subunit by the N-end rule pathway is essential for chromosome stability. Nature 410:955–959
Read MA, Brownell JE, Gladysheva TB, Hottelet M, Parent LA, Coggins MB, Pierce JW, Podust VN, Luo RS, Chau V, Palombella VJ (2000) Nedd8 modification of cul-1 activates $SCF^{\beta-TrCP}$-dependent ubiquitination of IkBa. Mol Cell Biol 20:2326–2333
Reiss Y, Kaim D, Hershko A (1988) Specificity of binding of NH_2-terminal residue of proteins to ubiquitin-protein ligase: use of amino acid derivatives to characterize specific binding sites. J Biol Chem 263:2693–2698
Rodgers KK, Bu Z, Fleming KG, Schatz DG, Engelman DM, Coleman JE (1996) A zinc-binding domain involved in the dimerization of RAG1. J Mol Biol 260:70–84
Rogers S, Wells R, Rechsteiner M (1986) Amino acid sequences common to rapidly degraded proteins: the PEST hypothesis. Science 234:364–368
Rothofsky ML, Lin SL (1997) *CROC-1* encodes a protein which mediates transcriptional activation of the human *FOS* promoter. Gene 195:141–149
Rouillon A, Barbey R, Patton EE, Tyers M, Thomas D (2000) Feedback-regulated degradation of the transcriptional activator Met4 is triggered by the SCF(Met30) complex. EMBO J 19:282–294
Russell SJ, Reed SH, Huang W, Friedberg EC, Johnston SA (1999) The 19S regulatory complex of the proteasome functions independently of proteolysis in nucleotide excision repair. Mol Cell 3:687–695
Sadis S, Atienza C, Finley D (1995) Synthetic signals for ubiquitin-dependent proteolysis. Mol Cell Biol 15:4086–4094
Salama SR, Hendricks KB, Thorner J (1994) G1 cyclin degradation: the PEST motif of yeast Cln2 is necessary, but not sufficient, for rapid protein turnover. Mol Cell Biol 14:7953–7966
Salvat C, Aquaviva C, Jariel-Encontre I, Ferrara P, Pariat M, Steff AM, Carillo S, Piechaczyk M (1999) Are there multiple proteolytic pathways contributing to c-Fos, c-Jun and p53 protein degradation in vivo? Mol Biol Rep 26:45–51
Sancho E, Vilá MR, Sánchez-Pulido L, Lozano JJ, Paciucci R, Nadal M, Fox M, Harvey C, Bercovich B, Loukili N, Ciechanover A, Lin SL, Sanz F, Estivill X, Valencia A, Thomson TM (1998) Role of UEV-1, an inactive variant of the E2 ubiquitin-conjugating enzymes, in in vitro differentiation and cell cycle behavior of HT-29-M6 intestinal mucosecretory cells. Mol Cell Biol 18:576–589
Saurin AJ, Borden KLB, Boddy MN, Freemont PS (1996) Does this have a familiar RING? Trends Biochem Sci 21:208–214
Schauber C, Chen L, Tongaonkar P, Vega I, Madura K (1998a) Sequence elements that contribute to the degradation of yeast G alpha. Genes Cells 3:307–319
Schauber C, Chen L, Tongaonkar P, Vega I, Lambertson D, Potts W, Madura K (1998b) Rad23 links DNA repair to the ubiquitin/proteasome pathway. Nature 391:715–718
Scheffner M, Huibregtse JM, Vierstra RD, Howley PM (1993) The HPV-16 E6 and E6-AP complex functions as a ubiquitin-protein ligase in the ubiquitination of p53. Cell 75:495–505
Scheffner M, Nuber U, Huibregtse JM (1995) Protein ubiquitination involving an E1-E2-E3 enzyme ubiquitin thioester cascade. Nature 373:81–83

Schimke RT (1973) Control of enzyme levels in mammalian tissues. Adv Enzymol 37:135–187

Schwab M, Lutum AS, Seufert W (1997) Yeast Hct1 is a regulator of Clb2 cyclin proteolysis. Cell 90: 683–693

Schwartz A, Ciechanover A (1999) The ubiquitin-proteasome pathway: involvement in the pathogenesis of human diseases. Annu Rev Med 50:57–74

Schwarz SE, Rosa JL, Scheffner M (1998) Characterization of human hect domain family members and their interaction with UbcH5 and UbcH7. J Biol Chem 273:12148–12154

Seol JH, Feldman RM, Zachariae W, Shevchenko A, Correll CC, Lyapina S, Chi Y, Galova M, Claypool J, Sandmeyer S, Nasmyth K, Shevchenko A, Deshaies R (1999) Cdc53/cullin and the essential Hrt1 RING-H2 subunit of SCF define a ubiquitin ligase module that activates the E2 enzyme Cdc34. Genes Dev 13:1614–1626

Seufert W, Jentsch S (1990) Ubiquitin-conjugating enzymes UBC4 and UBC5 mediate selective degradation of short-lived and abnormal proteins. EMBO J 9:543–550

Seufert W, Jentsch S (1992) In vivo function of the proteasome in the ubiquitin pathway. EMBO J 11:3077–3080

Seufert W, McGrath JP, Jentsch S (1990) *UBC1* encodes a novel member of an essential subfamily of yeast ubiquitin-conjugating enzymes involved in protein degradation. EMBO J 9:4535–4541

Shirayama M, Toth A, Galova M, Nasmyth K (1999) APC^{CDC20} promotes exit from mitosis by destroying the anaphase inhibitor Pds1 and cyclin Clb5. Nature 402:203–207

Shirayama M, Zachariae W, Ciosk R, Nasmyth K (1998) The Polo-like kinase Cdc5p and the WD-repeat protein Cdc20p/fizzy are regulators and substrates of the anaphase promoting complex in *Saccharomyces cerevisiae*. EMBO J 17:1336–1349

Singer JD, Gurian-West M, Clurman B, Roberts JM (1999) Cullin-3 targets cyclin E for ubiquitination and controls S phase in mammalian cells. Genes Dev 13:2375–2387

Skowyra D, Craig KL, Tyers M, Elledge SJ, Harper JW (1997) F-box proteins are receptors that recruit phosphorylated substrates to the SCF ubiquitin-ligase complex. Cell 91:209–219

Skowyra D, Koepp DM, Kamura T, Conrad MN, Conaway RC, Conaway JW, Elledge SJ, Harper JW (1999) Reconstitution of G1 cyclin ubiquitination with complexes containing SCF^{Grr1} and Rbx1. Science 284:662–665

Sommer T, Wolf DH (1997) Endoplasmic reticulum degradation: reverse protein flow of no return. FASEB J 11:1227–1233

Spence J, Sadis S, Haas AL, Finley D (1995) A ubiquitin mutant with specific defects in DNA repair and multiubiquitination. Mol Cell Biol 15:1265–1273

Spencer E, Jiang J, Chen ZJ (1999) Signal-induced ubiquitination of IkBa by the F-box protein Slimb/b-TrCP. Genes Dev 13:284–294

Staub O, Dho S, Henry P, Correa J, Ishikawa T, McGlade J, Rotin D (1996) WW domains of Nedd4 bind to the proline-rich PY motifs in the epithelial Na$^+$ channel deleted in Liddle's syndrome. EMBO J 15:2371–2380

Stebbins CE, Kaelin WG, Pavletich NP (1999) Structure of the VHL-ElonginC-ElonginB complex: implications for VHL tumor suppressor function. Science 284:455–461

Sudakin V, Ganoth D, Dahan A, Heller H, Hershko J, Luca FC, Ruderman JV, Hershko A (1995) The cyclosome, a large complex containing cyclin-selective ubiquitin ligase activity, targets cyclins for destruction at the end of mitosis. Mol Biol Cell 6:185–197

Sudol M (1996) Structure and function of the WW domain. Prog Biophys Mol Biol 65:113–132

Swanson R, Locher M, Hochstrasser M (2001) A conserved ubiquitin ligase of the nuclear envelope/endoplasmic reticulum that functions in both ER-associated and Matalpha2 repressor degradation. Genes Dev 15:2660–2674

Tanaka K, Suzuki T, Chiba T (1998) The ligation systems for ubiquitin and ubiquitin-like proteins. Mol Cells 8:503–512

Tang AH, Neufeld TP, Kwan E, Rubin GM (1997) PHYL acts to down-regulate TTK88, a transcriptional repressor of neuronal cell fates, by a SINA-dependent mechanism. Cell 90:459–467

Thrower JS, Hoffman L, Rechsteiner M, Pickart CM (2000) Recognition of the polyubiquitin proteolytic signal. EMBO J 19:94–102

Tongaonkar P, Chen L, Lambertson D, Ko B, Madura K (2000) Evidence for an interaction between ubiquitin-conjugating enzymes and the 26S proteasome. Mol Cell Biol 20:4691–4698

Travers KJ, Patil CK, Wodicka L, Lockhart DJ, Weissman JS, Walter P (2000) Functional and genomic analyses reveal an essential coordination between the unfolded protein response and ER-associated degradation. Cell 101:249–258

and the β rings are located internally (VOGES et al. 1999). The complex has the structure 7α; 7β; 7β, 7α. The crystal structure of the budding yeast 20S proteasome demonstrated that the interior of the cylinder contains three chambers (GROLL et al. 1997). Proteolysis has been shown to occur in the central chamber, which is formed by the β rings. Three of the seven β subunits are catalytic subunits and are responsible for the three different proteolytic activities that have been identified for the eukaryotic proteasome (SEEMULLER et al. 1995; BAUMEISTER et al. 1998).

In higher eukaryotes, added diversity is shown by the fact that, under certain conditions, the three catalytic β subunits, β1 or Y; β2 or Z; and β5 or X, can be replaced by the low molecular mass protein (LMP)2, multicatalytic endopeptidase complex-like (MECL)-1 and LMP7 subunits. Replacement of the β catalytic subunits changes the peptidase cleavage pattern obtained. It is thought that this change in cleavage pattern is important for antigenic processing. In agreement with this hypothesis the LMP2, MECL-1 and LMP7 subunits are all induced by γ-interferon (IFN) a known regulator of the immune response. Proteasomes containing these subunits have been called immunoproteasomes (FRUH et al. 1994; ROCK and GOLDBERG 1999).

Two regulatory complexes have been shown to associate with the 20S particle. The 19S regulatory complex (also called PA700) and the PA28 (also called 11S). The 19S complex is composed of up to 18 different subunits (GLICKMAN et al. 1998; HOLZL et al. 2000). The 19S complex is responsible for a number of different functions such as the recognition of the polyubiquitin tagged substrates, recycling of the ubiquitin and unfolding the protein substrate followed by translocation to the 20S complex for degradation. When a 19S complex is associated with either end of the 20S particle the structure is called the 26S proteasome.

The 11S regulatory complex is made up of two subunits, termed PA28α and PA28β. Both subunits are induced by γ-IFN and when associated with the 20S complex changes activate the proteolysis of short peptides as well as changing the cleavage pattern (DUBIEL et al. 1992; MA et al. 1992).

The PA28 complex has been implicated in antigenic processing and mice in which the gene for the β subunit has been disrupted are deficient in antigenic processing (PRECKEL et al. 1999). The PA28 regulatory complex has not been identified in yeast.

Recent work in a number of different cell systems has indicated that the proteasome occupies a number of discrete cellular compartments. Differential localization is a common way to regulate gene expression in biology. The evidence reviewed in this report indicates that the proteasome can also be localized to discrete cellular compartments under different cell conditions.

2 Localization in Yeast

2.1 Enrichment of Proteasomes at the Nuclear Periphery

In both budding and fission yeast two complementary studies have investigated the intracellular localization of proteasomes (WILKINSON et al. 1998; ENENKEL et al.

Schimke RT (1973) Control of enzyme levels in mammalian tissues. Adv Enzymol 37:135–187

Schwab M, Lutum AS, Seufert W (1997) Yeast Hct1 is a regulator of Clb2 cyclin proteolysis. Cell 90: 683–693

Schwartz A, Ciechanover A (1999) The ubiquitin-proteasome pathway: involvement in the pathogenesis of human diseases. Annu Rev Med 50:57–74

Schwarz SE, Rosa JL, Scheffner M (1998) Characterization of human hect domain family members and their interaction with UbcH5 and UbcH7. J Biol Chem 273:12148–12154

Seol JH, Feldman RM, Zachariae W, Shevchenko A, Correll CC, Lyapina S, Chi Y, Galova M, Claypool J, Sandmeyer S, Nasmyth K, Shevchenko A, Deshaies R (1999) Cdc53/cullin and the essential Hrt1 RING-H2 subunit of SCF define a ubiquitin ligase module that activates the E2 enzyme Cdc34. Genes Dev 13:1614–1626

Seufert W, Jentsch S (1990) Ubiquitin-conjugating enzymes UBC4 and UBC5 mediate selective degradation of short-lived and abnormal proteins. EMBO J 9:543–550

Seufert W, Jentsch S (1992) In vivo function of the proteasome in the ubiquitin pathway. EMBO J 11:3077–3080

Seufert W, McGrath JP, Jentsch S (1990) *UBC1* encodes a novel member of an essential subfamily of yeast ubiquitin-conjugating enzymes involved in protein degradation. EMBO J 9:4535–4541

Shirayama M, Toth A, Galova M, Nasmyth K (1999) APC^{CDC20} promotes exit from mitosis by destroying the anaphase inhibitor Pds1 and cyclin Clb5. Nature 402:203–207

Shirayama M, Zachariae W, Ciosk R, Nasmyth K (1998) The Polo-like kinase Cdc5p and the WD-repeat protein Cdc20p/fizzy are regulators and substrates of the anaphase promoting complex in *Saccharomyces cerevisiae*. EMBO J 17:1336–1349

Singer JD, Gurian-West M, Clurman B, Roberts JM (1999) Cullin-3 targets cyclin E for ubiquitination and controls S phase in mammalian cells. Genes Dev 13:2375–2387

Skowyra D, Craig KL, Tyers M, Elledge SJ, Harper JW (1997) F-box proteins are receptors that recruit phosphorylated substrates to the SCF ubiquitin-ligase complex. Cell 91:209–219

Skowyra D, Koepp DM, Kamura T, Conrad MN, Conaway RC, Conaway JW, Elledge SJ, Harper JW (1999) Reconstitution of G1 cyclin ubiquitination with complexes containing SCF^{Grr1} and Rbx1. Science 284:662–665

Sommer T, Wolf DH (1997) Endoplasmic reticulum degradation: reverse protein flow of no return. FASEB J 11:1227–1233

Spence J, Sadis S, Haas AL, Finley D (1995) A ubiquitin mutant with specific defects in DNA repair and multiubiquitination. Mol Cell Biol 15:1265–1273

Spencer E, Jiang J, Chen ZJ (1999) Signal-induced ubiquitination of IkBa by the F-box protein Slimb/b-TrCP. Genes Dev 13:284–294

Staub O, Dho S, Henry P, Correa J, Ishikawa T, McGlade J, Rotin D (1996) WW domains of Nedd4 bind to the proline-rich PY motifs in the epithelial Na$^+$ channel deleted in Liddle's syndrome. EMBO J 15:2371–2380

Stebbins CE, Kaelin WG, Pavletich NP (1999) Structure of the VHL-ElonginC-ElonginB complex: implications for VHL tumor suppressor function. Science 284:455–461

Sudakin V, Ganoth D, Dahan A, Heller H, Hershko J, Luca FC, Ruderman JV, Hershko A (1995) The cyclosome, a large complex containing cyclin-selective ubiquitin ligase activity, targets cyclins for destruction at the end of mitosis. Mol Biol Cell 6:185–197

Sudol M (1996) Structure and function of the WW domain. Prog Biophys Mol Biol 65:113–132

Swanson R, Locher M, Hochstrasser M (2001) A conserved ubiquitin ligase of the nuclear envelope/endoplasmic reticulum that functions in both ER-associated and Matalpha2 repressor degradation. Genes Dev 15:2660–2674

Tanaka K, Suzuki T, Chiba T (1998) The ligation systems for ubiquitin and ubiquitin-like proteins. Mol Cells 8:503–512

Tang AH, Neufeld TP, Kwan E, Rubin GM (1997) PHYL acts to down-regulate TTK88, a transcriptional repressor of neuronal cell fates, by a SINA-dependent mechanism. Cell 90:459–467

Thrower JS, Hoffman L, Rechsteiner M, Pickart CM (2000) Recognition of the polyubiquitin proteolytic signal. EMBO J 19:94–102

Tongaonkar P, Chen L, Lambertson D, Ko B, Madura K (2000) Evidence for an interaction between ubiquitin-conjugating enzymes and the 26S proteasome. Mol Cell Biol 20:4691–4698

Travers KJ, Patil CK, Wodicka L, Lockhart DJ, Weissman JS, Walter P (2000) Functional and genomic analyses reveal an essential coordination between the unfolded protein response and ER-associated degradation. Cell 101:249–258

Treier M, Staszewski LM, Bohmann D (1994) Ubiquitin-dependent c-jun degradation in vivo is mediated by the d domain. Cell 78:787–798

Tsvetkov LM, Yeh KH, Lee SJ, Sun H, Zang H (1999) p27(Kip1) ubiquitination and degradation is regulated by the SCF(Skp2) complex through phosphorylated Thr187 in p27. Curr Biol 9:661–664

Tyers M, Jorgensen P (2000) Proteolysis and the cell cycle: with this RING I do thee destroy. Curr Opinion Genet Dev 10:54–64

Ulrich HD, Jentsch S (2000) Two RING finger proteins mediate cooperation between ubiquitin-conjugating enzymes in DNA repair. EMBO J 19:3388–3397

van Nocker S, Sadis S, Rubin DM, Glickman M, Fu H, Coux O, Wefes I, Finley D, Vierstra R (1996) The multiubiquitin-chain-binding protein Mcb1 is a component of the 26S proteasome in *Saccharomyces cerevisiae* and plays a nonessential, substrate-specific role in protein turnover. Mol Cell Biol 16:6020–6028

Varshavsky A (1996) The N-end rule: Functions, mysteries, uses. Proc Natl Acad Sci USA 93:12142–12149

Varshavsky A (1997) The N-end rule pathway of protein degradation. Genes Cells 2:13–28

Visintin R, Prinz S, Amon A (1997) CDC20 and CDH1: a family of substrate specific activators of APC-dependent proteolysis. Science 278:460–463

Wang G, Yang J, Huibregtse JM (1999) Functional domains of the Rsp5 ubiquitin-protein ligase. Mol Cell Biol 19:342–352

Whaley JM, Naglich J, Gelbert L, Hsia YE, Lamiell JM, Green JS, Collins D, Neumann HP, Laidlaw J, Li FP, et al. (1994) Germ-line mutations in the von Hippel-Lindau tumor-suppressor gene are similar to somatic von Hippel-Lindau aberrations in sporadic renal cell carcinoma. Am J Hum Genet 55:1092–1102

Wilhovsky S, Gardner R, Hampton R (2000) HRD gene dependence of endoplasmic reticulum-associated degradation. Mol Biol Cell 11:1697–1708

Wilkinson KD (1997) Regulation of ubiquitin-dependent processes by deubiquitinating enzymes. FASEB J 11:1245–1256

Winston JT, Strack P, Beer-Romero P, Chu CY, Elledge SJ, Harper JW (1999) The SCF b-TrCP-ubiquitin ligase complex associates specifically with phosphorylated destruction motifs in IkBa and b-catenin and stimulates IkBa ubiquitination in vitro. Genes Dev 13:270–283

Xie Y, Varshavsky A (1999) The E2-E3 interaction in the N-end rule pathway: the RING-H2 finger of E3 is required for the synthesis of multiubiquitin chains. EMBO J 18:6832–6844

Xie Y, Varshavsky A (2000) Physical association of ubiquitin ligases and the 26S proteasome. Proc Natl Acad Sci USA 97:2497–2502

Yaglom J, Linskens MH, Sadis S, Rubin DM, Futcher B, Finley D (1995) p34^{Cdc28}-mediated control of Cln3 cyclin degradation. Mol Cell Biol 15:731–741

Yang Y, Fang S, Jensen JP, Weissman AM, Ashwell JD (2000) Ubiquitin protein ligase activity of IAPs and their degradation in proteasomes in response to apoptotic stimuli. Science 288:874–877

Yaron A, Hatzubai A, Davis M, Lavon I, Amit S, Manning AM, Andersen JS, Mann M, Mercurio F, Ben-Neriah Y (1998) Identification of the receptor component of the IkBa-ubiquitin ligase. Nature 396:590–594

Yashiroda H, Kaina D, Toh-e A, Kikuchi Y (1998) The PY-motif of Bul1 protein is essential for growth of *Saccharomyces cerevisiae* under various stress conditions. Gene 225:39–46

Yashiroda H, Oguchi T, Yasuda Y, Toh-e A, Kikuchi Y (1996) Bul1, a new protein that binds to the Rsp5 ubiquitin ligase in *Saccharomyces cerevisiae*. Mol Cell Biol 16:3255–3263

Yokouchi M, Kondo T, Houghton A, Bartkeiwicz M, Horne WC, Zhang H, Yoshimura A, Baron R (1999) Ligand-induced ubiquitination of the epidermal growth factor receptor involves the interaction of the c-Cbl RING finger and UbcH7. J Biol Chem 274:31707–31712

Zachariae W, Schwab M, Nasmyth K, Seufert W (1998b) Control of cyclin ubiquitination by CDK-regulated binding of Hct1 to the anaphase promoting complex. Science 282:1721–1724

Zachariae W, Shevchenko A, Andrews PD, Ciosk R, Galova M, Stark MJ, Mann M, Nasmyth K (1998a) Mass spectrometric analysis of the anaphase-promoting complex from yeast: identification of a subunit related to cullins. Science 279:1216–1219

Zachariae W, Shin T-H, Galova M, Obermaier B, Nasmyth K (1996) New subunits of the anaphase promoting complex of *Saccharomyces cerevisiae*. Science 274:1201–1204

Zhou P, Howley PM (1998) Ubiquitination and degradation of the substrate recognition subunits of SCF ubiquitin-protein ligases. Mol Cell 2:571–580

The Intracellular Localization of the Proteasome

C. Gordon

1 Introduction	175
2 Localization in Yeast	176
2.1 Enrichment of Proteasomes at the Nuclear Periphery	176
2.2 Possible Mechanism for Increased Concentration of Proteasomes at the Nuclear Periphery	177
2.3 Localization in Live Cells During Mitosis and Meiosis	179
3 Localization in Mammalian Cells	180
3.1 Immunoproteasome Localization	180
3.2 Localization of Proteasomes to the Aggresome	181
4 Summary	182
References	183

1 Introduction

Proteasomes are large multiprotein complexes which constitute the main non-lysosomal (non-vacuolar in yeast) mechanism to degrade intracellular proteins. One major function of proteasomes is to degrade proteins that have been targeted for destruction by the ubiquitin pathway. By determining the levels of key regulatory proteins the proteasome plays a critical role in many different biological processes, such as cell cycle proliferation, development, apoptosis and antigen presentation (Hershko and Ciechanover 1998; Voges et al. 1999). The proteasome is also responsible for degrading other proteins independently of the ubiquitin pathway. Misfolded proteins and ornithine decarboxylase are examples of proteins that do not require the addition of a polyubiquitin signal for proteolysis by the proteasome to occur (Strickland et al. 2000; Coffino 2001).

The 20S proteasome is composed of 14 different subunits. These are grouped into seven α subunits and seven β subunits based on their homology to the simpler version of the complex found in the Archaebacterium *Thermoplasma acidophilum*. The seven α and seven β subunits form ring structures. These rings stack one on top of each other to form a cylinder. The α rings are found on the outside of the complex

---

MRC Human Genetics Unit, Western General Hospital, Crewe Road, Edinburgh EH4 2XU, UK

and the β rings are located internally (VOGES et al. 1999). The complex has the structure 7α; 7β; 7β, 7α. The crystal structure of the budding yeast 20S proteasome demonstrated that the interior of the cylinder contains three chambers (GROLL et al. 1997). Proteolysis has been shown to occur in the central chamber, which is formed by the β rings. Three of the seven β subunits are catalytic subunits and are responsible for the three different proteolytic activities that have been identified for the eukaryotic proteasome (SEEMULLER et al. 1995; BAUMEISTER et al. 1998).

In higher eukaryotes, added diversity is shown by the fact that, under certain conditions, the three catalytic β subunits, β1 or Y; β2 or Z; and β5 or X, can be replaced by the low molecular mass protein (LMP)2, multicatalytic endopeptidase complex-like (MECL)-1 and LMP7 subunits. Replacement of the β catalytic subunits changes the peptidase cleavage pattern obtained. It is thought that this change in cleavage pattern is important for antigenic processing. In agreement with this hypothesis the LMP2, MECL-1 and LMP7 subunits are all induced by γ-interferon (IFN) a known regulator of the immune response. Proteasomes containing these subunits have been called immunoproteasomes (FRUH et al. 1994; ROCK and GOLDBERG 1999).

Two regulatory complexes have been shown to associate with the 20S particle. The 19S regulatory complex (also called PA700) and the PA28 (also called 11S). The 19S complex is composed of up to 18 different subunits (GLICKMAN et al. 1998; HOLZL et al. 2000). The 19S complex is responsible for a number of different functions such as the recognition of the polyubiquitin tagged substrates, recycling of the ubiquitin and unfolding the protein substrate followed by translocation to the 20S complex for degradation. When a 19S complex is associated with either end of the 20S particle the structure is called the 26S proteasome.

The 11S regulatory complex is made up of two subunits, termed PA28α and PA28β. Both subunits are induced by γ-IFN and when associated with the 20S complex changes activate the proteolysis of short peptides as well as changing the cleavage pattern (DUBIEL et al. 1992; MA et al. 1992).

The PA28 complex has been implicated in antigenic processing and mice in which the gene for the β subunit has been disrupted are deficient in antigenic processing (PRECKEL et al. 1999). The PA28 regulatory complex has not been identified in yeast.

Recent work in a number of different cell systems has indicated that the proteasome occupies a number of discrete cellular compartments. Differential localization is a common way to regulate gene expression in biology. The evidence reviewed in this report indicates that the proteasome can also be localized to discrete cellular compartments under different cell conditions.

2 Localization in Yeast

2.1 Enrichment of Proteasomes at the Nuclear Periphery

In both budding and fission yeast two complementary studies have investigated the intracellular localization of proteasomes (WILKINSON et al. 1998; ENENKEL et al.

1998). In both studies immunofluoresence microscopy using either specific antibodies raised against proteasomal subunits or proteasomal subunits tagged with the green fluorescent protein (GFP) epitopes were used to identify where the 19S, 20S and 26S complexes were located in the cell. In the case of the GFP analysis the wild-type gene was replaced with the GFP tagged form by homologous recombination. The tagged gene was expressed from the native promoter and ensured that the GFP signal obtained was due to the authentic location and was not an artefact caused by over-expression of the GFP-tagged protein. Both studies identified the nuclear periphery as the major site of proteasome localization in vegetatively growing yeast cells (but see RUSSELL et al. 1999 for an alternative localization in budding yeast). In the budding yeast studies careful cell fractionation experiments estimated that up to 80% of proteasomes were found in the endoplasmic reticulum (ER)/nuclear membrane fraction. Similar cell fractionation experiments in fission yeast estimated that a similar proportion of proteasomes was associated with intact nuclei (C.R.M. Wilkinson and C. Gordon, unpublished results). It was not possible from the above experiments to determine whether the 26S proteasomes were localized to the outer or inner side of the nuclear membrane. However, immunogold electron microscopy (EM) staining carried out in fission yeast demonstrated convincingly that the proteasomes were found predominately at the nuclear periphery inside the nucleus.

What could be the function of concentrating the proteasomes inside the nucleus at the nuclear periphery? One possibility is that this is the site of assembly for proteasomes in the cell. Another, possibility is that this is where a pool of proteasomes are stored in a latent state ready to be transported to different intracellular compartments when an increase in proteolysis is required. Evidence against such a resting pool of proteasomes is the fact that if fission yeast cells are stressed either by heat shock or by the addition of the arginine amino acid analogue canavanine to cell cultures the localization of proteasomes remains at the nuclear periphery. As such treatments result in a dramatic increase in the level of ubiquitin-dependant proteolysis this suggests that the proteasomes at the nuclear periphery cannot represent a latent pool of proteasomes (C. Gordon, unpublished results). A third explanation could be that the nuclear periphery is the major site of proteolysis in yeast cells. Consistent with this third hypothesis the major site of cleavage of the proteasome-specific peptide substrate Cbz–Leu–Leu–Glu–β-napthylamide co-localized with the GFP proteasome signal in budding yeast. We therefore feel that the cytology and cell fractionation experiments highlight the nuclear periphery as the major site of proteolysis in vegatively dividing yeast cells (ENENKEL et al. 1998).

2.2 Possible Mechanism for Increased Concentration of Proteasomes at the Nuclear Periphery

A simple model to account for the observed discrete localization is that some intracellular receptor(s) interacts with one or more 26S proteasomal subunits. Three recent papers have shed light on what components could be involved in localization to the nuclear periphery. The first evidence came from a thorough characterization of the *cut8–563* fission yeast temperature sensitive (ts) mutant strain. The *cut8–563* mutant was originally isolated in a screen to look for mutants that showed aberrant nuclear segregation (HIRANO et al. 1986). The mutant cells were still able to undergo cytokinesis splitting the unsegregated nuclei to give the 'cut' (*c*ell *u*ntimely *t*orn) phenotype. The *cut8$^+$* gene was cloned by positional cloning and was shown to encode a 262 amino acid protein with no obvious protein domains (SAMEJIMA and YANAGIDA 1994). Although evolutionary conserved homologues could be identified in other fungi no homologues could be identified in vertebrates. Deletion of the *cut8$^+$* gene resulted in viable cells that were ts for growth. The key observation was that in a *cut8* ts mutant at the restrictive temperature, or in *cut8* null strain, the proteasomes were delocalized and were no longer concentrated at the nuclear periphery (TATEBE and YANAGIDA 2000). In addition, the fission yeast cyclin B Cdc13 protein and the Cut2 protein, both known cell cycle substrates of the ubiquitin pathway, were partially stabilized when the proteasome was delocalized. The addition of the multi-ubiquitin degradation signal to the substrates occurred normally, indicating that the stabilization occurred as a consequence of the interference of some factor downstream of the anaphase-promoting complex E3 ubiquitin ligase. It was proposed that the stabilization occurred as a direct result of delocalization of the proteasomes from the nuclear periphery and implied that the enrichment of 26S proteasomes found at the nuclear periphery, which involved the Cut8 protein, was required for efficient proteolysis of substrates targeted by the ubiquitin pathway. The second piece of the puzzle came from an elegant study on mutants in the importin α homologue, *SRP1*, in budding yeast (TABB et al. 2000).

By genetic criteria, Nomura and colleagues demonstrated that the budding yeast importin α homologue, *SRP1*, has at least two separable functions (TABB et al. 2000). Two mutant alleles of the *SRP1* gene were isolated, *srp1-31* and *srp1-49*, which had clear and separable phenotypes. While defects in nuclear localization signal (NLS) binding and protein import could be demonstrated for the *srp1-31* allele no such defect in NLS transport could be observed for the *srp1-49* mutation. Instead the *srp1-49* mutant strain showed a wide range of pleiotropic phenotypes such as arrest in mitosis, altered microtubule morphology and defective nucleolar morphology. This implied that Srp1 has a function(s) in addition to its role in NLS-dependent protein import. Interestingly, both mutant alleles showed intragenic complementation, clearly demonstrating that genetically the Srp1 function effected by the *srp1-49* mutation is different from that of the *srp1-31* mutation. Two high copy suppressors that specifically rescued the *srp1-49* mutation were isolated. These suppressors were in the budding yeast *STS1* and *RPN11* genes. The *STS1* gene is the budding yeast orthologue of the fission yeast *Cut8$^+$* gene described above. The *RPN11* gene encodes a non-ATPase subunit of the 19S regulatory complex (GLICKMAN et al. 1998). The *srp1-49* strain displayed defects in the degradation of model substrates of the ubiquitin/proteasome pathway, which were not observed with the *srp1-31* strain. The pleiotropic phenotypes observed in the *srp1-49* strain were probably the result of this defect in proteolysis as has been observed with other mutants in components of the ubiquitin/proteasome pathway. Evidence from

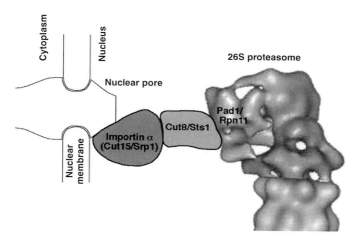

Fig. 1. Mechanism for enrichment to the nuclear periphery in yeast. The proteasome is tethered to the nuclear pore by interacting indirectly with the importin α subunit via the Cut8/Sts1 protein (see text for details)

two-hybrid experiments and in vitro pull-downs indicated that the Sts1 (Cut8) protein could interact directly with both the Srp1 and Rpn11 proteins. A simple model to account for the accumulation of proteasomes at the nuclear periphery is shown in Fig. 1. In this model, the 26S proteasome is tethered to the nuclear periphery by an interaction with the Srp1 protein via the Rpn11 proteasome subunit. This interaction occurs via the Sts1 (Cut8) protein. The localization of the fission yeast *SRP1* orthologue *Cut15$^+$* protein provides further evidence in favour of such a hypothesis (MATSUSAKA et al. 1998). The GFP staining observed in wild-type fission yeast cells expressing a Cut15–GFP-tagged protein displayed a remarkable similarity to that observed with the GFP-tagged proteasome subunits described earlier.

The model provides a framework within which to proceed with further experiments. If the model is correct the *cut15* mutant strain should delocalize the proteasome staining in a manner similar to that observed with the *cut8* mutant strain. In addition, a similar delocalization might also be seen in a *pad1 (rpn11)* mutant strain (PENNEY et al. 1998). The *cut15* mutant strain should also be defective in ubiquitin-dependent proteolysis. Finally, it is important to identify any other factors that are involved, in addition to Cut15/Srp1, Cut8/Sts1 and Pad1/Rpn11, in the discrete pattern of proteasome localization.

2.3 Localization in Live Cells During Mitosis and Meiosis

In fission yeast the distribution of proteasome localization was followed in time lapse films of fission yeast undergoing the mitotic and meiotic cycles (WILKINSON et al. 1998). The main conclusions of these studies were that during the mitotic cycle proteasomes were found at the nuclear periphery. However, during late anaphase at the end of nuclear division a concentrated spot of GFP signal was observed to remain between the separating nuclei. Such an enriched GFP signal was also found between separating nuclei during the meiotic divisions. Although it is not known if this concentration of proteasomes between the separating nuclei is functionally significant, one intriguing possibility is that proteolytic activity might be required for dismantling the mitotic spindle. Consistent with this hypothesis, the proteasome has been localized to the mitotic spindle in cells using specific antibodies to the 20S complex (AMSTERDAM et al. 1993).

The GFP tagged fission yeast strain was also used to investigate the localization of the 26S proteasome during the sexual cycle. The main results from this analysis was that during karyogamy, horsetail movement and the first meiotic division the GFP signal appeared to be no longer concentrated at the nuclear periphery but was found over the whole nucleus. The signal was not found in the nucleolus. A dramatic reorganization of GFP signal occurred at the end of meiosis II. The GFP signal concentrated between the separating nuclei before dispersing at the end of anaphase. The reason for this relocalization of proteasomes remains elusive. One possibility is that delocalized proteasome activity is necessary for the increased proteolysis required during spore formation.

3 Localization in Mammalian Cells

In mammalian cells the structure of the proteasome is complicated by the presence of two regulator complexes which interact with the 20S core catalytic particle. In addition to the 19S regulatory complex (PA700) found in yeast and described above, a second complex called 11S or PA28 has been characterized.

The distribution of the proteasomes has been investigated in a number of different cell types using immunogold EM, immunofluoresence microscopy and cell fractionation (Reviewed in RIVETT 1998). It is clear from these investigations that proteasomes are found in both the nucleus, the cytoplasm and associated with the ER.

3.1 Immunoproteasome Localization

Two recent reports have indicated that immunoproteasomes are localized to discrete intracellular compartments within the cell. However, the results are confusing as one group found immunoproteasomes concentrated at the ER whereas the other found them greatly enriched in subnuclear PML (*pro*myelocytic *l*eukemia oncoprotein) bodies (also called PODs or ND10 bodies; BROOKS et al. 2000; FABUNMI et al. 2000a). Surprisingly, the group that identified the ER localization of the LMP2 and LMP7 immunoproteasome specific subunits could not show co-localization with the PA28 regulator at the ER. In contrast, the second report showed enrichment of both immunoproteasomes and the PA28 regulator in PML bodies.

Further work is required to determine which one of these, or both, represents an accurate picture of where immunoproteasomes are found within cells.

3.2 Localization of the Proteasome to the Aggresome

A number of recent reports have demonstrated that under certain conditions the proteasome localizes to the centrosome region and that this area is an important site of proteolysis. These sites have been called 'proteasome centres' or 'aggresomes' (WOJCIK et al. 1996; JOHNSTON et al. 1998).

When investigating the turnover of the cystic fibrosis transmembrane conductance regulator (CFTR) and presenilin-1 integral membrane proteins, Kopito and co-workers found that the misfolded forms of the protein formed aggregates. Furthermore, the size of these aggregates could be increased by treating cells with proteasome inhibitors and/or by saturation of the ubiquitin/proteasome pathway by over-expressing mutant forms of the CFTR and presenillin genes, which resulted in misfolded proteins. When the localization of these aggregates was investigated by immunofluoresence using GFP-tagged proteins, it was found that the GFP signal accumulated as a discrete spot next to, but outside, the nuclear membrane. Kopito proposed that the structure was formed in response to the presence of slowly degraded (or undegradable) protein aggregates and called it an aggresome (JOHNSTON et al. 1998).

By immunofluoresence microscopy it was demonstrated that aggresomes did not co-localize with Golgi markers but with γ-tubulin, a known marker of the centrosome microtubule organizing centre (MTOC). Aggresome-containing cells appeared to have a normal distribution of microtubules, when stained with an anti-α tubulin antibody, implying that aggresomes did not interfere with MTOC function. However, if cells were treated with microtubule-destabilizing drugs, such as nocodazole, under conditions which had previously resulted in aggresome formation, no aggresomes were observed. This demonstrated that an intact microtubule network was required for aggresome formation. Over-expression of p50/dynamitin interferes with the dynactin complex and significantly interfered with aggresome formation. This implicated the microtubule-dependant minus end motor dynein in the transport of misfolded proteins to the MTOC/aggresome. Disruption of the actin microfilament network had no effect on aggresome formation. In contrast, aggresome formation resulted in dramatic reorganization of the intermediate filament vimentin. Immunofluoresence and EM demonstrated that the vimentin redistributed to form a cage-like structure surrounding the aggresome. This cage-like structure is thought to stop the release of protein aggregates into the cytoplasm (JOHNSTON et al. 1998; GARCIA-MATA et al. 1999).

This work was followed by several different reports that demonstrated that a number of different cytosolic proteins were capable of forming structures identical to the aggresome in a range of different cell types (ANTON et al. 1999; GARCIA-MATA et al. 1999; JOHNSTON et al. 2000). Therefore, it appears that aggresome formation is a general cellular response to the formation of insoluble protein aggregates in mammalian cells and is caused by an inability to degrade misfolded proteins by the ubiquitin/proteasome pathway.

This work was extended to identify the composition of aggresomes. Immunofluoresence microscopy has shown that aggresomes are a complex mixture of different components of the ubiquitin/proteasome pathway. Aggresomes contain high concentrations of the molecular chaperones Hdj1, Hdj2, Hsc70, Tcp1 (GARCIA-MATA et al. 1999; WIGLEY et al. 1999; WAELTER et al. 2001). In addition, the 20S proteasome and the 19S and PA28 regulators all co-localize with aggresomes (WIGLEY et al. 1999; WAELTER et al. 2001). This implies that the 26S and PA28 forms of the 20S complex are present. Surprisingly, experiments using microtubule-destabilizing drugs demonstrated that the transport of proteasomes to aggresomes was microtubule independent. Therefore, the mechanism of how proteasomes are transported to aggresomes appears to be different from that of the misfolded proteins and remains elusive. However, the 26S complexes present in the aggresome appear to be active because if one enriches for aggresomes by cell fractionation experiments the 26S proteasome complexes which co-purify are able to degrade a ubiquitinylated substrate in an ATP-dependent manner (WIGLEY et al. 1999; FABUNMI et al. 2000b). Ubiquitin has been found in aggresomes formed by expressing CFTR and influenza virus nucleoprotein but not those formed by expressing GFP-250 and the enzyme superoxide dismutase (JOHNSTON et al. 1998; ANTON et al. 1999; GARCIA-MATA et al. 1999; JOHNSTON et al. 2000). Perhaps the presence or absence of ubiquitin reflects a difference in how different proteins are dealt with by the cell. Alternatively, it could represent a difference in the activity of ubiquitin hydrolases present in different aggresomes.

The size of the aggresome was found to increase dramatically after the addition of proteasome inhibitors. As the aggresome structure became large it was found to recruit the cytosolic pools of proteasomes (WIGLEY et al. 1999). Recently, it was demonstrated that known substrates of the ubiquitin/proteasome pathway were stabilized in vivo after aggresome formation (BENCE et al. 2001). Presumably this inhibition of the ubiquitin/proteasome pathway was due to components of the pathway being sequestered by the aggresome and no longer being able to function in their normal context in the ubiquitin/proteasome pathway. It has been postulated that aggresome formation plays a key role in the initiation of cell death in the pathogenicity of many neurodegenerative diseases, such as Parkinson's and Huntingdon's diseases. This key observation, that the ubiquitin/proteasome pathway is impaired in aggresome-containing cells provides an explanation of why the protein aggregation found in a number of different neurodegenerative diseases results in cell death.

4 Summary

In summary, localization of proteasomes does appear to be important in the regulation of proteolysis. In yeast, a discrete localization is observed at the nuclear periphery for cells undergoing mitotic growth. This localization is clearly important as degradation by the ubiquitin/proteasome pathway is impaired in mutants that mislocalize proteasomes. In mammalian cells, proteasomes are present throughout the cell. However, the proteasome does appear to be enriched at the MTOC upon aggresome formation. The inhibition of the ubiquitin/proteasome pathway in aggresome-containing cells could provide an explanation for the pathogenicity of a number of neurodegenerative diseases.

References

Amsterdam A, Pitzer F, Baumeister W (1993) Changes in the intracellular localisation of proteasomes in immortalised ovarian granulosa cells during mitosis associated with a role in cell cycle control. Proc Natl Acad Sci USA 90:99–103
Anton LC, Schubert U, Bacik I, Princiotta MF, Wearsch PA, Gibbs J, Day PM, Realini C, Rechsteiner MC, Bennink JR, Yewdell JW (1999) Intracellular loclisation of proteasomal degradation of a viral antigen. J Cell Biol 146:113–124
Baumeister W, Walz J, Zuhl F, Seemuller E (1998) The proteasome: a paradigm of a self-compartmentalizing protease. Cell 92:367–380
Bence NF, Sampat RM, Kopito RR (2001) Impairment of the ubiquitin-proteasome system by protein aggregation. Science 292:1552–1555
Brooks P, Murray RZ, Mason GGF, Hendil KB, Rivett JA (2000) Association of immunoproteasomes with the endoplasmic reticulum. Biochem J 352:611–615
Coffino P (2001) Regulation of cellular polyamines by antizyme. Nature Rev Mol Cell Biol 2:188–194
Dubiel W, Pratt G, Ferrell K, Rechsteiner M (1992) Purification of an 11S regulator of the multicatalytic protease. J Biol Chem 274:22369–22377
Enenkel C, Lehman A, Kloetzel P-M (1998) Subcellular distribution of proteasomes implicates the major location of protein degradation in the nuclear envelope-ER network in yeast. EMBO J 17:6144–6154
Fabunmi RP, Wigley WC, Thomas PJ, DeMartino GN (2000a) Interferon gamma regulates accumulation of the proteasome activator PA28 and immunoproteasomes at nuclear PML bodies. J Cell Sci 114:29–36
Fabunmi RP, Wigley WC, Thomas PJ, DeMartino GN (2000b) Activity and regulation of the centrosome-associated proteasome. J Biol Chem 275:409–413
Fruh K, Gossen M, Wang K, Bujard H, Peterson PA, Yang Y (1994) Displacement of housekeeping proteasome subunits by MHC-encoded LMPs: a newly discovered mechanism for modulating the multicatalytic proteinase complex. EMBO J 13:3236–3244
Garcia-Mata R, Bebok Z, Sorscher EJ, Sztul ES (1999) Characterisation and dynamics of aggresome formation by a cytosolic GFP-chimera. J Cell Biol 146:1239–1254
Glickman MH, Rubin DM, Fried VA, Finley D (1998) The regulatory particle of the *Saccharomyces cerevisiae* proteasome. Mol Cell Biol 18:3149–3162
Groll M, Ditzel L, Lowe J, Stock D, Bochtler M, Bartunik HD, Huber R (1997) Structure of the 20S proteasome from yeast at 2.4Å resolution. Nature 386:463–471
Hershko A, Ciechanover A (1998) The ubiquitin system. Annu Rev Biochem 67:425–479
Hirano T, Funahashi S, Uemura T, Yanagida M (1986) Isolation and characterization of *Schizosaccharomyces pombe cut* mutants that block nuclear division but not cytokinesis. EMBO J 5:2973–2979
Holzl H, Kapelari B, Kellerman J, Seemuller E, Sumegi M, Udvardy A, Medalia O, Sperling J, Muller SA, Engel A, Baumeister W (2000) The regulatory complex of the *Drosphila melnogaster* 26S proteasomes: subunit composition and localisation of a deubiquitylating enzyme. J Cell Biol 150:119–129
Johnston JA, Ward CL, Kopito RR (1998) Aggresomes: a cellular response to misfolded proteins. J Cell Biol 143:1883–1898
Johnston JA, Dalton MJ, Gurney ME, Kopito RR (2000) Formation of high molecular weight complexes of mutant Cu, Zn-superoxide dismutase in a mouse model for famial amyotrophic lateral sclerosis. Proc Natl Acad Sci USA 97:12571–12576

Ma C-P, Slaughter CA, DeMartino GN (1992) Identification, purification and characterisation of a protein activator (PA28) of the 20S proteasome. J Biol Chem 267:10515–10523

Matsusaka T, Imamoto N, Yoneda Y, Yanagida M (1998) Mutations in fission yeast Cut15, an importin alpha homolog, lead to mitotic progression without chromosome condensation. Curr Biol 8:1031–1041

Penney M, Wilkinson CRM, Wallace M, Javerzat J-P, Ferrell K, Seeger M, Dubiel W, McKay S, Allshire R, Gordon C (1998) The $pad1^+$ gene encodes a subunit of the 26S proteasome in fission yeast. J Biol Chem 273:23938–23945

Preckel T, Fung-Leung W-P, Cai Z, Vitiello A, Salter-Cid L, Winqvist O, Wolfe TG, Von Herrath M, Angulo A, Ghazal P, Lee J-D, Fourie AM, Wu Y, Pang J, Ngo K, Peterson PA, Fruh KYY (1999) Impaired immunoproteasomal assembly and immune responses in $PA28^{-/-}$ mice. Science 286:2162–2165

Rivett, JA (1998) Intracellular distribution of proteasomes. Curr Opin Immunol 10:110–114

Rock KL, Goldberg AL (1999) Degradation of cell proteins and the generation of MHC class I-presented peptides. Annu Rev Immunol 17:739–779

Russell SJ, Steger KA, Johnston SA (1999) Subcellular localization, stoichiometry, and protein levels of 26S proteasome subunits in yeast. J Biol Chem 274:21943–21852

Samejima I, Yanagida M (1994) Identification of $cut8^+$ and $cek1^+$, a novel protein kinase gene which complement a fission yeast mutation that blocks anaphase. Mol Cell Biol 14:6361–6371

Seemuller E, Lupas A, Stock D, Lowe J, Huber R, Baumeister W (1995) Proteasome from *Thermoplasma acidophilum*: a threonine protease. Science 268:579–582

Strickland E, Hakala K, Thomas PJ, DeMartino GN (2000) Recognition of misfolding proteins by PA700, the regulatory subcomplex of the 26 proteasome. J Biol Chem 275:5565–5572

Tabb MM, Tongaonkar P, Vu L, Nomura M (2000) Evidence for separable functions of Srp1p, the yeast homolog of importin α (karyopherin α): role for Srp1p and Sts1p in protein degradation. Mol Cell Biol 20:6062–6073

Tatebe H, Yanagida M (2000) Cut8, essential for anaphase, controls localization of 26S proteasome, facilitating destruction of cyclin and Cut2. Curr Biol 10:1329–1338

Voges D, Zwickl P, Baumeister W (1999) The 26S proteasome: a molecular machine designed for controlled proteolysis. Annu Rev Biochem 68:1015–1068

Waelter S, Boeddrich A, Lurz R, Scherzinger E, Leuder G, Lehrach H, Wanker EE (2001) Accumulation of mutant Huntingtin fragments in aggresome-like inclusion bodies as a result of insufficient protein degradation. Mol Biol Cell 12:1393–1407

Wigley WC, Fabunmi RP, Lee MG, Marino CR, Muallem S, DeMartino GN, Thomas P (1999) Dynamic association of proteasomal machinery with the centrosome. J Cell Biol 145:481–490

Wilkinson CRM, Wallace M, Morphew M, Perry P, Allshire R, Javerzat J-P, McIntosh JR, Gordon C (1998) Localization of the 26S proteasome during mitosis and meiosis in fission yeast. EMBO J 17:6465–6476

Wojcik C, Schroeter D, Wilk S, Lamprecht J, Paweletz N (1996) Ubiquitin-mediated proteolysis centers in HeLa cells: indication from studies of an inhibitor of the chymotrypsin-like activity of the proteasome. Eur J Cell Biol 71:311–318

Proteasome Inhibitors: Complex Tools for a Complex Enzyme

M. Bogyo[1] and E.W. Wang[2]

1	Introduction	185
2	Initial Inhibitors Studies – Characterizing the Catalytic Mechanism of the Proteasome	186
3	Synthetic Inhibitors of the Proteasome	188
3.1	Synthetic Reversible Inhibitors	188
3.2	Synthetic Covalent Inhibitors	190
4	Bi-Functional Synthetic Inhibitors – Rational Design Based on Structure	193
5	Natural Product Inhibitors of the Proteasome	195
5.1	Small Molecule Inhibitors	195
5.2.	Protein Inhibitors of the Proteasome	198
6	Proteasome Inhibitors as Affinity Labels	199
7	Kinetic Studies of the Proteasome Using Inhibitors	201
7.1	Studies of Catalytic Mechanism	201
7.2	Analysis of Substrate Specificity	202
8	Proteasome Inhibitors as Therapeutic Agents	203
9	Summary	204
References		204

1 Introduction

Over the last two decades it has become abundantly clear that the proteasome is the pivotal component in cytosolic catabolism. Since its initial purification and biochemical characterization, this multi-component enzyme has been found to be essential for the regulation of fundamentally important processes such as cell division, cell death, signal transduction, and immune surveillance (for reviews see Coux et al. 1996; Goldberg and Rock 1992; Rivett 1993). Central to understanding any enzyme is the need to perturb its function in a highly controlled manner. This has been achieved through recent advances in the development and

---

[1] Department of Biochemistry and Biophysics, University of California, San Francisco, 513 Parnassus Avenue, San Francisco, USA
[2] Department of Pathology, Harvard Medical School 200 Longwood Avenue, Boston, MA 02115, USA

use of both natural and synthetic inhibitors that are capable of blocking the proteolytic activity of the proteasome. These advances are the focus of this chapter that highlights the use of diverse classes of inhibitors to probe the mechanism of the proteasome as well as to identify its physiological significance in the cell.

2 Initial Inhibitor Studies – Characterizing the Catalytic Mechanism of the Proteasome

The proteasome was first identified as a high molecular weight protease complex that resolved into a series of low molecular weight protein species upon denaturation (DAHLMANN et al. 1985; MCGUIRE and DEMARTINO 1986; ORLOWSKI and WILK 1981; TANAKA et al. 1986; WILK and ORLOWSKI 1980). It was subsequently purified and found to catalyze the hydrolysis of amide bonds adjacent to a variety of amino acids (ORLOWSKI and WILK 1981; WILK and ORLOWSKI 1980). Classification of its broad substrate specificity into three categories based on the nature of the amino acid found in the P1 position adjacent to the scissile amide bond soon followed. These three activities: chymotrypsin-like, trypsin-like and post-glutamyl peptide hydrolyzing (PGPH) activity, were established based on their similarity to well characterized proteolytic enzymes (ORLOWSKI and WILK 1981; WILK and ORLOWSKI 1980). Soon after the initial characterization of its biochemical properties, attention shifted to understanding the proteasome's enzymatic mechanism.

A common means of analyzing a novel proteolytic enzyme makes use of small molecule inhibitors whose reactivity towards an attacking nucleophile of a protease is well defined. Thus, the identification of a class of reagents capable of potent inhibition of a target protease can aid in the characterization of its underlying mechanism. At the time of initial identification of the proteasome, proteases were classified into four main groups based on their catalytic mechanism, and virtually all could be placed into aspartic, metallo-, cysteine, or serine protease families. However, initial biochemical studies of the proteasome quickly indicated that it did not fit into any of these classifications and thus, a new family must be established.

Rapid purification schemes that took advantage of the proteasome's large size made it possible to perform biochemical analysis of the proteasome using a variety of classical protease inhibitors. Initial studies by Orlowski and Wilk in the early 1980s established the utility of peptide aldehydes, identifying the natural product leupeptin as an inhibitor of the proteasome's trypsin-like activity (ORLOWSKI and WILK 1981; WILK and ORLOWSKI 1980, 1983a). Not surprisingly, this compound, with a basic arginine in the P1 position, showed little activity against the remaining two activities of the proteasome that preferred cleavage after acidic and hydrophobic P1 amino acids. The synthesis of a tri-peptide aldehyde (Cbz–Gly–Gly–

Leucinal) based on the sequence of a known peptide fluorogenic substrate resulted in a potent inhibitor of the chymotrypsin-like activity of the proteasome (WILK and ORLOWSKI 1983a). While these initial inhibitor studies provided useful new reagents for use in biochemical studies, the reactivity of peptide aldehydes towards both hydroxyl and thiol nucleophiles provided little information about the proteasome's catalytic mechanism.

Subsequently, a multitude of studies were performed using diverse sets of easily accessible, class-specific inhibitors (CARDOZO et al. 1992; DAHLMANN et al. 1985; McGUIRE and DeMARTINO 1986; WILK and ORLOWSKI 1980, 1983a). Chelating agents indicated that the proteasome did not belong to the metalloprotease family and pepstatin ruled out an aspartic protease mechanism (DAHLMANN et al. 1985; McGUIRE and DeMARTINO 1986; RIVETT 1985; WAGNER et al. 1986). However, it was determined that organic mercurials, known to be highly reactive towards thiol groups, profoundly inhibited the proteasome (DAHLMANN et al. 1985; McGUIRE and DeMARTINO 1986; RIVETT 1985; WAGNER et al. 1986; WILK and ORLOWSKI 1980; ZOLFAGHARI et al. 1987). Moreover, additional thiol-reactive compounds such as N-ethyl maleimide, and iodo-acetic acid could partially block multiple proteasomal activities, with preferential inhibition of the trypsin-like activity (DAHLMANN et al. 1985; McGUIRE and DeMARTINO 1986; RIVETT 1985; WAGNER et al. 1986; WILK and ORLOWSKI 1980; ZOLFAGHARI et al. 1987). The proteasome was also found to be resistant to inactivation by classical serine protease inhibitors such as di-isopropyl fluorophosphate (DFP) and phenylmethanesulfonyl fluoride. Thus, the proteasome was first classified as a thiol protease and the name 'macropain' was suggested to propose a link to the papain family of cysteine proteases (DAHLMANN et al. 1985; RIVETT 1985; WAGNER et al. 1986).

Controversy over the classification of the proteasome's catalytic mechanism arose when several groups reported that prolonged exposure of the proteasome to high concentrations of DFP resulted in inhibition of its chymotrypsin-like activity (ISHIURA et al. 1986; NOJIMA et al. 1986; WAGNER et al. 1986). Furthermore, POWERS and ORLOWSKI found that several structurally related isocoumarin derivatives, known to act as class-specific inhibitors of serine proteases, potently inhibited the proteasome (HARPER et al. 1985; KAM et al. 1988; ORLOWSKI and MICHAUD 1989). These findings prompted classification of the proteasome as a serine protease.

The cloning of several proteasome subunits from mammals and yeast (EMORI et al. 1991; FUJIWARA et al. 1989; SORIMACHI et al. 1990) unfortunately provided little information to assist in the classification of the proteasome's catalytic mechanism as none of the subunits showed homology to any known proteases. It was not until the proteasome was crystallized in a complex with a peptide aldehyde inhibitor that the true active site nucleophile was revealed to be the N-terminal threonine residue found on the catalytic β-type subunits (Fig. 1; LOWE et al. 1995; SEEMULLER et al. 1995; STOCK et al. 1995). Thus, the proteasome constitutes a new family of proteases that requires a free N-terminal threonine for activity.

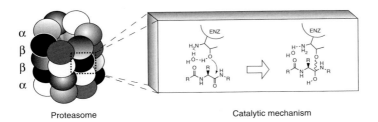

Proteasome   Catalytic mechanism

Fig. 1. Structure and catalytic mechanism of the proteasome. Cartoon representation (*left*) of the core 20S proteasome complex showing the topology of the multiple α and β subunits. The central rings of seven catalytic β subunits are repeated twice in the complex. These β subunits utilize an N-terminal threonine reside to initiate attack on a protein substrate (*right*). Catalysis requires coordination of a water molecule that serves as a charge-relay between the free amino group and the attacking hydroxyl nucleophile

3 Synthetic Inhibitors of the Proteasome

3.1 Synthetic Reversible Inhibitors

Potency and specificity are two features often considered to be most critical when designing protease inhibitors. Failure to achieve either of these traits in inhibitor design can adversely influence our understanding of a given protease, as illustrated by the initial synthetic compounds used to block the proteasome. These compounds all lacked specificity, as they were originally designed to block non-proteasomal proteases. However, with the proteasome's catalytic mechanism established and its three-dimensional structure determined, synthetic chemistry could be used to further refine inhibitory compounds leading to greater potency and enhanced selectivity for the proteasome.

The first synthetic inhibitors designed to target the proteasome were peptide aldehydes as they were relatively easy to synthesize and previous studies indicated that small peptidic substrates could serve as a template for inhibitor design. As a result, numerous peptide sequences have been synthesized as aldehydes (HARDING et al. 1995; IQBAL et al. 1995; VINITSKY et al. 1994; WILK and FIGUEIREDO-PEREIRA 1993) and several have proved widely useful. Peptide aldehydes such as leupeptin and calpain inhibitors I and II, as well as several closely related compounds, such as MG-132 (Cbz–Leu–Leu–leucinal) and MG-115 [Cbz–Leu–Leu–norvalinal; developed by Proscript (formerly Myogenics)], are frequently used to block proteasome activity both in vitro and in vivo (HARDING et al. 1995; LEE and GOLDBERG 1996; PALOMBELLA et al. 1994; ROCK et al. 1994). Additionally, the tetra peptide aldehyde Z–IE(OtBu)AL–H developed by Wilk and co-workers is among the commercially available proteasome inhibitors (WILK and FIGUEIREDO-PEREIRA 1993). Most of these compounds primarily inhibit the chymotrypsin-like activity of the proteasome but are capable of modifying all three primary catalytic β-subunits at high concentrations (GROLL et al. 1997). One of the drawbacks to these compounds is their reactivity towards both serine and cysteine proteases through formation of hemi-acetals or hemi-thioacetals with either hydroxyl or thiol nucleophiles (Fig. 2a). Specificity for the proteasome can only be achieved by design of peptide sequences optimized for proteasome binding. In fact, very few examples of selective peptide aldehydes have ever been documented (WILK and FIGUEIREDO-PEREIRA 1993; WILK and ORLOWSKI 1983b) as they usually exhibit broad specificity. Another drawback to the use of the aldehyde electrophile is its reactivity with free thiols and its instability in aqueous solution. Regardless of these shortcomings, the initial contributions of peptide aldehyde inhibitors laid the foundation for subsequent generations of proteasome inhibitors.

The first step towards designing compounds with increased selectivity for the proteasome was to take advantage of electrophiles that react specifically with a hydroxyl nucleophile (ADAMS and STEIN 1996; GARDNER et al. 2000; IQBAL et al. 1996; MCCORMACK et al. 1997). Peptide boron esters and acids are potent inhibitors of serine proteases that form reversible covalent interactions with the active site hydroxyl (Fig. 2b). These compounds lack reactivity towards cysteine proteases as a consequence of poor overlap between orbitals of the non-bonding electrons on sulfur with those of the vacant d-orbitals on boron, resulting in a weak sulfur–boron bond. Furthermore, these derivatives are less reactive to circulating

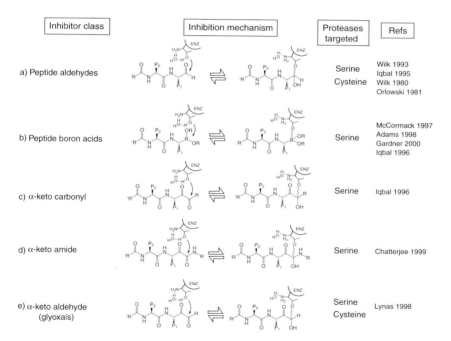

Fig. 2a–e. General structures and mode of action of reversible synthetic proteasome inhibitors. Partial list of compounds found to inhibit the proteasome in a reversible manner. Schematic representations of inhibition mechanism shows site of initial attack by the active site N-terminal threonine found on multiple proteasomal β-subunits (*arrows*), and the resulting transient covalent adduct

nucleophiles in aqueous solutions than their aldehyde counterparts. Chemists and biochemists at Proscript (now Millennium Pharmaceuticals) and Cephalon found that changing the electrophile from an aldehyde to a boron acid or ester created compounds with reduced cross-reactivity towards cysteine proteases and dramatically increased potency for the proteasome (ADAMS and STEIN 1996; IQBAL et al. 1996; MCCORMACK et al. 1997). Rivett and co-workers also explored the potency and selectivity of a number of peptide boron acids and esters including a boron ester derivative of the MG-132 aldehyde described above (GARDNER et al. 2000). Further refinement of these lead compounds has resulted in the generation of di-peptide boron acids that are capable of inhibiting the proteasome at picomolar concentrations (ADAMS et al. 1998; ADAMS and STEIN 1996; MCCORMACK et al. 1997). This high degree of potency for the proteasome essentially results in selective inhibitors. For example, the di-peptide boron ester PS-341 requires 20,000 times higher concentrations to inhibit other abundant serine proteases (ADAMS et al. 1998). These compounds are also highly bio-available and are currently being pursued in clinical studies as potential anti-inflammatory agents (see below; ADAMS and STEIN 1996).

Additional classes of peptide-based electrophiles that reversibly target the proteasome's active site threonine have been explored. Most of these compounds contain a short di- or tri-peptide recognition element fused to a C-terminally modified amino acid (often an aliphatic residue such a leucine). Examples include the peptide α-keto carbonyls (IQBAL et al. 1996; Fig. 2c), α-keto amides (CHATTERJEE et al. 1999; Fig. 2d), and α-keto aldehydes (LYNAS et al. 1998; Fig. 2e). All contain a highly electrophilic carbonyl carbon that forms a stable acetal or ketal linkage to the threonine hydroxyl when bound in the active sites of the proteasome (Fig. 2c–e). Moreover, compounds containing an α-keto amide at their C terminus have the potential for extension of inhibitor structures into the S' region of the target protease, located directly C-terminal to the site of amide bond hydrolysis (Fig. 2d). Peptides containing these extended binding elements were developed with the hope of gaining additional specificity and potency towards the proteasome's multiple active sites (LYNAS et al. 1998). Unfortunately, all of these reversible inhibitors suffer from the same limitations associated with the peptide aldehydes including broad specificity and instability in solution. To overcome these problems compounds must be generated that are potent enough to require a low dosage regime thus eliminating cross-reactivity and other toxic effects.

3.2 Synthetic Covalent Inhibitors

Another major class of synthetic proteasome inhibitors are compounds that inactivate the catalytic threonine nucleophile by irreversible covalent adduct formation (Fig. 3). Often referred to as suicide substrates, these inhibitors have found widespread use in biochemical studies of the proteasome. Furthermore, the covalent nature of these compounds allows protease activity to be traced using suitably labeled inhibitors (see Sect. 6).

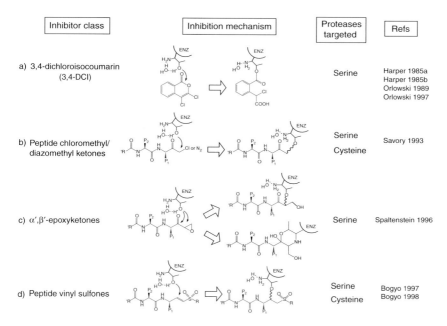

Fig. 3a–d. General structures and mode of action of irreversible synthetic proteasome inhibitors. Partial list of compounds that act as suicide substrates for the proteasome. Schematic diagrams show site of initial attack by the threonine hydroxyl and the resulting covalent adduct. In the case of the α′,β′ epoxyketones, multiple inhibition mechanisms are proposed

The general serine protease inhibitor 3,4 dichloroisocoumarin (3,4-DCI) was one of the first compounds found to act as a potent irreversible inhibitor of the proteasome (HARPER and POWERS 1985; KAM et al. 1988; ORLOWSKI and MICHAUD 1989). While many isocoumarin analogs possessing a variety of hydrophobic appendages have been synthesized, only 3,4-DCI shows appreciable activity against the proteasome (HARPER et al. 1985; HARPER and POWERS 1985; KAM et al. 1988; ORLOWSKI and MICHAUD 1989). 3,4-DCI functions as a masked acid chloride that binds in the active site near the base-activated hydroxyl side chain of threonine to form a covalent ester linkage (Fig. 3a). Inhibition is achieved by modification of one or more of the active sites of the proteasome. While 3,4-DCI initially found widespread use as a proteasome inhibitor, complications arose when it was discovered that, in contrast to its potent inactivation of the chymotrypsin-like activity of the proteasome, it simultaneously resulted in activation of other activities (CARDOZO et al. 1992; ORLOWSKI et al. 1993). Furthermore, 3,4-DCI treatment of proteasomes lead to the accelerated processing of select protein substrates (CARDOZO et al. 1992; ORLOWSKI et al. 1993; PEREIRA et al. 1992). Many studies have attempted to ascertain the reason for the proteasome's varied response to 3,4-DCI, yet a detailed biochemical understanding of this phenomenon is lacking. Furthermore, the utility of this class of reagents as proteasome inhibitors is limited by their broad reactivity with many serine-type protease as well as their instability in solution (POWERS and KAM 1994). These reasons prompted many research groups to turn their attention to other classes of covalent proteasome inhibitors.

Chloromethyl ketones comprise a distinct class of commonly used covalent irreversible serine protease inhibitors. The diazomethyl ketones – close relatives of the chloromethyl ketones – were initially thought to be reactive only towards cysteine proteases, but were later found to react with serine proteases as well (SAVORY et al. 1993). The function of both classes of peptide electrophiles is mechanistically similar to that of peptide aldehydes and boron acids. The chloride or diazo groups adjacent to the ketone moiety create a highly electrophilic site that is capable of reacting with activated nucleophiles (Fig. 3b). The ease with which these peptide-based inhibitors can be synthesized prompted the development of this class of electrophiles as covalent inhibitors of the proteasome (SAVORY et al. 1993). However, the low potency of these compounds towards the proteasome necessitated high concentrations of inhibitor to elicit appreciable inhibition, thus, limiting their use as proteasome inhibitors.

Many other classes of inhibitors initially designed to target serine proteases have been successfully converted into proteasome inhibitors. α',β' epoxyketone electrophiles have been incorporated into peptides sequences that have been optimized for binding to the proteasome (SPALTENSTEIN et al. 1996). Conversion of the potent, tri-peptide aldehyde inhibitor of the chymotrypsin-like activity of the proteasome, Cbz–Ile–Ile–Phe–H, to the corresponding α',β' epoxyketone produced a covalent inhibitor with a 40-fold improved potency (SPALTENSTEIN et al. 1996). The α',β'-epoxyketones, like the α-keto amides and aldehydes, have two electron-deficient carbon atoms that are susceptible to attack by the proteasome's threonine hydroxyl. This feature creates the potential for either reversible ketal formation with the carbonyl or irreversibly ether formation through ring opening of the strained epoxide moiety (Fig. 3c). Attack at the carbonyl carbon places the proteasome's terminal amino group in close proximity to the highly electrophilic epoxide ring. Subsequent ring opening by the amino group would result in the formation of a stable six-membered ring (Fig. 3c). Evidence for this unusual 'double attack' as the primary mechanism for inhibition of the proteasome by α',β' epoxyketones was supported by recent studies of the natural product epoxomicin, which relies on the same electrophilic group for inhibition of the proteasome (see Sect. 5).

Peptide vinyl sulfones are electrophiles initially designed as cysteine protease inhibitors (BROMME et al. 1996; PALMER et al. 1995) that act by formation of a covalent linkage with an active site nucleophile via a Michael addition (Fig. 3d). The vinyl sulfone electrophile was reported to be resistant to attack by virtually all serine proteases (BROMME et al. 1996; PALMER et al. 1995). The strong preference for a thiol nucleophile was believed to result from the 'soft' basic property of the thiol group that favors attack at the unsaturated carbon–carbon double bond (BROMME et al. 1996; PALMER et al. 1995). However, several peptide vinyl sulfones were found to inhibit the proteasome through covalent bond formation with the active site threonine hydroxyl (Fig. 3d; BOGYO et al. 1997, 1998). Remarkably, an analog of the potent tri-peptide aldehyde MG-132, when converted to a vinyl sulfone, covalently inhibited all three activities of the proteasome. Replacement of the carboxylbenzoyl (cbz) N-terminal capping group with a nitrophenol moiety produced a compound with increased potency that was easily modified by radioactive iodine. This class of electrophilic peptide proved to be valuable in affinity labeling and mechanistic studies of the proteasome (BOGYO et al. 1997, 1998). These aspects will be discussed in greater detail in Sect. 6.

4 Bi-Functional Synthetic Inhibitors – Rational Design Based on Structure

Determination of the three-dimensional structure of the proteasome provided the single greatest breakthrough in our understanding of its complex biochemical mechanism. The first structure was obtained for the archaebacterial form of the proteasome (SEEMULLER et al. 1995, LOWE et al. 1995) and revealed a complex comprised of a single α- and β-type subunit each repeated 14 times, to create a highly symmetrical core complex. In contrast, the core of the yeast proteasome is made up of seven distinct α- and β-type subunits each repeated twice in the complex (Fig. 1; GROLL et al. 1997). While the information from these structural studies was valuable for understanding mechanism and topology of the complex, they also paved the way for development of new classes of proteasome inhibitors the design of which is based on the detailed maps of the proteasome's inner cavity.

Initial structure-aided studies were aimed at creating inhibitors that could specifically target a single active site of the proteasome (LOIDL et al. 1999a). The β2 subunit of the yeast proteasome was shown to be responsible for the trypsin-like activity of the yeast proteasome (DICK et al. 1998). It also possesses a unique feature in that a portion of its substrate-binding pocket lies in close proximity to a cysteine residue of the neighboring β3 subunit (GROLL et al. 1997). Thus, the β2 active site depends on contacts created by multiple subunits and could potentially be targeted by reagents with two reactive electrophiles. Such β2-specific reagents would represent a new class of inhibitors specific for the trypsin-like activity of the yeast proteasome. Using the cysteine-reactive maleimide group, Moroder and co-workers synthesized a series of bi-functional peptide aldehydes (LOIDL et al. 1999a; Fig. 4a). The structure of the β2 active site provided a guide for design of peptide scaffolds that placed the maleimide group in the S3 pocket of the β2 subunit, proximal to the free thiol of the β3 subunit. Replacement of the P3 acetyl-leucine residue with a maleoyl-β-alanine residue of the peptide aldehyde, Ac-LLnL–H, converted this compound from a potent, reversible inhibitor of the chymotrypsin-like activity, into a specific, covalent inhibitor of the trypsin-like activity. This dramatic change in specificity was the direct result of a double covalent attack by threonine and cysteine in the active site of the β2 subunit (Fig. 4a). Further refinement of the P1 and P2 positions, to incorporate residues optimal for the

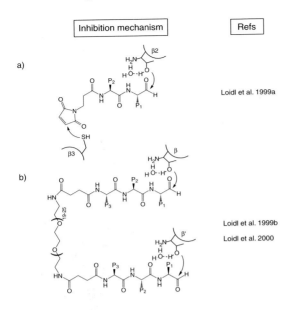

Fig. 4a,b. Synthetic bivalent inhibitors of the proteasome. Structures of two classes of inhibitors that target multiple nucleophiles on multiple proteasomal β-subunits. In the case of class (**a**) selective inhibition of the β2 proteasomal subunit is achieved by adduct formation between the active site threonine (β2) and a side-chain cysteine of an adjacent non-catalytic β-subunit (β3). Compounds of class (**b**) target two active site threonines on different catalytic β-subunits in the core 20S particle

trypsin-like activity, produced compounds with dramatic increases in potency. Thus, structural studies proved essential to the development of this novel, highly selective, and highly potent class of proteasome inhibitors.

The topology of the active sites of the proteasome was also used to generate inhibitors that could span multiple active sites (LOIDL et al. 1999b, 2000). The proteasome core structure contains two stacks of β-subunits. Each active site is repeated twice and thus can be targeted twice by a single compound that possesses reactive groups separated by the appropriate distance. An ethylene glycol polymer was selected as a scaffold for inhibitor design because it is composed of monomers that can be linked to create spacers of variable lengths, it contains no hydrolyzable peptide bonds, and it is highly soluble in water. Compounds were synthesized by fusing potent peptide aldehyde sequences end-to-end between a series of ethylene glycol monomers (Fig. 4b). The distance between two active site threonine residues was calculated and was used to determine the number of monomers required. The resulting compound, containing two identical peptide aldehydes specific for the chymotrypsin-like activity of the proteasome, was found to have a 100-fold increased potency towards the chymotrypsin-like activity as compared to the activity of the monomeric peptide aldehyde. Similarly, combining peptide sequences intended to target the trypsin-like activity of the proteasome resulted in enhanced potency towards the trypsin-like activity. The versatility of the technique was also demonstrated by production of hetero-bi-functional compounds that contain one chymotrypsin-like specific aldehyde and one trypsin-like specific aldehyde. These compounds are potent inhibitors of both activities (LOIDL et al. 1999b).

Collectively, these studies highlight the importance of detailed structural information for inhibitor design. They also demonstrate how information from structural studies of complex, multi-component enzymes such as the proteasome can help to define mechanisms for controlling potency and selectivity of synthetic inhibitors.

5 Natural Product Inhibitors of the Proteasome

5.1 Small Molecule Inhibitors

While chemists have expended a great deal of energy developing compounds that target enzymes such as the proteasome, nature often creates molecules that are far more specific and potent than anything made by the hands of a chemist. Unfortunately, it is often difficult or impossible to identify which natural products target a desired enzyme simply by inspection of structure or analysis of biochemical activity. In most cases compounds are singled out based on their biological effects in a pre-defined assay. Only after isolation of an active compound coupled with detailed biochemical analysis can the mode of action of the natural product be determined. Several new classes of proteasome inhibitors were identified through the synthesis of chemically modified versions of natural products that were then used to identify their biological targets (FENTEANY et al. 1995; MENG et al. 1999a, 1999b). Another class of natural product inhibitors of the proteasome was uncovered by direct screening of microbacterial extracts for inhibition of proteasome activity (KOGUCHI et al. 1999, 2000a,b; KOHNO et al. 2000).

The *Streptomyces* metabolite lactacystin was initially identified by its ability to inhibit cell cycle progression and induce neurite outgrowth in neuronal cell lines (OMURA et al. 1991a,b). Chemists have produced synthetic versions of it, but its five chiral centers made its synthesis far from trivial (Fig. 5a). Initial structure/activity studies using synthetic analogs proved critical for determining lactacystin's mode of action (FENTEANY et al. 1994). These studies showed that modification by deletion or inversion of steriochemistry of the hydroxyl or methyl groups attached to the lactam ring at carbons C6 and C7 resulted in complete loss of activity (FENTEANY et al. 1994). Hydrolysis of the thioester at carbon C4 also abolished activity, while replacement of the N-acetyl cysteine residue with other thiol-containing groups that leave the thioester moiety intact had no effect (FENTEANY et al. 1994). Similarly, the β-lactone derivative of lactacystin formed by an intra-molecular lactonization retained full activity, suggesting the importance of the thioester as an electrophilic site for attack by its target enzyme.

Derivatives of lactacystin in which a single hydrogen was replaced with its radioactive counterpart, tritium (FENTEANY et al. 1995), showed that lactacystin modified predominantly one polypeptide in crude extracts, identified as the X (β5) subunit of the proteasome. Thus, lactacystin targets the proteasome by covalent modification of the active site threonine hydroxyl (Fig. 5a). Subsequent studies found that lactacystin blocks multiple activities of the proteasome through modification of all of the major catalytically active β subunits (BOGYO et al. 1997; CRAIU

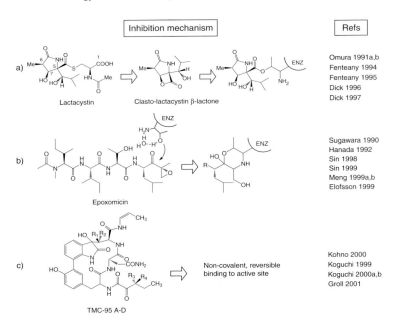

Fig. 5a–c. Natural product inhibitors of the proteasome. **a** The natural product lactacystin spontaneously reacts in basic aqueous solution to form the intra-molecular lactone (clasto-lactacystin β-lactone) which then reacts with the active site threonine of the proteasome. **b** Epoxomicin initially forms a covalent bond between the proteasome's N-terminal threonine hydroxyl at its C-terminal ketone carbonyl (not shown). This primary adduct formation is followed by a second attack by the terminal free amino group resulting in the formation of the stable six-membered ring adduct shown. **c** Cyclic peptide natural products TMC-95 A–D, from re-arrangement of the variable R groups (R1–R4)

et al. 1997). Other studies focused on kinetic analysis of the inhibition of the proteasome by lactacystin (DICK et al. 1996, 1997). These studies found that lactacystin acts by spontaneous lactonization between its thioester and the hydroxyl group at carbon C6 to form the highly reactive lactone, clasto-lactacystin β-lactone (DICK et al. 1996). This lactone species is the primary agent responsible for alkylation of the proteasome's active site. Moreover, lactonization is necessary for inhibition of the proteasome in intact cells, as lactacystin itself is not membrane permeable (DICK et al. 1997). Consequently, clasto-lactacystin β-lactone is now widely used and is commercially available.

The lack of structural resemblance of lactacystin to synthetic peptide-based compounds currently being used as protease inhibitors implied that it was highly specific for the proteasome. Several studies confirmed the selective nature of lactacystin, leading many to consider it as the gold standard by which to judge specificity of proteasome inhibitors (CRAIU et al. 1997; DICK et al. 1997; OSTROWSKA et al. 1997). Since these initial studies of lactacystin's specificity, it has been tested against a wide range of protease targets and was eventually found to be cross-reactive with cathepsin A. This lysosomal serine protease was effectively inhibited by low micromolar concentrations of lactacystin (OSTROWSKA et al. 1997) and may have escaped detection as a target owing to its lack of enzymatic activity in non-acidic assay conditions. Although it remains one of the most useful inhibitors currently available, the high cost of obtaining this compound from commercial sources financially limits its frequent use.

A second class of natural product proteasome inhibitors was isolated in screens of actinomycete fermentation broths for agents that blocked tumor growth (HANADA et al. 1992; SUGAWARA et al. 1990). Epoxomicin and eponemycin are modified peptides that contain an α',β'-epoxyketone group at their C terminus attached to an aliphatic P1 amino acid (Fig. 5b). Eponemycin was also identified in assays of blood vessel formation as a potent inhibitor of angiogenesis (OIKAWA et al. 1991), instigating a search for its biological targets.

At the core of both epoxomicin and eponemycin is an easily synthesized peptide backbone. This characteristic allowed the rapid facile synthesis of derivatives of epoxomicin in which the N-terminal acetate group was replaced with an aliphatic biotin linker (MENG et al. 1999a,b; SIN et al. 1998, 1999). This biotin-labeled epoxomicin derivative has identical activity to the parent natural product and was used to identify target proteins by affinity blotting. Moreover, the biotin molecule permitted simple affinity purification of targets using immobilized avidin. These studies identified four catalytic β-subunits of the proteasome as the primary targets of epoxomicin (X, LMP-7, Z, and MECL-1; MENG et al. 1999b). β-subunit modification correlated directly with inactivation of all three proteolytic activities of the proteasome and with anti-inflammatory activity in vivo (MENG et al. 1999b). Additional derivatives of epoxomicin were synthesized by optimizing the peptide backbone sequence to increase potency and selectivity for the chymotrypsin-like activity of the proteasome. Several of these derivatives showed as much as fivefold increased activity when compared to epoxomicin. Therefore simple chemical manipulation of the core peptide backbone sequence can be used to alter the selectivity and potency of this new class of proteasome inhibitors (ELOFSSON et al. 1999).

Epoxomicin contains an α',β'-epoxyketone group at its C terminus such that when bound in the active site, two electrophilic carbon atoms are found in close proximity to the proteasome's nucleophilic threonine. The crystal structure of the epoxomicin-inhibited yeast proteasome was used to establish its mode of inhibition (GROLL et al. 2000). Unlike other electrophilic peptide proteasome inhibitors that specifically target the hydroxyl nucleophile, epoxomicin reacts covalently with both the hydroxyl and the free amino groups of the N-terminal threonine, to produce a highly stable six-member ring (Fig. 5b). This unusual mode of inhibition provides an explanation for the extreme potency and selectivity of epoxomicin, as other inhibitors rarely possess an N-terminal nucleophile that can form this type of double adduct. To date, no other cellular targets of epoxomicin are known, reinforcing its utility as a new tool for studies of proteasome function.

A third class of compounds very recently identified in the fermentation broth of *Apiospora monagnei* was selected for its ability to directly inhibit the proteasome (KOGUCHI et al. 1999, 2000a,b; KOHNO et al. 2000). Four compounds (TMC-95

A–D; Fig. 5c) were isolated that have potent activity against the proteasome and have nearly identical structures as determined by a combination of nuclear magnetic resonance, infra-red, and mass spectrometry (KOHNO et al. 2000). The active species were modified cyclic peptides formed by a covalent link between a highly oxidized tryptophan side chain and the meta-carbon of a nearby tyrosine. The C terminus of the peptides contain a highly electrophilic α′-keto-carbonyl group, similar to that observed for several other synthetic inhibitors (IQBAL et al. 1996). While this class of natural products contains the electron-deficient α-keto-carbonyl group found on other previously characterized proteasome inhibitors, structural analysis of the yeast 20S proteasome bound to TMC-95A shows that it is not involved in direct covalent interaction with the active site hydroxyl (Groll et al. 2001). TMC-95A and its derivatives therefore represent the first class of truly non-covalent, selective inhibitors of the proteasome. These molecules will probably be the focus of much attention in the future as they provide a scaffold for the design of new classes of potentially therapeutically important proteasome inhibitors.

5.2 Protein Inhibitors of the Proteasome

In addition to the multitude of small molecule and peptide-based compounds, inhibitors can also take the form of large macromolecular protein structures. Most protein inhibitors of the proteasome were identified through screens of fractionated crude cellular extracts applied to purified proteasomes, which are then assayed for hydrolysis activity. In addition to endogenous proteins that modulate proteasome activity, several examples of macromolecular and exogenous protein inhibitors have been identified.

The paucity of endogenous inhibitors can be attributed to the limited number of large protease complexes that exist in the cell. The proteasome is one example of a large, complex protease that is comprised of a core particle, containing catalytically active β-subunits, that is able to form complexes with different multi-component cap complexes to facilitate protein breakdown. Some regulatory complexes facilitate activity of the 20S core and by analogy, similar mechanisms may exist that negatively modulate proteasome function (COUX et al. 1996). Furthermore, there are examples of endogenously synthesized protein inhibitors of nearly all classes of proteolytic enzymes (BARRETT et al. 1998). Studies aimed at finding mediators for negative regulation of proteasomal proteolysis identified two high molecular weight protein complexes that bound to the proteasome and inhibited proteolysis of both protein and small fluorogenic substrates (LI et al. 1991; LI and ETLINGER 1992). Denaturation of these 240- and 200-kDa complexes revealed that each consists of a single monomeric species with molecular weights of 40 and 50kDa, respectively. Biochemical studies established that the 40-kDa monomer was conjugated to ubiquitin and formed an ATP-stabilized complex with the 26S proteasome. The 40-kDa monomer was later determined to be δ-aminolevulinic acid dehydratase, the second enzyme in the pathway of heme synthesis (GUO et al. 1994). This intriguing discovery suggests that the inhibitory component of the proteasome was the result of gene sharing. Moreover, the 240-kDa purified proteasome complex has dehydratase activity that when blocked has no effect on the ability of the complex to inhibit proteasomal proteolysis (GUO et al. 1994). The exact role of these large protein complexes is not clear but further studies will define their significance in bulk protein turnover.

A single monomeric protein of 31kDa was also identified as having an inhibitory influence on hydrolysis of both protein and small peptide substrates by the proteasome (CHU-PING et al. 1992). This protein, isolated from red blood cells and given the name PI31, forms multimers under non-denaturing conditions and associates directly with the central core of the 20S proteasome. Recently PI31 was cloned and recombinantly expressed (MCCUTCHEN-MALONEY et al. 2000; ZAISS et al. 1999). When applied in vitro to purified 20S proteasomes recombinant PI31 effectively blocks hydrolysis of several small peptide substrates. The sequence of this protein has no significant homology to any known protein family. It contains a proline-rich region at its C terminus and mutational studies suggest that this region contains the active domain (MCCUTCHEN-MALONEY et al. 2000). Furthermore, truncation of the proline-rich domain results in C-terminal fragments with extended secondary structure that may form direct contacts with the proteasome core to modulate its activity. Studies performed in the presence of activator protein complexes such as the PA28 and PA700 caps indicate that PI31 competes for complex formation with these activators (ZAISS et al. 1999) suggesting an important regulatory role in the ubiquitin–proteasome pathway for protein breakdown.

Infectious pathogens provide an ideal opportunity to view natural mechanisms designed to block proteasome function. Viruses for example, may benefit from down-regulation of proteasome function, allowing endogenously synthesized viral proteins to evade processing into peptide fragments, a necessary process to elicit an immune response. The human immunodeficiency virus (HIV) Tat protein can function as a transcription factor and was shown to be stimulated by two of the ATPase subunits of the 26S proteasome. This observation prompted further biochemical studies on the interaction of Tat with the proteasome (TSUBUKI et al. 1994). These studies lead to the finding that Tat was able to bind and inhibit the 20S proteasome core, thereby preventing the formation of complexes with the PA28 subunits, a requirement for antigen presentation (SEEGER et al. 1997). Tat also exerts a small activation effect on the assembled 26S proteasome necessary for recognition and clearance of ubiquitinated protein substrates. Therefore inhibition by Tat leads to a specific blockade of the antigen presentation pathway without perturbing other critical functions of the proteasome.

6 Proteasome Inhibitors as Affinity Labels

Small molecules are often designed to target a single enzyme thereby allowing analysis of its function through inhibitor studies. A lack of specificity usually plagues such designs, where absolute specificity of the reagent is often difficult to determine and inhibitors may effectively block unidentified enzymes. Small molecule inhibitors that are capable of covalently attaching themselves to their targets in an activity-dependent manner can be used as affinity labeling reagents to circumvent these difficulties. Affinity labeling techniques utilize the small molecule as a probe rather than an inhibitor. Upon modification, a target protein bound to a labeled probe can be resolved and labeling intensity used to determine the activity of that target protease. Multiple targets can be assessed simultaneously by this approach, including species that may not have been identified previously. Thus, labeling provides a direct indication of the global reactivity of the inhibitor in total cellular extracts. Furthermore, indirect visualization of binding by unlabeled compounds can be accomplished through competition with labeled probes for the active site of a given target of interest. Several groups have taken advantage of chemically tagged suicide inhibitors to create several classes of useful new affinity probes for studying proteasome function.

Both lactacystin and epoxomicin are covalent, specific inhibitors of the proteasome that have been chemically synthesized, and therefore can be easily converted into a labeled form. For lactacystin, attachment of a tritium atom in place of hydrogen yielded an affinity label that could identify its cellular target (FENTEANY et al. 1995). Similarly, epoxomicin was chemically converted into an affinity label by attachment of a biotin moiety (SIN et al. 1999). Both of these labeled compounds were crucial for identification of the proteasome as their primary protein target (FENTEANY et al. 1995; MENG et al. 1999b). In addition to the obvious utility of these reagents for target identification, both classes of affinity probes can also be used to rapidly monitor proteasome activity under different physiological conditions. Tritium-labeled lactacystin however, is difficult and expensive to synthesize, so its full potential as a probe will probably never be realized. On the other hand, biotin-epoxomicin is relatively easily produced and is likely to find increased use as a reagent for specific monitoring of proteasome activity.

In addition to the natural product affinity labels, several other groups have generated labeled versions of known synthetic inhibitors of the proteasome. A ^{14}C-labeled analog of the coumarin inhibitor 3,4-DCI was synthesized (ORLOWSKI et al. 1997) and used to identify the proteasome subunits that it modifies. Pre-treating proteasomes with peptide aldehyde inhibitors prior to labeling with probe allowed assignment of subunits targeted by these reagents. Similar studies made use of labeled peptide diazo-methylketones and chloro-methylketones (REIDLINGER et al. 1997), but unlike the results from earlier studies (BOGYO et al. 1997), subunits that do not posses the catalytic N-terminal threonine residue were found to be labeled. This unexpected result is probably due to non-specific alkylations resulting from long incubations with high concentrations of purified enzymes using highly reactive labeled electrophiles. Consequently, careful consideration must be made when choosing compounds and conditions for affinity labeling studies of the proteasome.

Peptide vinyl sulfones have also been developed as affinity labeling probes of the proteasome (BOGYO et al. 1997, 1998; NAZIF et al. 2000; KESSLER et al. 2001). These compounds are simple peptide structures in which the C-terminal carboxylic acid is replaced by the electrophilic vinyl sulfone electrophile. Simple attachment of a nitro-phenol or phenol moiety to a number of possible sites on the peptide backbone creates compounds that could be labeled with radioactive iodine. A variety of peptide sequences were used to produce probes that covalently label each of the six catalytically active β-subunits of the proteasome. Resolution of labeled proteins using either one- or two-dimensional gel electrophoresis permitted visualization of both overall levels of proteasomal activity as well as activity of individual subunits. This method therefore represents a considerable advantage over commonly used fluorogenic peptide substrates that provide information regarding only overall levels of proteasome activity.

These affinity labeling techniques highlight an often overlooked feature of covalent protease inhibitors, namely that specificity and potency are not always essential factors to be considered when designing new reagents. A compound having modest activity for a desired target that can be easily synthesized and modified for label attachment, for some applications, can be more effective than more potent counterparts that lack these features.

7 Kinetic Studies of the Proteasome Using Inhibitors

7.1 Studies of Catalytic Mechanism

The complexity resulting from the proteasome's multiple active sites makes conventional strategies of kinetic analysis ineffective. Most biochemical studies of the proteasome therefore rely heavily on inhibitors to help decipher these multiple proteolytic events. This section will discuss some of the uses for proteasome inhibitors in kinetic studies of peptide hydrolysis.

The use of the isocoumarin compound 3,4-DCI as a proteasome inhibitor led to the identification of additional proteasomal proteolytic activities distinct from the chymotrypsin-like, trypsin-like and PGPH activities (CARDOZO et al. 1992; ORLOWSKI et al. 1993; PEREIRA et al. 1992). The hydrolysis of certain protein substrates was accelerated in the presence of this isocoumarin inhibitor (PEREIRA et al. 1992), and products generated by DCI-treated proteasomes resulted in a majority of cleavages after branched aliphatic amino acids. This distinct DCI-resistant activity was given the name branched chain amino acid preferring (BrAAP) activity (ORLOWSKI et al. 1993). Several small fluorogenic substrates designed to mimic the polypeptides produced from DCI-treated proteasomes include Cbz–Gly–Pro–Ala–Leu–Gly–p-aminobenzoate and Cbz–Gly–Pro–Ala–Leu–Ala–p-aminobenzoate. These substrates are cleaved by the BrAAP activity (after the leucine residue), and therefore can be used to effectively monitor this activity (ORLOWSKI et al. 1993). A fifth activity known as small neutral amino acid preferring (SNAAP), which cleaves the same BrAAP-like substrates but has cleavage preference for Gly–Gly and Ala–Gly bonds was also identified. In contrast to the BrAAP activity, SNAAP is sensitive to DCI and a variety of other thiol reagents, but like BrAAP activity is insensitive to treatment with the peptide aldehyde Z–LLF–H (ORLOWSKI et al. 1993).

Only three catalytic subunits have been found to exist in the proteasome core (GROLL et al. 1997). As five proteolytic activities have now been observed and no evidence exists for additional catalytically active proteasomal subunits, these newly defined BrAAP and SNAAP activities may result from hydrolysis by combined activity of the three active subunits. A series of elegant kinetic experiments analyzed BrAAP activity in the presence of peptide aldehydes that contained a P1 branched aliphatic or aromatic residue (MCCORMACK et al. 1998). The branched aliphatic P1 aldehydes exhibited simple inhibition kinetics with respect to the BrAAP substrate whereas the aromatic substrates revealed a bi-phasic or partial inhibition of the BrAAP activity. Simple kinetic inhibition of the BrAAP activity correlated directly with compounds that show similar activity against both the chymotrypsin-like and PGPH activities, and the bi-phasic or partial inhibitors of BrAAP inhibited specifically the chymotrypsin-like activity. These findings combined with mutational studies in yeast (DICK et al. 1998) indicate that the BrAAP and SNAAP activities are not distinct activities but rather a combination of the chymotrypsin-like and PGPH activities of the proteasome.

7.2 Analysis of Substrate Specificity

Traditionally, the amino acid reside found at the site of hydrolysis defines each proteolytic activity of the proteasome. However, this classification appears to be oversimplified. The ability of peptide aldehyde inhibitors containing hydrophobic P1 residues to block all three of the major proteolytic activities of the proteasome suggests that substrate specificity is regulated by multiple factors. A series of detailed kinetic and biochemical studies using inhibitors of varied peptide sequences are beginning to define the specificity elements used by the proteasome to determine how protein substrates are processed.

Examination of the catalytic mechanism for the proteasome was accomplished by kinetic analysis using several peptide reporters. A model based on these studies proposed that the 20S proteasome is a dynamic structure with multiple conformers having least two cooperative sites for hydrolysis of the chymotrypsin-like substrate (STEIN et al. 1996). More extensive analysis revealed a non-linear dependence between steady-state velocity and inhibitor concentration, as the result of the peptide aldehyde inhibitor, Ac-LLnL-H, binding at multiple active sites in the complex. This study also suggested a model for substrate specificity in which active sites can bind substrates with diverse P1 residues leading to hydrolysis of a single substrate by more than one proteasomal active site. This model also proposed that the P1 residue of a substrate has a relatively minor role in defining cleavage sites on a substrate.

Modified peptides represent valuable tools for determining inhibitor specificity. Relatively large numbers of sequences can be synthesized to incorporate a desired electrophile creating substrates that can then be used to directly monitor binding to the proteasome's active sites or to indirectly monitor inhibition of proteolysis. Studies of extended peptide aldehydes showed that alteration in the P4 and P5 positions of inhibitors had, in some cases, a more profound effect on inhibitor potency than changes to the P1 position. These results suggest that positions distal to the site of amide bond hydrolysis represent a second critical binding determinant (T. Akopian, B. Gilbert, R.R. Rando, and A.L Goldberg, unpublished results). The importance of this distal P4 binding site was confirmed by direct labeling analysis using peptide vinyl sulfones (BOGYO et al. 1998). Addition of an aromatic P4 residue to the tri-leucine-containing core peptide vinyl sulfone dramatically changed its subunit labeling profile indicating the importance of this position for recognition by the proteasome's multiple active sites (BOGYO et al. 1998). Furthermore, these studies performed with peptide vinyl sulfones showed that changes in modification of individual catalytic β-subunits can be correlated with changes in inhibitory peptide sequences. Thus information can be obtained regarding primary sequence specificity of each catalytic subunit. Similar findings using tetra-peptide α',β'-epoxyketones further established that regulation of substrate specificity requires positions distal to the P1 site (ELOFSSON et al. 1999).

While these studies represent a step towards the characterization of proteasomal substrate processing, a more systematic approach is needed to better define absolute substrate specificity. Extending this affinity labeling approach to include inhibitors that address the contributions of each of the possible 20 amino acids to binding of each of the three proteasomal active sites has recently been accomplished (NAZIF et al. 2000). These studies will help lead to a better understanding of how the proteasome is able to perform the highly controlled process of protein breakdown. Furthermore, information from this study has led to the design of inhibitors that target a single subunit of the proteasome. These reagents are likely to be of particular importance for biochemical studies of proteasome function.

8 Proteasome Inhibitors as Therapeutic Agents

A review of proteasome inhibitors is incomplete without mention of the possible uses of proteasome inhibitors as therapeutic agents. The proteasome is an enticing target for chemical intervention especially in view of its essential role in cellular physiology. Processing and activation of the transcription factor NF-κB, a proteasome substrate that has implications in inflammation, makes it an ideal system to assess the role of proteasome inhibitors as anti-inflammatory agents (PALOMBELLA et al. 1994).

Proteasome-mediated cyclin degradation is required for initiation of mitosis, and this poses yet another role for inhibitors as anti-cancer agents (GLOTZER et al. 1991). It is not surprising that the biotech industry has taken a keen interest in inhibitor efficacy towards inflammation and cancer. Proscript (now Millennium

Pharmaceuticals) is applying peptide boron acid proteasome inhibitors to models for arthritis and delayed type hypersensitivity (ADAMS and STEIN 1996). Preliminary experiments using oral delivery of inhibitors has yielded promising results. However, questions still remain about the benefit of targeting an enzyme that is central to so many processes required for cell survival. Continued studies should uncover any benefits to proteasome inhibition as a means for therapeutic intervention.

9 Summary

As the dominant protease dedicated to protein turnover, the proteasome shapes the cellular protein repertoire. Our knowledge of proteasome regulation and activity has improved considerably over the past decade. Novel inhibitors, in particular, have helped to advance our understanding of proteasome biology. They range from small peptide-based structures that can be modified to vary target specificity, to large macromolecular inhibitors that include proteins. While these reagents have played an important role in establishing our current knowledge of the proteasome's catalytic mechanism, many questions remain. Rapid advances in the synthesis and identification of new classes of proteasome inhibitors over the last 10 years serve as a positive indicator that many of these questions will soon be resolved. The future lies in designing compounds that can function as drugs to target processes involved in disease progression. It may only be a short while before the products of such research have safe application in a practical setting. Structural and combinatorial chemistry approaches are powerful techniques that will bring us closer to these goals.

References

Adams J, Stein R (1996) Novel Inhibitors of the proteasome and their therapeutic use in inflammation. Annu Rep Med Chem 31:279–288

Adams J, Behnke M, Chen S, Cruickshank AA, Dick LR, Grenier L, Klunder JM, Ma YT, Plamondon L, Stein, RL (1998). Potent and selective inhibitors of the proteasome: dipeptidyl boronic acids. Bioorg Med Chem Lett 8:333–338

Barrett AJ, Rawlings ND, Woessner JF (1998) Handbook of proteolytic enzymes. Academic, London

Bogyo M, McMaster JS, Gaczynska M, Tortorella D, Goldberg AL, Ploegh H (1997) Covalent modification of the active site threonine of proteasomal β subunits and the Escherichia coli homolog HslV by a new class of inhibitors. Proc Natl Acad Sci USA 94:6629–6634

Bogyo M, Shin S, McMaster JS, Ploegh HL (1998) Substrate binding and sequence preference of the proteasome revealed by active-site-directed affinity probes. Chem Biol 5:307–320

Bromme D, Klaus JL, Okamoto K, Rasnick D, Palmer JT (1996) Peptidyl vinyl sulphones: a new class of potent and selective cysteine protease inhibitors: S2P2 specificity of human cathepsin O2 in comparison with cathepsins S and L. Biochem J 315:85–89

Cardozo C, Vinitsky A, Hidalgo MC, Michaud C, Orlowski M (1992) A 3,4-dichloroisocoumarin-resistant component of the multicatalytic proteinase complex. Biochemistry 31:7373–7380

Chatterjee S, Dunn D, Mallya S, Ator MA (1999) P'-extended α-ketoamide inhibitors of proteasome. Bioorg Med Chem Lett 9:2603–2606

Chu-Ping M, Slaughter CA, DeMartino GN (1992) Purification and characterization of a protein inhibitor of the 20S proteasome (macropain). Biochim Biophys Acta 1119:303–311

Coux O, Tanaka K, Goldberg AL (1996) Structure and functions of the 20S and 26S proteasomes. Annu Rev Biochem 65:801–847

Craiu A, Gaczynska M, Akopian T, Gramm CF, Fenteany G, Goldberg AL, Rock KL (1997) Lactacystin and clasto-lactacystin β-lactone modify multiple proteasome β-subunits and inhibit intracellular protein degradation and major histocompatibility complex class I antigen presentation. J Biol Chem 272:13437–13445

Dahlmann B, Kuehn L, Rutschmann M, Reinauer H (1985) Purification and characterization of a multicatalytic high-molecular mass proteinase from rat skeletal muscle. Biochem J 228:161–170

Dick LR, Cruikshank AA, Grenier L, Melandri FD, Nunes SL, Stein RL (1996) Mechanistic studies on the inactivation of the proteasome by lactacystin: a central role for *clasto*-lactacystin β-lactone. J Biol Chem 271:7273–7276

Dick LR, Cruikshank AA, Destree AT, Grenier L, McCormack TA, Melandri FD, Nunes SL, Palombella VJ, Parent LA, Plamondon L, Stein RL (1997) Mechanistic studies on the inactivation of the proteasome by lactacystin in cultured cells. J Biol Chem 272:182–188

Dick TP, Nussbaum AK, Deeg M, Heinemeyer W, Groll M, Schirle M, Keilholz W, Stevanovic S, Wolf DH, Huber R, Rammensee HG, Schild H (1998) Contribution of proteasomal beta-subunits to the cleavage of peptide substrates analyzed with yeast mutants. J Biol Chem 273:25637–25646

Elofsson M, Splittgerber U, Myung J, Mohan R, Crews CM (1999) Towards subunit-specific proteasome inhibitors: synthesis and evaluation of peptide α',β'-epoxyketones. Chem Biol 6:811–822

Emori Y, Tsukahara T, Kawasaki H, Ishiura S, Sugita H, Suzuki K (1991) Molecular cloning and functional analysis of three subunits of yeast proteasome. Mol Cell Biol 11:344–353

Fenteany G, Standaert RF, Reichard GA, Corey EJ, Schreiber SL (1994) A β-lactone related to lactacystin induces neurite outgrowth in a neuroblastoma cell line and inhibits cell cycle progression in an osteosarcoma cell line. Proc Natl Acad Sci USA 91:3358–3362

Fenteany G, Standaert RF, Lane WS, Choi S, Corey EJ, Schreiber SL (1995) Inhibition of proteasome activities and subunit-specific amino-terminal threonine modification by lactacystin. Science 268:726–731

Fujiwara T, Tanaka K, Kumatori A, Shin S, Yoshimura T, Ichihara A, Tokunaga F, Aruga R, Iwanaga S, Kakizuka A et al. (1989) Molecular cloning of cDNA for proteasomes (multicatalytic proteinase complexes) from rat liver: primary structure of the largest component (C2). Biochemistry 28:7332–7340

Gardner RC, Assinder SJ, Christie G, Mason GG, Markwell R, Wadsworth H, McLaughlin M, King R, Chabot-Fletcher MC, Breton JJ, Allsop D, Rivett AJ (2000) Characterization of peptidyl boronic acid inhibitors of mammalian 20S and 26S proteasomes and their inhibition of proteasomes in cultured cells. Biochem J 2:447–454

Glotzer M, Murray AW Kirschner, MW (1991) Cyclin is degraded by the ubiquitin pathway. Nature 349:132–138

Goldberg AL, Rock KL (1992) Proteolysis, proteasomes and antigen presentation. Nature 357:375–379

Groll M, Ditzel L, Lowe J, Stock D, Bochtler M, Bartunik HD, Huber R (1997) Structure of 20S proteasome from yeast at 2.4Å resolution. Nature 386:463–471

Groll M, Kim KB, Kairies N, Huber R, Crews CM (2000) Crystal structure of epoxomicin: 20S proteasome reveals a molecular basis for selectivity of α',β'-epoxyketone proteasome inhibitors. J Am Chem Soc 122:1237–1238

Groll M, Koguchi Y, Huber R, Kohno J (2001) Crystal structure of the 20S proteasome: TMC-95A complex: a non-covalent proteasome inhibitor. J Mol Biol 311:543–548

Guo GG, Gu M, Etlinger JD (1994) 240-kDa proteasome inhibitor (CF-2) is identical to δ-aminolevulinic acid dehydratase. J Biol Chem 269:12399–12402

Hanada M, Sugawara K, Kaneta K, Toda S, Nishiyama Y, Tomita K, Yamamoto H, Konishi M, Oki T (1992) Epoxomicin, a new antitumor agent of microbial origin. J Antibiot (Tokyo) 45:1746–1752

Harding CV, France J, Song R, Farah JM, Chatterjee S, Iqbal M, Siman R (1995) Novel dipeptide aldehydes are proteasome inhibitors and block the MHC-I antigen-processing pathway. J Immunol 155:1767–1775

Harper JW, Powers JC (1985) Reaction of serine proteases with substituted 3-alkoxy-4-chloro-isocoumarins and 3-alkoxy-7-amino-4-chloroisocoumarins: new reactive mechanism-based inhibitors. Biochemistry 24:7200–7213

Harper JW, Hemmi K, Powers JC (1985) Reaction of serine proteases with substituted isocoumarins: discovery of 3,4-dichloroisocoumarin, a new general mechanism based serine protease inhibitor. Biochemistry 24:1831–1841

Iqbal M, Chatterjee S, Kauer JC, Das M, Messina P, Freed B, Biazzo W, Siman R (1995) Potent inhibitors of proteasome. J Med Chem 38:2276–2277

Iqbal M, Chatterjee S, Kauer JC, Mallamo JP, Messina PA, Reiboldt A, Siman R (1996) Potent α-ketocarbonyl and boronic ester derived inhibitors of proteasome. Bioorg Med Chem Lett 6:287–290

Ishiura S, Yamamoto T, Nojima M, Sugita H (1986) Ingensin, a fatty acid-activated serine proteinase from rat liver cytosol. Biochim Biophys Acta 882:305–310

Kam CM, Fujikawa K, Powers JC (1988) Mechanism-based isocoumarin inhibitors for trypsin and blood coagulation serine proteases: new anticoagulants. Biochemistry 27:2547–2557

Kessler BM, Tortorella D, Altun M, Kisselev AF, Fiebiger E, Hekking BG, Ploegh HL, Overkleeft HS (2001) Extended peptide-based inhibitors efficiently target the proteasome and reveal overlapping specificities of the catalytic beta-subunits. Chem Biol 8:913–929

Koguchi Y, Kohno J, Suzuki S, Nishio M, Takahashi K, Ohnuki T, Komatsubara S (1999) TMC-86A, B and TMC-96, new proteasome inhibitors from *Streptomyces* sp. TC 1084 and *Saccharothrix* sp. TC 1094. I. Taxonomy, fermentation, isolation, and biological activities. J Antibiot (Tokyo) 52:1069–1076

Koguchi Y, Kohno J, Nishio M, Takahashi K, Okuda T, Ohnuki T, Komatsubara S (2000a) TMC-95 A, B, C, and D, novel proteasome inhibitors produced by *Apiospora montagnei Sacc.* TC 1093. Taxonomy, production, isolation, and biological activities. J Antibiot (Tokyo) 53:105–109

Koguchi Y, Kohno J, Suzuki S, Nishio M, Takahashi K, Ohnuki T, Komatsubara S (2000b) TMC-86 A, B and TMC-96, new proteasome inhibitors from *Streptomyces* sp. TC 1084 and *Saccharothrix* sp. TC 1094. II. Physico-chemical properties and structure determination. J Antibiot (Tokyo) 53:63–65

Kohno J, Koguchi Y, Niskio M, Nakao K, Kuroda M, Shimizu R, Ohnuki T, Komatsubara S (2000) Structures of TMC-95A-D: novel proteasome inhibitors from Apiospora montagnei Sacc. TC 1093. J Org Chem 65:990–995

Lee DH, Goldberg AL (1996) Selective inhibitors of the proteasome-dependent and vacuolar pathways of protein degradation in Saccharomyces cerevisiae. J Biol Chem 271:27280–27284

Li XC, Gu MZ, Etlinger JD (1991) Isolation and characterization of a novel endogenous inhibitor of the proteasome. Biochemistry 30:9709–9715

Li XS, Etlinger JD (1992) Ubiquitinated proteasome inhibitor is a component of the 26S proteasome complex. Biochemistry 31:11964–11967

Loidl G, Groll M, Musiol HJ, Ditzel L, Huber R, Moroder L (1999a) Bifunctional inhibitors of the trypsin-like activity of eukaryotic proteasomes. Chem Biol 6:197–204

Loidl G, Groll M, Musiol HJ, Huber R, Moroder L (1999b) Bivalency as a principle for proteasome inhibition. Proc Natl Acad Sci USA 96:5418–5422

Loidl G, Musiol HJ, Groll M, Huber R, Moroder L (2000) Synthesis of bivalent inhibitors of eucaryotic proteasomes. J Pept Sci 6:36–46

Lowe J, Stock D, Jap B, Zwickl P, Baumeister W, Huber R (1995) Crystal structure of the 20S proteasome from the archaeon *T. acidophilum* at 3.4Å resolution. Science 268:533–539

Lynas JF, Harriott P, Healy A, McKervey MA, Walker B (1998) Inhibitors of the chymotrypsin-like activity of proteasome based on di- and tri-peptidyl α-keto aldehydes (glyoxals). Bioorg Med Chem Lett 8:373–378

McCormack T, Baumeister W, Grenier L, Moomaw C, Plamondon L, Pramanik B, Slaughter C, Soucy F, Stein R, Zuhl F, Dick L (1997) Active site-directed inhibitors of Rhodococcus 20S proteasome. Kinetics and mechanism. J Biol Chem 272:26103–26109

McCormack TA, Cruikshank AA, Grenier L, Melandri FD, Nunes SL, Plamondon L, Stein RL, Dick LR (1998) Kinetic studies of the branched chain amino acid preferring peptidase activity of the 20S proteasome: development of a continuous assay and inhibition by tripeptide aldehydes and *clasto*-lactacystin β-lactone. Biochemistry 37:7792–7800

McCutchen-Maloney SL, Matsuda K, Shimbara N, Binns DD, Tanaka K, Slaughter CA, DeMartino G (2000) cDNA cloning, expression, and functional characterization of PI31, a proline-rich inhibitor of the proteasome. J Biol Chem 275:18557–18565

McGuire MJ, DeMartino GN (1986) Purification and characterization of a high molecular weight proteinase (macropain) from human erythrocytes. Biochim Biophys Acta 873:279–289

Meng L, Kwok BH, Sin N, Crews CM (1999a) Eponemycin exerts its antitumor effect through the inhibition of proteasome function. Cancer Res 59:2798–2801

Meng L, Mohan R, Kwok BH, Elofsson M, Sin N, Crews CM (1999b) Epoxomicin, a potent and selective proteasome inhibitor, exhibits in vivo antiinflammatory activity. Proc Natl Acad Sci USA 96:10403–10408

Nazif T, Bogyo M (2000) Global analysis of proteasomal substrate specificity using positional-scanning libraries of covalent inhibitors. Proc Natl Acad Sci USA 2001 98:2967–2972

Nojima M, Ishiura S, Yamamoto T, Okuyama T, Furuya H, Sugita H (1986) Purification and characterization of a high-molecular-weight protease, ingensin, from human placenta. J Biochem (Tokyo) 99:1605–1611

Oikawa T, Hasegawa M, Shimamura M, Ashino H, Murota S, Morita I (1991) Eponemycin, a novel antibiotic, is a highly powerful angiogenesis inhibitor. Biochem Biophys Res Commun 181:1070–1076

Omura S, Fujimoto T, Otoguro K, Matsuzaki K, Moriguchi R, Tanaka H, Sasaki Y (1991a) Lactacystin, a novel microbial metabolite, induces neuritogenesis of neuroblastoma cells. J Antibiot (Tokyo) 44:113–116

Omura S, Matsuzaki K, Fujimoto T, Kosuge K, Furuya T, Fujita S, Nakagawa A (1991b) Structure of lactacystin, a new microbial metabolite which induces differentiation of neuroblastoma cells. J Antibiot (Tokyo) 44:117–118

Orlowski M, Wilk S (1981) A multicatalytic protease complex from pituitary that forms enkephalin and enkephalin containing peptides. Biochem Biophys Res Commun 101:814–822

Orlowski M, Michaud C (1989) Pituitary multicatalytic proteinase complex. Specificity of components and aspects of proteolytic activity. Biochemistry 28:9270–9278

Orlowski M, Cardozo C, Michaud C (1993) Evidence for the presence of five distinct proteolytic components in the pituitary multicatalytic proteinase complex. Properties of two components cleaving bonds on the carboxyl side of branched chain and small neutral amino acids. Biochemistry 32:1563–1572

Orlowski M, Cardozo C, Eleuteri AM, Kohanski R, Kam CM, Powers JC (1997) Reactions of [^{14}C]-3,4-dichloroisocoumarin with subunits of pituitary and spleen multicatalytic proteinase complexes (proteasomes). Biochemistry 36:13946–13953

Ostrowska H, Wojcik C, Omura S, Worowski K (1997) Lactacystin, a specific inhibitor of the proteasome, inhibits human platelet lysosomal cathepsin A-like enzyme. Biochem Biophys Res Commun 234:729–732

Palmer JT, Rasnick D, Klaus JL, Bromme D (1995) Vinyl sulfones as mechanism-based cysteine protease inhibitors. J Med Chem 38:3193–3196

Palombella VJ, Rando OJ, Goldberg AL, Maniatis T (1994) The ubiquitin-proteasome pathway is required for processing the NF-κB1 precursor protein and the activation of NF-κB. Cell 78:773–785

Pereira ME, Nguyen T, Wagner BJ, Margolis JW, Yu B, Wilk S (1992) 3,4-dichloroisocoumarin-induced activation of the degradation of β-casein by the bovine pituitary multicatalytic proteinase complex. J Biol Chem 267:7949–7955

Powers JC, Kam CM (1994) Isocoumarin inhibitors of serine peptidases. Methods Enzymol 244:442–457

Reidlinger J, Pike AM, Savory PJ, Murray RZ, Rivett AJ (1997) Catalytic properties of 26S and 20S proteasomes and radiolabeling of MB1, LMP7, and C7 subunits associated with trypsin-like and chymotrypsin-like activities. J Biol Chem 272:24899–24905

Rivett AJ (1985) Purification of a liver alkaline protease which degrades oxidatively modified glutamine synthetase. Characterization as a high molecular weight cysteine proteinase. J Biol Chem 260:12600–12606

Rivett AJ (1993) Proteasome: multicatalytic proteinase complexes. Biochem J 291:1–10

Rock KL, Gramm C, Rothstein L, Clark K, Stein R, Dick L, Hwang D, Goldberg AL (1994) Inhibitors of the proteasome block the degradation of most cell proteins and the generation of peptides presented on MHC class I molecules. Cell 78:761–771

Savory PJ, Djaballah H, Angliker H, Shaw E, Rivett AJ (1993) Reaction of proteasomes with peptidylchloromethanes and peptidyldiazomethanes. Biochem J 296:601–605

Seeger M, Ferrell K, Frank R, Dubiel W (1997) HIV-1 tat inhibits the 20S proteasome and its 11S regulator-mediated activation. J Biol Chem 272:8145–8148

Seemuller E, Lupas A, Stock D, Lowe J, Huber R, Baumeister W (1995) Proteasome from *Thermoplasma acidophilum*: a threonine protease. Science 268:579–582

Sin N, Meng L, Auth H, Crews CM (1998) Eponemycin analogues: syntheses and use as probes of angiogenesis. Bioorg Med Chem 6:1209–1217

Sin N, Kim KB, Elofsson M, Meng L, Auth H, Kwok BH, Crews CM (1999) Total synthesis of the potent proteasome inhibitor epoxomicin: a useful tool for understanding proteasome biology. Bioorg Med Chem Lett 9:2283–2288

Sorimachi H, Tsukahara T, Kawasaki H, Ishiura S, Emori Y, Sugita H, Suzuki K (1990) Molecular cloning of cDNAs for two subunits of rat multicatalytic proteinase. Existence of N-terminal conserved and C-terminal diverged sequences among subunits. Eur J Biochem 193:775–781

Spaltenstein A, Leban JJ, Huang JJ, Reinhardt KR, Viveros OH, Sigafoos J, Crouch R (1996) Design and synthesis of novel protease inhibitors. Tripeptide α',β'-epoxyketones as nanomolar inactivators of the proteasome. Tet Lett 37:1343–1346

Stein RL, Melandri F, Dick L (1996) Kinetic characterization of the chymotryptic activity of the 20S proteasome. Biochemistry 35:3899–3908

Stock D, Ditzel L, Baumeister W, Huber R, Lowe J (1995) Catalytic mechanism of the 20S proteasome of Thermoplasma acidophilum revealed by X-ray crystallography. Cold Spring Harb Symp Quant Biol 60:525–532

Sugawara K, Hatori M, Nishiyama Y, Tomita K, Kamei H, Konishi M Oki T (1990) Eponemycin, a new antibiotic active against B16 melanoma. I. Production, isolation, structure and biological activity. J Antibiot (Tokyo) 43:8–18

Tanaka K, Ii K, Ichihara A, Waxman L, Goldberg AL (1986) A high molecular weight protease in the cytosol of rat liver. I. Purification, enzymological properties, and tissue distribution. J Biol Chem 261:15197–15203

Tsubuki S, Saito Y, Kawashima S (1994) Purification and characterization of an endogenous inhibitor specific to the Z-Leu-Leu-Leu-MCA degrading activity in proteasome and its identification as heatshock protein 90. FEBS Lett 344:229–233

Vinitsky A, Cardozo C, Sepp-Lorenzino L, Michaud C, Orlowski M (1994) Inhibition of the proteolytic activity of the multicatalytic proteinase complex (proteasome) by substrate-related peptidyl aldehydes. J Biol Chem 269:29860–29866

Wagner BJ, Margolis JW, Abramovitz AS (1986) The bovine lens neutral proteinase comprises a family of cysteine-dependent proteolytic activities. Curr Eye Res 5:863–868

Wilk S, Orlowski M (1980) Cation-sensitive neutral endopeptidase: isolation and specificity of the bovine pituitary enzyme. J Neurochem 35:1172–1182

Wilk S, Orlowski M (1983a) Evidence that pituitary cation-sensitive neutral endopeptidase is a multicatalytic protease complex. J Neurochem 40:842–849

Wilk S, Orlowski M (1983b) Inhibition of rabbit brain prolyl endopeptidase by *n*-benzyloxycarbonyl-prolyl-prolinal, a transition state aldehyde inhibitor. J Neurochem 41:69–75

Wilk S, Figueiredo-Pereira ME (1993) Synthetic inhibitors of the multicatalytic proteinase complex (proteasome). Enzyme Protein 47:306–313

Zaiss DM, Standera S, Holzhutter H, Kloetzel P, Sijts AJ (1999) The proteasome inhibitor PI31 competes with PA28 for binding to 20S proteasomes. FEBS Lett 457:333–338

Zolfaghari R, Baker CR Jr, Canizaro PC, Amirgholami A, Behal FJ (1987) A high-molecular-mass neutral endopeptidase-24.5 from human lung. Biochem J 241:129–135

Subject Index

A

AAA ATPase   2, 3, 6–8, 10–15, 34, 35
- archaeal and bacterial   34, 35
Ac-LLnL-H   193, 202
actinomycete   2, 3, 9, 10
affinity probes   193, 199, 200, 203
allostery   82
δ-aminolevulinic acid dehydratase   198
amphipathic helix   158
anaphase-promoting complex (APC)   8, 16, 17
- phosphorylation   153, 157
- regulators   153, 157
- substrates   153
- subunits   144, 145, 152
angiogenesis   197
anti-cancer agents   203
antigen presentation   83, 84, 199
- evolution   110
antizyme (AZ)   164, 165
APC (*see* anaphase-promoting complex)
apoptosis inhibitors   142, 146
Arabidopsis   7, 17
ARC   10, 35
archaebacterial proteasomes   193
arginine finger   15
armadillo repeat motif (ARM)   8, 17
arthritis   204
ATPases of the AAA subfamily   2, 3, 6–8, 10–15
19S ATPase   27, 35
19S ATPase regulator   73, 78, 79
avidin   197
AZ (*see* antizyme)

B

base subcomplex   3, 11–13, 15, 16, 18
biotin label   197, 200
biotin-epoxomicin   197, 200
bleomycin hydrolase   33
budding yeast   176–179
- RPN11 protein   178, 179
- SRP1 protein   179
- STS1 protein   179
BUL1,2   147

C

C2 domain   147
calpain inhibitor   188
19S cap   3, 8, 11, 13, 14, 17
cathepsin A   197
c-Cbl   145, 155
CDC4   150, 157
CDC5   140, 153
CDC20   153, 157
CDC48   34, 35, 162, 163
CDC53 (*see* cullins)
CDH1   153, 157
CDK (*see* cyclin-dependent kinase)
cell cycle   138–140, 145, 149–153, 156, 157
centrosome   181
Cephalon   190
c-fos   140, 143
CFTR (cystic fibrosis transmembrane conductance regulator)   181
chaperones   5, 7, 159, 163, 165, 182
chaperonins   7
c-jun   140, 165
clasto-lactacystin beta-lactone   196
Clp   2, 3, 6, 7, 14
ClpA   34
ClpAP   2
ClpP   6
ClpX   14, 34
ClpXP   2
coiled coil   11
conformational changes   80, 81
conjugation   46, 64
COP9   45, 58, 59
- signalosome   3, 8, 17, 18
covalent
- adducts   191, 192, 200
- modification   191, 192
CP (20S proteolytic core of the proteasome)   49–52
CTL (*see* cytotoxic T-lymphocytes)
cullins   144, 149–152
cyclic peptides   198

Subject Index

cyclin(s)   139, 140, 155, 145, 150–153
– degradation   203
cyclin-dependent kinase (CDK)   139, 150, 151
– inhibitor   144, 149–151
cyclosome   8, 16
cystic fibrosis transmembrane conductance regulator (CFTR)   181
cytotoxic T-lymphocytes (CTL)
– responses   94
– – to hepatitis B virus   94
– – to human immunodeficiency virus   94
– – in inbread mice   94

D

3,4-DCI (see 3,4 dichloroisocoumarin)
defective ribosomal products (DRIPs)   95
Deg1   158
delayed type hypersensitivity   204
DER3 (see HDR1)
destruction box   153
deubiquitination   54, 64
DFP (see di-isopropyl fluorophosphate)
3,4 dichloroisocoumarin   191, 200
– 14C-labeling   200
Dictyostelium   8
di-isopropyl fluorophosphate (DFP)   187
DNA repair   143, 147, 156, 161, 163
double attack   192
DRIPs (defective ribosomal products)   95
dynein   181

E

E1 (see ubiquitin-activating enzyme)
E2 (see ubiquitin-conjugating enzyme)
E3 (see ubiquitin protein ligase)
E3α (see UBR1)
E6   143, 147, 155
E6-associated protein (E6-AP)   143, 144, 147, 156
eIF3   3, 8, 17, 18
endocytosis   149, 160–162
endoplasmatic reticulum (ER)-associated degradation   141, 156, 158, 159
eponemycin   197
epoxide   192
epoxomicin   196, 197, 200
Escherichia coli   3, 5
eukaryotes, subunit differentiation   7–9

F

F-box protein   144, 149–151, 153, 157
fission yeast   176–180
– Cut8 protein   178, 179
– Cut15 protein   179
– *pad1+* gene   179
fizzy-related   153

Frankia   9
FtsH   2, 6, 7, 14

G

Giardia   7, 8, 17
green fluorescent protein (GFP)   34, 177, 180
GRR1   150, 157

H

HCT1 (see CDH1)
HECT E3s   143, 145, 147–149, 156, 162, 164
hemi-acetal   189
hemi-thioacetal   189
HHR23A (see RAD23)
HIV (see human immunodeficiency virus)
HLA (see human leukocyte antigens)
HLA-B27   95
HRD1   156, 159
HRT1 (see RBX1)
Hsc73   31
HslU   2, 3, 7, 14
HslV   2, 3, 6, 7, 9, 25
HSP100   2, 3
human immunodeficiency virus (HIV)   94, 95, 151, 199
human leukocyte antigens (HLA)   94
hybrid proteasome   84
hydrophobic surfaces, recognition by the ubiquitin system   154, 157–159

I

IκBα phosphorylation, degradation   140, 150, 151, 163
immunoproteasome(s)   9, 33, 84, 85, 104
importin α 178
independent degradation   47
inflammation   203
influenza virus nucleoprotein   182
γ-interferon   3, 9–11, 25, 26, 33, 83, 84, 104
iodio-acetic acid   187

K

KEKE motif   16
KEN box   153
Ki antigen   10, 74

L

lactacystin   195–197, 200
– tritium labeling   195, 200
lactam ring   195
leucine aminopeptidase   33, 125
leucine rich repeat (LRR)   150
Leupeptin   186, 188
Lid   3, 11, 12, 16–18
Liddle's syndrome   147
LMP (low molecular weight polypeptide)   104

Lon 2, 6, 7, 14
low molecular weight polypeptide (LMP) 104
– deficient cells 105
– knockout mice 105
LRR (*see* leucine rich repeat)

M

major histocompatibility complex (MHC) 9, 33
– class I molecules 83, 93
– – allele-specific peptide motifs 94
– – peptide ligands 93, 98
– – processing and presentation pathway 92
maleolyl-β-alanine 193
mammalian cells, localization 180–182
MATα2 158
MCB1 (*see* RPN10)
Mdm2 155, 156
MET30 150, 157
Methanococcus jannaschii 24, 34, 35
MG-115 188
MG-132 188, 190, 193
MHC (*see* major histocompatibility complex)
Michael addition 192
microtubule organizing centre (MTOC) 181
mitosis 203
MMS2 143, 156, 161
MPN/JAB 3, 8, 18
MTOC (*see* microtubule organizing centre)
multiubiquitin chains 139, 160, 162
– geometry 143, 161, 162, 165
– hydrophobic patches 160
– recognition by the proteasome 160–162, 164, 165
Mycobacterium 2, 9, 10
Myogenics 188

N

Nedd4 147–149
N-end rule 153, 154, 164
N-ethyl maleimide (NEM) 187
NF-kappa-B transcription factor processing 203
nitro-phenol 201
non-ATPase subunits 12, 15–17
N-terminal
– nucleophile hydrolases 2–5
– threonine 187
Ntn hydrolases 2–5, 28
nuclear factor (NF)-κB 140, 150, 151
nuclear transport 163

O

ornithin decarboxylase (ODC) 164, 165

P

p53 140, 143, 147, 155, 156
p97 (*see also* CDC48) 34, 35
PA26 3, 11, 74
– activator 27
PA28 (proteasome activator 28; *see also* 11S regulator) 3, 10, 11, 16, 27, 60, 74, 114, 199
PA700 199
PAN 14, 34, 35
PCI/PINT 3, 8, 18
PEST motif 151
phenol 201
phenylmethanesulfonyl fluoride 187
phosphorylation as prerequisite for ubiquitylation 140, 146, 151, 155
PI31 199
P-loop NTPases 14, 15
PML (promyelocytic leukemia oncoprotein) 180
presenilin-1 181
processing, proteasomal 140, 163
processivity 85
product release 84, 85
promyelocytic leukemia oncoprotein (PML) bodies 180
Proscript 188, 190, 203
proteasomal
– functions, regulating 46–49
– processing 140, 163
proteasome(s)
– activator 28 (*see* PA28)
– affinity labeling 200, 203
– affinity purification 200
– archaebacterial 193
– binding 47, 54, 56, 64
– catalytic mechanism 186, 188, 201
– channel, gating 50, 51
– cleavage
– – site(s)
– – – altering 52
– – – usage in polypeptides and proteins 97, 99, 107, 114
– – specificity 96, 97
– cloning of subunits 187
– composition
– – changes 61
– – diversity 61–63
– – 11S composition 176
– – 19S composition 176
– – 20S composition 175
– generation of epitopes 108, 115, 119, 126
– hybrid 84
– hydrolysis activities
– – branched chain amino acid preferring (BrAAP) 201, 202

proteasome(s)
- - chymotrypsin-like activity   186, 193, 194, 197, 201
- - post-glutamyl peptide hydrolyzing activity   186, 201
- - small neutral amino acid preferring (SNAAP)   201, 202
- - trypsin-like activity   186, 193, 194, 201
- immunological functions   91–128
- immunoproteasome   104, 176, 180
- inhibitors   96
- - anti-inflammatory agents   190, 203
- - bi-functional peptide aldehydes   193
- - bi-functional synthetic inhibitors   193, 194
- - chloromethyl ketones   200
- - clinical studies   190
- - commercially available   188
- - cross-reactivity   190, 196
- - diazomethyl ketones   192, 200
- - endogenous protein inhibitors   198
- - α', β' epoxyketones   192, 197, 203
- - ethylene glycol scaffolds   194
- - extended binding elements   190
- - irreversible synthetic inhibitors   190, 191
- - isocoumarins   187
- - α-keto
- - - aldehydes   190, 192
- - - amides   190, 192
- - natural product inhibitors   195, 196
- - organic mercurials   187
- - peptide aldehydes   186, 188–190, 192, 194, 200, 202, 203
- - peptide α-keto carbonyls   190, 192
- - peptide boron esters   189, 190
- - peptide vinyl sulfones   192, 200, 201, 203
- - reversible synthetic inhibitors   188, 189
- - S′ spanning inhibitors   190
- - screening of microbial extracts   195
- - subunit specific inhibitors   203
- - as therapeutic agents   203, 204
- kinetic studies   196, 201, 202
- length of degradation products   101
- molecular
- - evolution   1–18
- - ruler   101
- peptidase specificities   96
- processive protein degradation   102
- structure aided inhibitor studies   193, 194, 197
- substrates   95
- - processing   48
- - specificity   186, 202
- - unfolding   49, 55
- three-dimensional structure   193, 194, 197

20S proteasome   23–36
- activation   80
- active sites   80
- activity   73–85
- α annulus   80, 81
- assembly   29–31
- catalytic mechanism   28
- degradation products   31–34
- inactive conformation   79
- inhibitors   82
- occurence   23–25
- processing   29–31
- proteolytic activity   31–34
- structural features   25–28
- structure   73, 79
- subunit composition   23–25
20S proteasome/11S regulator crystal structure   80
26S proteasome   27, 33
- regulating   43–66
20S proteolytic core complex   2–10, 49–52, 198
PS-341   190
puromycin-sensitive aminopeptidase   33
PY motif   147

R
RAD6   154, 156, 161
RAD23   147, 163
radioactive iodine labeling   193, 201
RBX1   150, 151
RecA   3, 14, 15
11S regulator (see also PA28)   3, 10, 11, 16, 73–85, 114
- binding affinity   81
- biological roles   83, 85
- crystal structure   76, 78
- functional domains   78
- heptamer   75–77
- hexamer   75
- oligomeric state   75
- stochastic model   76
- tissue distribution   74
19S regulator   23, 27, 34
19S regulatory particle of the proteasome (RP)   52–54
regulatory   74
- complexes   27, 57–61
- particle   54–57
Rhodococcus   3, 6, 9
- erythropolis   25, 30, 31, 35
RING finger
- domain   145, 146
- proteins
- - dimerization   146
- - downregulation   146, 156, 157
- - as ubiquitin ligases   149–156

ROC1 (see RBX1)
RP (19S regulatory particle of the proteasome) 52–54
RPN10   161–163
RSP5   147–149, 156

S
S5a (see RPN10)
SCF complex
- regulation of activity   157
- substrates   150, 151
- subunits   144, 149, 150
self-compartmentalization   5–7
self-tolerance   94, 110, 112
sensor-1   15
sevenfold symmetry   81
SH2 domain   152, 155
Siah-1   155, 156, 163
SKP1   144, 150, 151, 157
SOCS box   151, 152
soft base nucleophiles   192
ssr A   34
Streptomyces   9
stress response   138, 141, 142, 147, 157–159
substrate binding pocket   193
suicide inhibitors   190, 191, 200
superoxide dismutase   182

T
TAP (transporter associated with antigen processing)   94
Tap transporter   9
tat protein   199
TFIIH   8, 16
Thermoplasma   2–5, 13
- *acidophilum*   24–30, 32–35
thimet oligopeptidase   33
thioester   195
TMC-95   196, 197
TPR repeat   150, 152
transcriptional regulators
- activation by the proteasome   140, 147, 163
- degradation by the proteasome   138, 140, 150, 154, 155, 165
transporter associated with antigen processing (TAP)   94
- peptide sequence specificity   94
- transported peptides   94
tricorn   33, 34
tripeptidylpeptidase II   34
Trypanosoma   8, 11, 17
two-dimensional gel electrophoresis   201

U
UBC4   142, 150, 152, 155, 157, 161
UBC5   142, 150, 152, 157, 161
UBC6   158, 159
UBC7   157–159
UBC13   143, 156, 161
UbcH5   155
UbcH7   155
ubiquitin
- activation   138, 141
- functions unrelated to the proteasome   143, 147, 149, 156, 160–163
- fusion degradation, pathway and components   161–164
- hydrolases (see deubiquitination)
- lysine residues   138, 139, 143–146
- protein ligase (E3)   138, 139, 143–146
- – association with the proteasome   164
- – downregulation   156, 157
- recognition by the proteasome   160, 161
ubiquitin-activating enzyme (E1)   138, 141, 146
ubiquitin-conjugating enzyme (UBC, E2)   138, 141, 142, 146
- association with the proteasome   164
- dimerization   142, 143, 161
- variants (UEVs)   142, 143, 161
ubiquitin-related proteins   142, 150, 163
UBR1   144, 145, 153, 154, 156, 161, 164
Ump1   34
unfoldase activity   3, 15
unfolded protein response (UPR)   141, 159

V
VAT   34, 35
VCB complex (see *von-Hippel-Lindau* tumor suppressor protein)
VCP (see CDC48)
VHL (see *von-Hippel-Lindau* tumor suppressor protein)
vimentin   181
viral proteins as adaptors for E3s   147, 150, 151
von Willebrand factor type A domain (VWA)   8, 16
von-Hippel-Lindau tumor suppressor protein (VHL)   144, 151, 152

W
WD40 motif   150, 153
WW motif   147

Y
yeast
- localization   176–179
- proteasomes   193

Z
Z-IE(OtBu)AL-H   188
Z-LLF-H   202

Current Topics in Microbiology and Immunology

Volumes published since 1989 (and still available)

Vol. 225: **Vogt, Peter K.; Mahan, Michael J.** (Eds.): Bacterial Infection: Close Encounters at the Host Pathogen Interface. 1998. 15 figs. IX, 169 pp. ISBN 3-540-63260-3

Vol. 226: **Koprowski, Hilary; Weiner, David B.** (Eds.): DNA Vaccination/Genetic Vaccination. 1998. 31 figs. XVIII, 198 pp. ISBN 3-540-63392-8

Vol. 227: **Vogt, Peter K.; Reed, Steven I.** (Eds.): Cyclin Dependent Kinase (CDK) Inhibitors. 1998. 15 figs. XII, 169 pp. ISBN 3-540-63429-0

Vol. 228: **Pawson, Anthony I.** (Ed.): Protein Modules in Signal Transduction. 1998. 42 figs. IX, 368 pp. ISBN 3-540-63396-0

Vol. 229: **Kelsoe, Garnett; Flajnik, Martin** (Eds.): Somatic Diversification of Immune Responses. 1998. 38 figs. IX, 221 pp. ISBN 3-540-63608-0

Vol. 230: **Kärre, Klas; Colonna, Marco** (Eds.): Specificity, Function, and Development of NK Cells. 1998. 22 figs. IX, 248 pp. ISBN 3-540-63941-1

Vol. 231: **Holzmann, Bernhard; Wagner, Hermann** (Eds.): Leukocyte Integrins in the Immune System and Malignant Disease. 1998. 40 figs. XIII, 189 pp. ISBN 3-540-63609-9

Vol. 232: **Whitton, J. Lindsay** (Ed.): Antigen Presentation. 1998. 11 figs. IX, 244 pp. ISBN 3-540-63813-X

Vol. 233/I: **Tyler, Kenneth L.; Oldstone, Michael B. A.** (Eds.): Reoviruses I. 1998. 29 figs. XVIII, 223 pp. ISBN 3-540-63946-2

Vol. 233/II: **Tyler, Kenneth L.; Oldstone, Michael B. A.** (Eds.): Reoviruses II. 1998. 45 figs. XVI, 187 pp. ISBN 3-540-63947-0

Vol. 234: **Frankel, Arthur E.** (Ed.): Clinical Applications of Immunotoxins. 1999. 16 figs. IX, 122 pp. ISBN 3-540-64097-5

Vol. 235: **Klenk, Hans-Dieter** (Ed.): Marburg and Ebola Viruses. 1999. 34 figs. XI, 225 pp. ISBN 3-540-64729-5

Vol. 236: **Kraehenbuhl, Jean-Pierre; Neutra, Marian R.** (Eds.): Defense of Mucosal Surfaces: Pathogenesis, Immunity and Vaccines. 1999. 30 figs. IX, 296 pp. ISBN 3-540-64730-9

Vol. 237: **Claesson-Welsh, Lena** (Ed.): Vascular Growth Factors and Angiogenesis. 1999. 36 figs. X, 189 pp. ISBN 3-540-64731-7

Vol. 238: **Coffman, Robert L.; Romagnani, Sergio** (Eds.): Redirection of Th1 and Th2 Responses. 1999. 6 figs. IX, 148 pp. ISBN 3-540-65048-2

Vol. 239: **Vogt, Peter K.; Jackson, Andrew O.** (Eds.): Satellites and Defective Viral RNAs. 1999. 39 figs. XVI, 179 pp. ISBN 3-540-65049-0

Vol. 240: **Hammond, John; McGarvey, Peter; Yusibov, Vidadi** (Eds.): Plant Biotechnology. 1999. 12 figs. XII, 196 pp. ISBN 3-540-65104-7

Vol. 241: **Westblom, Tore U.; Czinn, Steven J.; Nedrud, John G.** (Eds.): Gastroduodenal Disease and Helicobacter pylori. 1999. 35 figs. XI, 313 pp. ISBN 3-540-65084-9

Vol. 242: **Hagedorn, Curt H.; Rice, Charles M.** (Eds.): The Hepatitis C Viruses. 2000. 47 figs. IX, 379 pp. ISBN 3-540-65358-9

Vol. 243: **Famulok, Michael; Winnacker, Ernst-L.; Wong, Chi-Huey** (Eds.): Combinatorial Chemistry in Biology. 1999. 48 figs. IX, 189 pp. ISBN 3-540-65704-5

Vol. 244: **Daëron, Marc; Vivier, Eric** (Eds.): Immunoreceptor Tyrosine-Based Inhibition Motifs. 1999. 20 figs. VIII, 179 pp. ISBN 3-540-65789-4

Vol. 245/I: **Justement, Louis B.; Siminovitch, Katherine A.** (Eds.): Signal Transduction and the Coordination of B Lymphocyte Development and Function I. 2000. 22 figs. XVI, 274 pp. ISBN 3-540-66002-X

Vol. 245/II: **Justement, Louis B.; Siminovitch, Katherine A. (Eds.):** Signal Transduction on the Coordination of B Lymphocyte Development and Function II. 2000. 13 figs. XV, 172 pp. ISBN 3-540-66003-8

Vol. 246: **Melchers, Fritz; Potter, Michael (Eds.):** Mechanisms of B Cell Neoplasia 1998. 1999. 111 figs. XXIX, 415 pp. ISBN 3-540-65759-2

Vol. 247: **Wagner, Hermann (Ed.):** Immunobiology of Bacterial CpG-DNA. 2000. 34 figs. IX, 246 pp. ISBN 3-540-66400-9

Vol. 248: **du Pasquier, Louis; Litman, Gary W. (Eds.):** Origin and Evolution of the Vertebrate Immune System. 2000. 81 figs. IX, 324 pp. ISBN 3-540-66414-9

Vol. 249: **Jones, Peter A.; Vogt, Peter K. (Eds.):** DNA Methylation and Cancer. 2000. 16 figs. IX, 169 pp. ISBN 3-540-66608-7

Vol. 250: **Aktories, Klaus; Wilkins, Tracy, D. (Eds.):** Clostridium difficile. 2000. 20 figs. IX, 143 pp. ISBN 3-540-67291-5

Vol. 251: **Melchers, Fritz (Ed.):** Lymphoid Organogenesis. 2000. 62 figs. XII, 215 pp. ISBN 3-540-67569-8

Vol. 252: **Potter, Michael; Melchers, Fritz (Eds.):** B1 Lymphocytes in B Cell Neoplasia. 2000. XIII, 326 pp. ISBN 3-540-67567-1

Vol. 253: **Gosztonyi, Georg (Ed.):** The Mechanisms of Neuronal Damage in Virus Infections of the Nervous System. 2001. approx. XVI, 270 pp. ISBN 3-540-67617-1

Vol. 254: **Privalsky, Martin L. (Ed.):** Transcriptional Corepressors. 2001. 25 figs. XIV, 190 pp. ISBN 3-540-67569-8

Vol. 255: **Hirai, Kanji (Ed.):** Marek's Disease. 2001. 22 figs. XII, 294 pp. ISBN 3-540-67798-4

Vol. 256: **Schmaljohn, Connie S.; Nichol, Stuart T. (Eds.):** Hantaviruses. 2001, 24 figs. XI, 196 pp. ISBN 3-540-41045-7

Vol. 257: **van der Goot, Gisou (Ed.):** Pore-Forming Toxins, 2001. 19 figs. IX, 166 pp. ISBN 3-540-41386-3

Vol. 258: **Takada, Kenzo (Ed.):** Epstein-Barr Virus and Human Cancer. 2001. 38 figs. IX, 233 pp. ISBN 3-540-41506-8

Vol. 259: **Hauber, Joachim, Vogt, Peter K. (Eds.):** Nuclear Export of Viral RNAs. 2001. 19 figs. IX, 131 pp. ISBN 3-540-41278-6

Vol. 260: **Burton, Didier R. (Ed.):** Antibodies in Viral Infection. 2001. 51 figs. IX, 309 pp. ISBN 3-540-41611-0

Vol. 261: **Trono, Didier (Ed.):** Lentiviral Vectors. 2002. 32 figs. X, 258 pp. ISBN 3-540-42190-4

Vol. 262: **Oldstone, Michael B. A. (Ed.):** Arenaviruses I. 2002, 30 figs. XVIII, 197 pp. ISBN 3-540-42244-7

Vol. 263: **Oldstone, Michael B. A. (Ed.):** Arenaviruses II. 2002, 49 figs. XVIII, 268 pp. ISBN 3-540-42705-8

Vol. 264/I: **Hacker, Jörg; Kaper, James B. (Eds.):** Pathogenicity Islands and the Evolution of Microbes. 2002. 34 figs. XVIII, 232 pp. ISBN 3-540-42681-7

Vol. 264/II: **Hacker, Jörg; Kaper, James B. (Eds.):** Pathogenicity Islands and the Evolution of Microbes. 2002. 24 figs. XVIII, 228 pp. ISBN 3-540-42682-5

Vol. 265: **Dietzschold, Bernhard; Richt, Jürgen A. (Eds.):** Protective and Pathological Immune Responses in the CNS. 2002. 21 figs. X, 278 pp. ISBN 3-540-42668-X

Vol. 266: **Cooper, Koproski (Eds.):** The Interface Between Innate and Acquired Immunity. 2002. 15 figs. XIV, 116 pp. ISBN 3-540-42894-1

Vol. 267: **Mackenzie John S.; Barrett Alan D. T.; Deubel, Vincent (Eds.):** Japanese Encephalitis and West Nile Viruses. 2002. 66 figs. X, 418 pp. ISBN 3-540-42783-X

Printing (Computer to Film): Saladruck Berlin
Binding: Stürtz AG, Würzburg

BEETHOVEN REMEMBERED

The Biographical Notes of Franz Wegeler and Ferdinand Ries

Foreword by Christopher Hogwood
Introduction by Eva Badura-Skoda

GREAT OCEAN PUBLISHERS
ARLINGTON, VIRGINIA

Translated from the German
Biographische Notizen über Ludwig van Beethoven (1838, 1845)
by Frederick Noonan

Notes by Dr. Alfred C. Kalischer
from his edition of the text (2nd edition, Berlin, 1906)
translated by Frederick Bauman and Tim Clark

Illustrations reproduced through the courtesy and with permission
of the sources cited in the list of Illustrations.

Book and cover design by M. Esterman.

All rights reserved. No part of this book may be used or reproduced in any
manner whatsoever without written permission from the publishers.

Copyright © 1987 by Great Ocean Publishers, Inc.
1823 North Lincoln Street
Arlington, VA 22207

First Printing
Library of Congress Cataloging in Publication Data

Wegeler, Franz Gerhard, 1765-1848.
 [Biographische Notizen über Ludwig van Beethoven. English]
 Beethoven remembered.

 Includes index.
  1. Beethoven, Ludwig van, 1770-1827. 2. Composers —
Austria — Biography. I. Ries, Ferdinand, 1784-1838. II. Title.
ML410.B4W3413   1987   780'.92'4 [B]   85-10043
ISBN 0-915556-15-4

Printed and bound in Canada

Contents

List of Illustrations vi

Foreword by Christopher Hogwood ix

Introduction by Eva Badura-Skoda xi

Part I by Franz Gerhard Wegeler (1838) 1

Part II by Ferdinand Ries (1838) 63

Supplement by Franz Gerhard Wegeler (1845) . . . 144

Notes by Alfred C. Kalischer 168

Index 193

Index of Musical Compositions 198

Illustrations

1 Beethoven's silhouette at age 16, by Joseph Neesen, 1786 — frontispiece of the original edition *(Beethoven-Haus, Bonn, H.C. Bodmer Collection)* — *frontispiece*

2 Sketch of Beethoven by J.D. Böhm, c. 1820 *(Beethoven-Haus, Bonn)* — viii

3 Anton Schindler (1798-1864) *(Bild-Archiv der Österreichischen Nationalbibliothek, Vienna)* — xv

4 Autograph letter from Ferdinand Ries to Carl Czerny, 1830 *(Archiv der Gesellschaft der Musikfreunde, Vienna)* — xix

5 Franz Gerhard Wegeler (1765-1848) *(Bild-Archiv der Österreichischen Nationalbibliothek, Vienna)* — 1

6 Beethoven's birthplace (Bonngasse), pencil drawing by R. Beissel, 1889 *(Bild-Archiv der Österreichischen Nationalbibliothek, Vienna)* — 12

7 The von Breuning family, silhouette, 1782 *(Beethoven-Haus, Bonn)* — 15

8 The von Breuning family house, Munsterplatz, Bonn, watercolor by M. Frickel *(Stadtarchiv und Wissenschaftliche Stadtbibliothek, Bonn)* — 16

9 Christian Gottlob Neefe (1748-1798), engraving by Liebe after a drawing by J.G. Rosenberg *(Archiv der Gesellschaft der Musikfreunde, Vienna)* — 18

10 Count Ferdinand von Waldstein (1762-1823), anonymous silhouette from Beethoven's scrapbook *(Bild-Archiv der Österreichischen Nationalbibliothek, Vienna)* — 19

11 Stephan von Breuning (1774-1827), portrait by Franz Gerhard von Kugelgen *(Beethoven-Haus, Bonn)* — 30

12 Beethoven's portrait (1814) engraved by Blasius Höfel from a pencil drawing by Louis Letronne *(Archiv der Gesellschaft der Musikfreunde, Vienna)* — 47

13 Excerpt from Beethoven's letter to Eleonore von Breuning, 2 Nov. 1793 *(from the original German edition)* — 54

14 Elector Maximilian Franz (1756-1801), anonymous oil portrait *(Beethoven-Haus, Bonn)* — 56

15 Ferdinand Ries (1784-1838), anonymous oil portrait *(Beethoven-Haus, Bonn)* — 63

16	Heiligenstadt c. 1821, watercolor by T.D. Raulino *(Historisches Museum der Stadt, Vienna)*	67
17	Nikolaus Simrock (1752-1833), lithograph by Weber *(Stadtarchiv und Wissenschaftliche Stadtbibliothek, Bonn)*	78
18	Prince Karl Lichnowsky (1756-1814) *(Hradec u Opavy, Czechoslovakia)*	92
19	Princess Maria Christiane Lichnowsky (1765-1841)*(Hradec u Opavy, Czechoslovakia)*	92
20	Friedrich Heinrich Himmel (1765-1814) *(Archiv der Gesellschaft der Musikfreunde, Vienna)*	97
21	Archduke Rudolph (1788-1831), engraving by Blasius Höfel after a painting by Adalbert Suchy *(Beethoven-Haus, Bonn)*	99
22	Johann van Beethoven (1740-1792), the composer's father, *(Bild-Archiv der Österreichischen Nationalbibliothek, Vienna)*	110
23	Excerpt from Beethoven's letter to Ferdinand Ries, 20 Jan. 1816 *(from the original German edition)*	122
24	Karl van Beethoven, Beethoven's nephew (1806-1858) *(Bild-Archiv der Österreichischen Nationalbibliothek, Vienna)*	124
25	Excerpt from Beethoven's letter to Ferdinand Ries, 25 April 1823 *(from the original German edition)*	136
26	Nikolaus Johann van Beethoven, Beethoven's youngest brother (1776-1848). Oil painting by Leopold Gross *(Bild-Archiv der Österreichischen Nationalbibliothek, Vienna)*	139
27	Ferdinand Ries, lithograph by L. Lehmann *(Archiv der Gesellschaft der Musikfreunde, Vienna)*	142
28	Musical score of Beethoven's "Die Klage", WoO 113 *(from the original German edition)*	143
29	Beethoven's birthplace (Bonngasse), photograph *(Bild-Archiv der Österreichischen Nationalbibliothek, Vienna)*	147
30	The Fischer house (Rheingasse), Beethoven family residence, drawing *(Bild-Archiv der Österreichischen Nationalbibliothek, Vienna)*	147
31	Miniature portrait of Beethoven on ivory by Christian Hornemann, 1803 *(Beethoven-Haus, Bonn, H.C.Bodmer Collection)*	160
32	Beethoven's study in the Schwarzspanierhaus, lithograph from a sepia drawing by J.N. Hochle, 1827 *(Historisches Museum der Stadt, Vienna)*	165
33	Musical score of Beethoven's "Empfindungen bei Lydiens Untreue", WoO 132 *(from the original German edition)*	166

Sketch of Beethoven by J.D. Böhm, c. 1820

Foreword
by Christopher Hogwood

A significant watershed has been passed recently in the relationship between music and musical biography. We now exist in the post-*Amadeus* period. In earlier times, marked by large volumes of collected correspondence and three-volume studies in depth, the attitude was "rearrange the composer's music if you like, but present the true account of the man." The new age, marked by cinematic afflatus and theatrical fictions, prefers an *Urtext* performance, but rearranges the life.

The change has gone unnoticed by much of the musical public, who were blissfully unaware of lacunae in the documentation of great historical figures, or drowsily assumed the existence of an omniscient college of musicologists monitoring and rationing the flow of propaganda necessary to maintain *The Four Seasons* and Albinoni's *Adagio* at the top of the annual charts.

It will come as something of a shock, therefore, to discover that one of the prime sources of information on Beethoven as seen by a contemporary and colleague has, until now, been unavailable to the English reader. Still more surprising will be the singularity of this most significant human document, produced at a time when it was unprecedented to write of Beethoven with insight, but without the tinge of mythology.

Although extracts from these *Notizen* have appeared in earlier biographies, this new translation of the complete account of Wegeler and Ries carries more immediacy — the immediacy we associate with Beethoven's own conversation books, uncensored by later adulation — and an informality which in itself is a remarkable literary accomplishment.

Ries in particular comes over as a warm and approachable human being, lacking the spite and side of a Schindler, and quite prepared to admit amusedly to his errors in the face of 'modern' music. Like so many of us who dare not admit it so gracefully, he too was fooled the first time he heard the "false entry" of the horn in the *Eroica*. But he also gives us nuggets of psychological insight — Beethoven's improvisations were at their best "especially when he was in a good mood or was irritated" — and pertinent information for the performing musician that can never be found in more portentous tomes — the sparing use of rubato, the effect of a slight rallentando during a crescendo, or the liberating information that Beethoven added octaves to the rondo theme in the C major piano concerto, "to make it more brilliant."

This is Beethoven the Man rather than Beethoven the Myth. To approach the personality behind the music via this unusual memoir can only add to every reader's listening pleasure.

Introduction
by Eva Badura-Skoda

The publication of the *Biographische Notizen über Ludwig van Beethoven* by Franz Gerhard Wegeler and Ferdinand Ries in an English translation is long overdue. Indeed, it is hard to believe that this book has not been published in its entirety for so long. Until now, only excerpts have appeared in the Beethoven books of H. E. Krehbiel and O. G. Sonneck. The value of the recollections of Wegeler is clearly beyond question. Wegeler was Beethoven's close friend during his difficult adolescent years in Bonn and during two of his happy first years in Vienna, a friend in whom Beethoven never lost confidence and who remained in contact with him throughout his life. Of even greater value than the recollections of Wegeler are those of the well-trained musician Ries, Beethoven's only acknowledged pupil besides Archduke Rudolf. Both authors' *Notizen* are comparable in their documentary value to the reports of Haydn's interviewers Griesinger and Dies; and like these they rightly continue to serve as an indispensable source of information for all biographers of the great composer.

Apart from the fact that the accounts of Wegeler and Ries have supplied and will always supply Beethoven biographers with documentary material, these recollections are fascinating reading for every music lover. As A. C. Kalischer pointed out in the foreword to his German edition, Robert

Schumann told a friend immediately after its appearance: "I shall lend you the book; one cannot stop reading it."

Twenty years ago when I started to teach music history at American universities, I was so surprised to discover that the *Notizen* of Wegeler and Ries had never been published in English that I commissioned a translation of it for my own use, not for publication. Besides, others in the field of Beethoven scholarship also were aware of this lack, including English and American colleagues who I thought were in a better position to supply the English speaking world with this urgently needed little book. Since many years have passed with the book still unavailable, I welcomed the invitation of Great Ocean Publishers to write an Introduction to the English edition which they were preparing for publication, especially since it includes the translation of the footnotes of the German Beethoven scholar Kalischer.

The commentaries given in Kalischer's footnotes (indicated by the bold superscripts in the text) help to rectify some memory slips of the two authors. There are some additional minor inaccuracies as the learned reader may notice, but they are of no real importance — diminishing neither the vividness of the picture of Beethoven's personality, nor the trustworthiness of the accounts of these two faithful friends and admirers of Beethoven's genius. The interested reader may be referred to the writings of Stephan Ley (*Wahrheit, Zweifel und Irrtum in der Kunde von Beethoven's Leben*, Wiesbaden 1955), to Clemens Brenneis (*"Das Fischhof-Manuskript: Zur Fruehgeschichte der Beethoven-Biographik,"* in *Zu Beethoven*, ed. by H. Goldschmidt,

Berlin/DDR, 1979), to the valuable edition of nearly 500 letters from and to Ries (*Ferdinand Ries: Briefe und Dokumente*, ed. by Cecil Hill, *Veroeffentlichungen des Stadtarchivs Bonn*, vol. 27, Bonn, 1982) and — last but certainly not least — to the highly relevant article of Alan Tyson in *Nineteenth Century Music* (vol. 7, 3 April 1984), *"Ferdinand Ries (1784-1838): The History of his Contribution to Beethoven Biography."* In this valuable study, which admirably summarizes the latest research results, Tyson explains for example how it could have happened that Ries erred with regard to the year of his arrival in Vienna (he did not come to Vienna prior to 1801), and that this and some other dating errors were committed *bona fide*. As the prefaces to both parts of the *Notizen* show, the authors' principal aim was to provide accurate data for future studies of Beethoven's life, and they were very concerned indeed about the truthfulness of their reports.

Wegeler can be trusted when he says in his preface: "I feel it necessary to mention that I have set down only what I know for certain." An eminent doctor with a scientific mind, *Rector Magnificus* of the University of Bonn at the age of 29, Wegeler was born in 1765, five years before Beethoven. His relationship to his greatly admired friend in his teenage years reminds one of that between Josef von Spaun and Schubert. After two years of studies in Vienna he returned to the Rhineland, but since he could not pursue his university career because of political reasons he settled in Koblenz and practiced there as a medical doctor, occupying in later years some official positions as well. He married Eleonore von Breuning (who might have been Beethoven's first real love). For more information on Wegeler the interested reader may be referred to the book of Stephan Ley, *Beethoven als Freund der Familie Wegeler-von Breuning*, Bonn, 1927.

In the article on Ries mentioned above, Tyson also explains why a span of ten years passed between Beethoven's death and the publication of the *Notizen* of Wegeler and Ries. This had to do with an assertion of Schindler, Beethoven's "factotum," secretary and self-appointed "friend," who claimed in a letter to Wegeler after Beethoven's and Stephan von Breuning's deaths in 1827 that Beethoven had commissioned him to write his biography together with Stephan von Breuning and Wegeler. Schindler's suggestion was that now, after von Breuning's death, Ries could take over his part in the joint venture of a Beethoven biography. Wegeler, who did not know Schindler well, at first agreed to this proposal. It was Ries who had no real interest in collaborating with Schindler, especially after recognizing that Schindler was less interested in conveying actual facts and impressions than in giving a censored picture of Beethoven according to his own limited understanding of his great master.

Since Schindler in later years made some derogatory, hostile remarks about Ries whom he obviously envied greatly (Ries, unlike Schindler, was a successful composer and performer and highly esteemed as Beethoven's pupil and friend), it seems necessary to say a few words about Schindler.

Anton Schindler was born in 1798 (not 1795 as E. Hueffler claimed) and died in 1864. He wanted the world to believe that he had been Beethoven's "private secretary without pay" almost continuously from the year 1816 (i.e., from the age of 18) on. This assertion is incorporated into most of the modern music dictionaries and encyclopedias when they state that Schindler was "in contact almost daily" with Beethoven for a full ten years. In reality Schindler appears not to have become Beethoven's "unpaid secretary" until 1819 if not later, perhaps even as late as 1822. The greatest number of the

Anton Schindler (1798-1864)

many authenticated errors in the first and second edition of his biography of Beethoven have been found to concern events which happened prior to 1820. From 1822 until May 1823 Schindler lived in the same house as Beethoven. Afterwards, however, their relationship broke off completely, and it was not until December 1826, when Beethoven returned from Gneixendorf to Vienna, already a sick man, that he became reconciled with Schindler and once again entrusted him with secretarial duties. Furthermore Schindler appears to have helped to a considerable extent in caring for Beethoven throughout the serious illness leading up to his death three months later.

Some years ago, when D. W. MacArdle's translation of Schindler's biography of Beethoven appeared in the United States, there were a number of Beethoven scholars who justifiably voiced their concern about the possible harm this new edition might do because the editor had neglected to emphasize sufficiently Schindler's questionable character and his tendency even to forge evidence if he considered it 'necessary' or advisable. This may have been the main reason why he had destroyed some 260 of the approximately 400 conversation notebooks of Beethoven which came into his possession after Beethoven's death. That several entries in the conversation notebooks were indeed forgeries of Schindler, written after 1840, has been proven only rather recently (1978); but the suspicion that Schindler is not trustworthy had been voiced since Nottebohm's times and even earlier; I myself wrote in 1975 in *Notes* (vol. 32, Sept. 1975, p. 47): "Schindler definitely was a neurotic and not always truthful person." Ferdinand Hiller, who as a student of Hummel was allowed to visit Beethoven, gave in his memoirs (p. 46 f.) a lively picture of this man:

> Schindler . . . had brought us to Beethoven, and from then on we saw him rather frequently. The fact that the great master was able to associate with this knight of the most woeful countenance day after day for a number of years, that he did not turn him out, can only be due to the fact that at that time contact with the out-

side world had become a matter of indifference to him and that he had need of an intelligent servant. I do not by any means deny that Schindler had a knowledge of music and certain intellectual abilities — but his personality was as spare as his figure and as dry as his facial features. It is certain that he was of service to Beethoven in many ways — but it is just as certain that no friendship in the world was ever exploited more cleverly and to greater advantage. In later years, on occasions when Schindler was visited by a music lover whom he wanted to honor or please particularly, he would appear wearing an unattractive dressing gown which Beethoven had once worn threadbare, thus presenting, without realizing it, a most accurate picture of his relation to the great man who had merely tolerated him.

Hiller also relates that Beethoven shortly before his death asked his friend Hummel to give a concert for Schindler's benefit, as he himself was unable to do so. And there is indeed no reason to doubt that Beethoven felt indebted to Schindler for the many services he had been unable or unwilling to pay him for, and he may have been very grateful for Schindler's apparently devoted care for him during the days of his final illness. Schindler's biography of Beethoven will always remain an interesting source of information on Beethoven, but far less reliable than the *Notizen* of Beethoven's real friends Wegeler and Ries.

Schindler's unkind remarks about Ferdinand Ries were made in later years and probably caused by Ries's lack of interest in collaborating with Schindler in the biography project. Only after Wegeler had freed himself from any feelings of obligation towards Schindler (he may have even started to doubt Schindler's claim that Beethoven wanted Schindler to write part of his biography) did Ries develop a serious interest in the project.

Years ago and some months before the scholarly world learned about Schindler's forged entries in Beethoven's conversation notebooks, I had commissioned Austria's most gifted and best-known graphologist Robert Muckenschnabel, who had studied 18th and 19th century handwritings in addition to modern ones, to analyze the handwriting of various friends of Beethoven and Schubert, among them Schindler and Ries. Prior to the invention of typewriters, the handwriting of cultivated persons was distinctively personal. Dr. Muckenschnabel, who did not know anything about the writers of the letters he had examined, wrote a shockingly negative character analysis of Schindler (published in my article *"Zum Charakterbild Anton Schindlers,"* Oesterreichische Musikzeitschrift, vol. 32, p. 241 f.) which, however, corroborates surprisingly the recently discovered evidence. The examination of Ries's handwriting (based on letters preserved in the Archiv der Gesellschaft der Musikfreunde in Wien and dated Dec. 26, 1825; Nov. 14, 1828; April 25, 1830 and March 6, 1837) resulted in the following character analysis:

> A talented individualist, who has a clear and realistic conception of his own abilities and an enthusiastic love of beauty and noble humanistic goals. He was a reliable, extraordinarily diligent and faithful person, faithful to his friends as well as to himself and his own principles. Sensitive but also easily hurt, grateful and devoted to everyone who had helped him, he was apparently too good-natured and sometimes therefore easily taken advantage of, which may later have been balanced by self-criticism, but

also criticism of others. He suffered perhaps from the fact that he could not do as much for others as he would have liked to, but he had the courage to speak the truth, which may have made him some enemies. Orderly, self-controlled and prudent, he disliked lack of clarity ("das Unklare"). His diligence helped him to fully develop his talent and to become a self-confident, creative and noble human being with high qualities and a cultured style of living.

Letter from Ferdinand Ries to Carl Czerny, 1830

No doubt, a graphologist can misjudge handwriting — as no human being can escape the common fate of erring from time to time — but Muckenschnabel's analysis is corroborated by many other sources.

Studies of physiognomy may provide an auxiliary tool, too, for historians who attempt to understand the character qualities of a "witness" from the past; and, indeed, the pictures of Ries and Schindler included in the present volume are "telling" portraits for any perceptive observer.

Ries's father had been a friend of the Beethoven family and, as a violinist in the court orchestra in Bonn, a colleague of Beethoven for many years. Ferdinand Ries learned to admire Beethoven's musical abilities probably even before becoming his student at the age of 17. During the following four years of apprenticeship with Beethoven in Vienna he had unrivalled opportunities for observing Beethoven at work, and through the commissions he was entrusted with such as copying music or proofreading, he gained insight into many of Beethoven's opinions and experiences. It is noteworthy that he and Wegeler made similar observations on certain aspects of Beethoven's character. For instance, Beethoven's hot temper and his easy suspiciousness often led to quarrels and misunderstandings with even his best friends, but would be typically followed by exaggerated feelings of remorse. On the other hand, naïveté was also typical of Beethoven. The inclusion of the priceless story of his childlike faith in Himmel's "news-report" of the invention of a "lantern for the blind" shows that both Wegeler and Ries must have had direct reports that the event had indeed happened, otherwise they would not have included it.

As interesting as the accounts of Beethoven's human reactions are, the real value of the *Notizen* to our knowledge of Beethoven's music lies in those contributions of Ries which allow us an insight into his teacher's musical taste and those few remarks which concern the compositional process and some major works of Beethoven. May the readers of this book share the authors' admiration and love for Beethoven and his music.

Fall 1984

Franz Gerhard Wegeler (1765-1848)

Foreword

Just as these memories of Beethoven's life, jointly collected by Ries and myself, were about to be published, the first report of Ries's illness reached me on the twelfth of January. Then on the following day I was stunned to hear the painful and unexpected news that at one o'clock in the afternoon my fine friend had passed away in the arms of his wife and my son. The shock was especially severe because only a short while before I had spent a week with him and then had received a long letter from him written in the best of spirits on the twenty-eighth of December. All our friends and acquaintances would have been justified in believing he would outlive me by thirty years. Alas, the physical aches of old age are by no means the most painful. What the loss of Ries means for art all Europe knows. He was one of those few who earn their reputation with profound works which continue to live beyond their time. But reflection on this noble life dedicated to art still engenders within me a strange melancholy feeling — sorrow, actually — since we were bound together for many years by the closest friendship. His devotion to me as an older friend was exceeded perhaps only by his well-founded love for his father. I accompanied him either in person or in spirit from his earliest childhood through the unforgettable years of apprenticeship with Beethoven, the first emergence of his brilliance in Germany, Russia and England, to the many triumphs which England, Germany, and France accorded the mature artist and conductor of great music.

As a person Ries was also a man of remarkable character. He was a noble man in the full sense of the word, a man without falsehood, a man warm and generous. Ries was indeed one of the very rarest of beings, the very pattern of faithfulness and love, as son, brother, husband, father, and friend. I was privileged to know intimately his faultless nature and his genuine, fervent love for the good and beautiful, the muses of his every thought. Recollections of him are among my tenderest, though most melancholy, memories in the evening of my long life. Ries first found himself through his great master and friend Beethoven, and to him he remained faithful in unchanging devotion and gratitude until his final breath. The recording of these notes, especially those dealing with their time spent together, was the most urgent business of the last months, indeed the last weeks and days, of his life. Alas, he had no premonition of how soon he was to follow his cherished master! — And how candidly, yet tenderly, he had depicted Beethoven and his relations with him. — Indeed, he could not have created a more beautiful memorial to his master than these unadorned stories. Moreover, he has woven himself into this picture with such an amiable frankness and modesty that the future biographer of Ries must consult above all Ries's own disclosures about Beethoven if he has any desire to paint a genuine portrait of the most faithful of pupils and masters.

But it is time to put an end to this involuntary outbreak of feeling. Old age is garrulous and it is not easy to fall silent when the subject is so great and the heart is so full.

Wegeler

Coblenz, May 1838

PART ONE

Preface

When an affectionate friendship forms between young people during seven or eight years in their own hometown and develops and deepens as they mature, when as grown men these friends share their lives in a foreign capital for nearly two years and since then, though separated, have remained in close contact for thirty years, then after the death of one — especially one whose outstanding achievement in the field of science or art has secured him lasting European fame — the surviving friend may consider himself justified, even obliged, not to withhold from the world any information which might contribute to a just appreciation of that man and artist.

These few words above characterize my relationship to Ludwig van Beethoven. I was born in Bonn in 1765. In 1782 I became acquainted with the twelve-year-old youth, already a composer, and I remained in constant and closest communication with him until September 1787 when I entered school in Vienna to complete my medical studies.[1] When I returned in October 1789 our same cordial relationship continued until Beethoven later departed for Vienna towards the end of 1792.

I followed in October 1794.* Here we were reunited once more with the same undiminished warmth of feeling and hardly a day passed when we did not meet.

In the middle of the year 1796 I returned to Bonn and we began to exchange letters. During that period of great stress this exchange could hardly be called very active. However, there was little necessity for a lively correspondence on either side since we were kept otherwise informed of each other's changing destinies — he through Simrock's business letters and through my letters to my brother-in-law Stephen von Breuning as well as to friends and colleagues of mine to whom I had introduced Beethoven; and I through the same sources and through letters from Ferdinand Ries.

The following notes will furnish the reader proof that an unbroken exchange of letters was not in the least necessary to sustain our affectionate relations.

I feel it necessary to mention that I have set down in the following pages only what I know for certain. Consequently a

---

* As Rector of the University of Bonn I had signed the decree of the Academic Senate intending to stop the spread of hospital fever by forbidding the students to visit those Frenchmen captured at Quesnoi, Landrecies, Condé, etc., then being transported to Austria. Alms intended for the prisoners were to be brought to them by certain clergymen. The *Moniteur* received this decree some ten or twelve days later and without a hearing branded me a rabid enemy of the Republic. Those were evil times! The people's representative in Bonn gave orders that he was to be addressed with the familiar *du* form. At that time *la queue de Robespierre* was hardly less dangerous than his head had been and it was advisable to save my own.

somewhat exaggerated insistence on providing proof of almost every statement may become apparent. This applies also to the contributions of my friend Ries. However, this also gives us the right to expect that these contributions will correct many a mistake and many unfounded or distorted details in the writings on Beethoven and may thus be considered as genuine source material in this respect.

Since I am no more than a poor dilettante as far as music is concerned, the reader will find among the statements on Beethoven as an artist only those which do not go beyond my sphere. Even when describing how Beethoven accompanied the *Lamentations* I called upon Ries's father for assistance.

Wegeler

Coblenz, May 1838

Ludwig van Beethoven's Family and His Birth

Since so many contradictory reports of Beethoven's origins, family, and date of birth have been circulated, it does not seem out of place to document these notes with accurate, though perhaps too detailed, proof.

The Cologne Electoral Court Calendar for 1760 furnishes on page nine the following information about grandfather Ludwig and father Johann van Beethoven:

Electoral Chamber, Orchestra and Court Music
. . .
Vocalists
Ludwig van Beethoven, Vocalist
. . .
Johann van Beethoven, Unpaid assistant

For the year 1761 the Court Calendar, page nine, contains the entry:

Kapellmeister
Vacant
Vocalists
Herr Ludwig van Beethoven
Johann van Beethoven, Unpaid assistant

In the Court Calendar for the year 1763, page fourteen, the former is entered as Kapellmeister and the latter as full vocalist.

The mother of our Ludwig was Maria Magdalena

Kewerich. The parish records of Ehrenbreitstein give the date of her baptism, which was usually the actual day of birth or certainly the day after.[2]

> Anno 1746, 20$^{ma.}$ Decembris, renata est *Maria Magdalena Kewerich*, Domini *Henrici Kewerich*, coqui primarii Em^{mi} et *Mariae Westorfs*[3], conjugum legitima filia.

From this we can see that she was born in Ehrenbreitstein near Coblenz, the daughter of the chef of the Elector of Trier who had his residence there. Her first marriage had been to the Elector's valet Laym:

> Anno 1763, 30$^{ma.}$ Januarii prævia dispensatione super omnibus denunciationibus de expressa licentia Em$^{mi.}$ sub vesperam in sacello apud R. R. P. P. Capucinos coram requisitis testibus . . . matrimonialiter copulati sunt praenobilis Dominus *Johannes Laym*, Em$^{mi.}$ Cubicularius et praenobilis virgo *Maria Magdalena Kewerich*, Vallensis. (i.e. from the valley).
> — Extract from the parish marriage register.

Laym died at the age of thirty, two years and ten months later:

> Anno 1763, 28va Novembris obiit . . . praenobilis Dominus *Johannes Laym*, Em$^{mi.}$ cubicularius etc.
> — Extract from the parish register of Ehrenbreitstein

This next extract is taken from the parish register of St. Remigius in Bonn.

> 12. November 1767.
> Copulavi *Johannem van Beethoven*, filium legitimum *Ludovici van Beethoven* et *Mariae Josephae Poll*,
> et
> *Mariam Magdalenam Keferich* viduam *Leym* ex

Ehrenbreitstein, filiam *Henrici Keferich* et *Annae Mariae Westorffs*.[4]

(The place in question is sometimes called Thal, sometimes Ehrenbreitstein and sometimes Thal-Ehrenbreitstein.)

The somewhat languid Bonn dialect, which is similar to that spoken in Cologne, turned Kewerich into Keferich and Laym into Leym.

To this marriage was born on 2 April 1769[5] Ludwig Maria whose godparents were grandfather Ludwig van Beethoven and Anna Maria Lohe, called Courtin.[6] This child died on the eighth of that month.

Our Ludwig was born on 17 December 1770[7] as is absolutely clear from the following document:

(Official stamp)
Administrative District of Cologne
District of Bonn.   Office of the Lord Mayor of Bonn.
Extract from the parish baptismal register of St. Remigius in Bonn, filed at the office of the Lord Mayor of Bonn.
Anno millesimo septingentesimo septuagesimo, die decima septima Decembris baptizatus est **Ludovicus**, Domini *Johannis van Beethoven* et *Helenae Keverichs*, conjugum filius legitimus: Patrini: Dominus *Ludovicus van Beethoven*, et *Gertrudis Müllers*, dicta Baums.

This extract certified correct:
Bonn, 28th June 1827.
Lord Mayor
  (Seal)   (Signed) Windeck.

Since *Lenchen* and *Lene* are diminutives for both *Magdalena* and *Helene*, the mother has obviously been mistakenly entered above as Helene instead of Magdalena.

For these extracts from the official records I am indebted partly to Father Geschwind, Vicar of Ehrenbreitstein, and partly to my childhood friend, Lord Mayor Windeck of Bonn, who spared no effort in seeking out everything that could in any way shed light on Beethoven's family and birth.

*
* *

The nonsense concocted by Fayolle and Choron that Beethoven was the off-spring of Friedrich Wilhelm II, King of Prussia[1] needs no refutation. The monarch was not in Bonn prior to Beethoven's birth nor did his mother ever leave the city during her married life.[2] This vulgar foolishness is surpassed only by the amusing ignorance of an English author who found the idea that Friedrich II should be Beethoven's father hilarious on the grounds that this king had of course died in 1740.[3]9

But in 1837 even this nonsense was, it seems difficult to believe, considerably outdone by W. van Marsdyk from Holland. This man, in a printed letter addressed *à Monsieur le*

---

[1] *Conversations-Lexicon*, fifth edition, p. 621.

[2] Beethoven expresses himself about this in his letter of 7 December 1826 [*sic*, see p. 40].

[3] "That Beethoven is a wonderful man, there can be no doubt; but if this prince were really his father, he is the greatest prodigy the world ever saw, or most likely will see again: for as Frederick II died in 1740, the period of Mad. Beethoven's gestation must in such a case have been exactly thirty years."

(*The Harmonicon*, November 1823.)

Bourgmestre de la ville de Bonn, claimed our Beethoven was Dutch. The grounds? "If he were really German, he would be called Ludwig von Rübengärten (Ludwig of the beet gardens). Beethoven had in common with Napoleon not only a general uncertainty about his birth date, but also *Lodewyk van Beethoven et Napoleone Buonaparte ont nombre égal de lettres!*"

Furthermore, old inhabitants of Zütphen remembered that Beethoven was born in a hostel built for travelling musicians which has since disappeared (though the site is now no doubt marked by a monument!). — Thus was Beethoven, recognized already as a hero on the very day of his birth, indeed "the greatest prodigy the world ever saw, or most likely will ever see again! — "

Has the reader had enough?

Quand l'absurde est outré, l'on lui fait trop d'honneur
De vouloir par raison combattre son erreur.

As for the house where Beethoven was born, it would seem in all probability to be the Graus house in the Bonngasse, the fourth on the right from the Judengässchen and bearing the number 515, opposite the present Post House. When a child is born, it is the custom for parents who have no relatives in the neighborhood to ask the neighbors to stand as baptismal witnesses. Ludwig Maria's godmother, Frau Courtin, was a neighbor from the house to the right; the godmother of our Ludwig, Frau Baums, was the next-door neighbor to the left (No. 516).

Furthermore Frau Mertens, née Lengersdorf, whose parental home was diagonally opposite the house where

Beethoven was born, remembers quite clearly that Beethoven was in fact born in this particular house.

Beethoven's birthplace

Two more sons were born after our Beethoven, the elder Caspar Anton Carl on 8 April 1774, the younger Nicolaus Johannes on 2 October 1776. — Caspar's godparents were the Minister Plenipotentiary von Belderbusch and the Abbess of Vylich, Countess von Satzenhofen. Caspar later earned his living as a piano teacher. Nicolaus studied pharmacy in the Court Pharmacy in Bonn. Both of them followed Ludwig to Vienna where they lived under the names of Carl and Johann.

Ludwig's grandfather died on 24 December 1773; his mother, 17 July 1787; his father, 18 December 1792; his brother Caspar or Carl in the last quarter of 1815. Johann is probably still alive and living in Vienna.

Beethoven's Education and First Schooling

Ludwig van Beethoven received as a young boy his first instruction in music from his father who was, as mentioned earlier, a tenor in the service of the Elector Max Friedrich residing in Bonn. The father was, however, not a pianist. His grandfather, the kapellmeister and bass singer, had once produced operas at the National Theater built by the Elector. Among the other singers at that theater were two sisters of the famous Salomon, who was born in Bonn in 1745 and died in London in 1815; the kapellmeister's son, father of our Beethoven; and a married woman named Drever, who was the sister of the old but still vigorous father Franz Ries, the elder. The grandfather was said to have been much applauded, especially in the operetta *L'Amore Artigiano (Love Among the Artisans)*[10] and in the *Deserteur von Monsigny*.

Little Louis clung with great affection to this grandfather, who as noted above, was also Beethoven's godfather, and even though he lost him so soon, he retained the most vivid, early impression of him. He was fond of talking about his grandfather with his childhood friends, and his pious and gentle mother, whom he loved much more than his very strict father, was often obliged to tell him about his grandfather. The portrait of him by the court painter Radoux was the only thing which he had sent to Vienna from Bonn and it gave him much pleasure his entire life. This grandfather was a short, sturdy man with extremely animated eyes and was highly respected as an artist.

Our Beethoven was also, as Ritter von Seyfried correctly describes him, "of sturdy build, medium height, strong-boned, full of vitality, a picture of strength."

Beethoven's education was neither particularly neglected nor especially good. He learned his reading, writing, arithmetic, and a little Latin in a public school where the present President of the County Court at Coblenz, Herr Wurzer, was a fellow pupil. Music he learned at home and he was constantly and strictly kept at it by his father. The family enjoyed no income other than the father's salary and consequently their circumstances were somewhat straitened. The father, not particularly distinguished for either his intellect or his morals, was rather strict since he hoped to find in his eldest son a help in the upbringing of the other children.

Ludwig made his first acquaintance with German literature, especially poetry, in the von Breuning family in Bonn where he also received his first introduction to social life and behavior. As this family will be mentioned repeatedly in the future, this would seem to be the proper place to describe its members and Beethoven's relations with them.

The family consisted of the mother, who was the widow of the Court Councilor von Breuning, three sons of about Beethoven's age, and one daughter. Beethoven taught the daughter as well as the youngest son, who had become an excellent pianist by the time he died in 1798 just after finishing his medical studies. The second son, Stephen von Breuning, became an Imperial Court Councilor in Vienna. He remained a loyal friend throughout Beethoven's entire life and died very soon after him. The third, Christoph von Breuning, is *Geheimer-Revisions-und Cassations-Rath* in Berlin. The daughter, Eleonora von Breuning, to whom he dedicated the *Variations No. 1*, is the wife of the author of these pages.

The von Breuning family, silhouette, 1782

An unforced atmosphere of culture reigned in the house in spite of an abundance of youthful high spirits. Christoph von Breuning experimented with little verses at an early age as did Stephen von Breuning much later, not without success. Friends of the house excelled in the sort of sociable entertainment that unites usefulness with pleasure.

If we also add that the family enjoyed a good measure of affluence, especially before the war, it is easy to understand that Beethoven's youthful exuberance first developed there.

Beethoven was soon treated as one of the children in the family and not only did he spend the greatest part of the day there but even many nights. He felt free there, he could move around with ease, and everything combined to make him cheerful and to develop his mind. Being five years older than Beethoven, I was able to observe and form a judgment on these things. The mother of the von Breuning family, who was

The von Breuning family house

born on 3 January 1750 and is living with me still, had great power over the boy, who was frequently stubborn and unsociable. Everything here can be confirmed by numerous passages in Beethoven's letters.

However, let us return to Beethoven's musical education.

Again, he received his first instruction from his father.

Later on he had much better schooling from a certain Pfeiffer, a music director and oboist subsequently engaged as a conductor in Düsseldorf for the band of the Bavarian Regiment. Pfeiffer was known as an excellent artist and extremely bright man. Beethoven owed the most to this teacher and was so grateful that through Simrock he provided him with financial support even from Vienna.

Whether he received instruction from the court organist van der Eder[11] after Pfeiffer's departure from Bonn, I do not know, but I think it most likely because I do not know of anyone else from whom he might have learned the technique of organ playing. When the famous organist Abbé Vogler played in Bonn, I was sitting at Beethoven's sickbed.

Neefe, the former musical director of the Grossmann Theater Company later employed as court organist and known as a composer, had little influence on the instruction of our Ludwig; indeed he complained about Neefe's excessively harsh criticism of his first attempts at composition.[12]

In 1785 Beethoven was engaged as organist of the Electoral Chapel by Elector Max Franz, the brother of Emperor Joseph II, where he alternated with Neefe in duties hardly strenuous. In making this appointment the Elector apparently was thinking only of Beethoven's need for support.

Ritter von Seyfried is mistaken when he states in his *Ludwig van Beethoven's Studies*, page four: "Now that young Beethoven was no less competent on the organ, the art-loving

Elector nominated him as Neefe's successor."[12a] Neefe and Beethoven were court organists at the same time.

Christian Gottlob Neefe (1748-1798)

In the Cologne Court for the year 1790, page sixteen, is the entry:

> Electoral chamber, chapel and court music
> Organists
> Christian Neefe
> Ludwig van Beethoven

The duties of an organist at court were slight. Furthermore the organ was small, in keeping with the size of the court chapel (the present Protestant church), and was not even visible to the congregation. Great skill in playing was not required; indeed it could hardly even be put to use consider-

---

[1] Neefe was a fine example of the attitude prevalent at the time, since he was the organist of a prince of the church although he was a Protestant himself.

ing the limitations of the instrument. Neefe was also in good health and not prevented from executing his duties by other business[1]. Thus, as already suggested, Beethoven's appointment was obviously intended only as a most tactfully extended support.

But who helped him to obtain this post? A man who, as far I know, has never been given the credit he deserves in writings about Beethoven.

Count Ferdinand von Waldstein (1762-1823)

Beethoven's first and in every respect most important patron, his Mæcenas, was Count Waldstein, Knight of the Teutonic Order and, what is most important in this connection, favorite and constant companion of the young Elector. Subsequently he became Commander of the Teutonic Order at Virnsberg and Exchequer to the Emperor of Austria. Not just a connoisseur, he was also a practicing musician. It was he who in every way possible supported our Beethoven and first appreciated his genius to the full. With his assistance the young artist developed the talent for extemporizing and improvising variations on a given theme. From him he often received financial support, bestowed with such consideration for his easily wounded feelings that Beethoven usually assumed they were small gratuities from the Elector. Beethoven's appointment as organist, his being sent by the Elector to Vienna, etc., were all the Count's doing. That Beethoven dedicated the great and important *Sonata in C Major, Opus 53,* to him in later years is proof of the undiminished gratitude of the mature man.

Count Waldstein was the man Beethoven had to thank that the first blossoming of his genius was not suppressed, and so we, too, are indebted to this Mæcenas for Beethoven's later glory.

But let us return to our organist.

In his new position Beethoven first casually proved his talent to the orchestra by the following incident: In the Catholic church the Lamentations of Jeremiah are sung on three days of Holy Week. These consist of short movements of four to six lines apiece performed as chorales, each to a particular rhythm. The vocal line consisted of four consecutive notes, for example *c, d, e, f,* with several words or even complete sentences changed on the third until a few notes at the end lead back to the key-note. Since the organ must remain silent during these three days, the singer receives only an improvised accompaniment from a pianist.

Once when it was Beethoven's turn to perform this duty, he asked the singer Heller, who was very secure indeed in his intonation, whether he might try to throw him off, and he used the rather rashly given permission to wander about so much in the accompaniment that the singer was completely bewildered and could no longer find the closing cadence, even though Beethoven kept striking the note to be chanted in the treble with his little finger.

Franz Anton Ries, then first violinist and musical director of the Electoral Chapel, is still fond of relating in detail how Kapellmeister Luchesi, who was present, was startled by Beethoven's playing. In his first flush of anger Heller complained to the Elector about Beethoven and although the clever young Elector, who could be quite mischievous himself at times, was amused by the whole story, he nevertheless commanded that a simpler accompaniment be played in future.

At that time Beethoven also became a Chamber Musician. Once he played in that capacity before the Elector in a small group with the elder Ries and the famous Bernhard Romberg[13] who is still alive today. The artists were sightreading a new trio by Pleyel. In the second part of the Adagio, they were not able to stay together although they did not break down completely. They kept on playing courageously and luckily reached the end at the same time. Afterwards they figured out that two bars had been left out of the piano part. The Elector was very intrigued by Pleyel's composition and had it repeated a week later when their secret was discovered, much to the Prince's delight.

*
**

When Haydn first returned from England, a breakfast was given for him by the Electoral Orchestra in Godesberg, a resort near Bonn. On this occasion Beethoven showed him a cantata. Haydn examined it very closely and then warmly encouraged the composer to pursue his studies. Later this cantata was supposed to be performed in Mergentheim, but several sections were so difficult for the wind instruments that some musicians declared they could not possibly play them. As a result, the performance was cancelled. As far as we know, this cantata was never printed.[14]

*
**

Beethoven's first compositions were the sonatas published in Speyer's *Blumenlese,* followed by the song *Wenn jemand eine Reise thut,* then the music to a *Ritter-ballet* performed by the high nobility during Carnival season but not yet published.[15] The piano score is at present in the hands of the music publisher Dunst in Frankfurt who will presumably include it in the edition of Beethoven's works he is now completing. This edition of the *Ritter-ballet* should also include a minstrel's song, a German song, a drinking song, etc. Since Beethoven had not indicated himself as composer of these songs, the compositions were long assumed to be the work of Count Waldstein, especially since the Count had put the ballet together in collaboration with the dancing master Habich from Aachen.

Next came the Variations on *Vieni amore,* a theme by Righini, dedicated to the Countess of Hatzfeld. These Variations gave rise to the following unusual incident. Because he had not yet heard any great or celebrated pianists, Beethoven knew nothing of the finer nuances of handling the instrument; his playing was rough and hard. While on a journey from Bonn to Mergentheim, where the Elector kept a residence as Grand Master of the Teutonic Order, Beethoven and the Elector's Orchestra stopped in Aschaffenburg. Ries, Simrock, and the two Rombergs took him to Sterkel who, in response to everyone's request, sat himself down to play. Sterkel's playing was very light, highly pleasing, and, as the elder Ries put it, somewhat ladylike. Beethoven stood beside him concentrating intensely. Then he was asked to play but only complied when Sterkel intimated that he doubted whether even the composer of the Variations could play them all the way through. Beethoven played not only these variations, as far as he could remember them (Sterkel could not find the music), but also a number of others no less difficult and, to the amazement of his listeners, he played everything in precisely the same pleasant manner with which Sterkel had impressed him. That is how easy it was for him to adapt his style of playing to someone else's.[16]

This journey, incidentally, on which the whole orchestra travelled in two yachts up the Rhine and the Main during the most beautiful season of the year, was an abundant source of the loveliest memories for Beethoven. Lux, the singer and famous comedian, was elected Great King of the

Journey. Parcelling out duties, he assigned Beethoven and Bernhard Romberg to the kitchen and, indeed, they did serve there. Beethoven's diploma promoting him through the kitchen ranks, awarded on the Heights of Rüdesheim, will probably have been found among his effects. I certainly saw it in his possession, lovingly kept, in 1796. A large seal in pitch affixed to the lid of a box and attached with a few unravelled strands of the ship's rigging gave this diploma a most ceremonious aspect.

From his earliest youth Beethoven had an extraordinary aversion to teaching. At times Frau von Breuning would try to force him to continue teaching in the house of the Austrian Ambassador, the Duke of Westphalia, which was opposite her own[1]. He would set out, *ut iniquae mentis asellus*[2]/[16a], because he knew he was being observed. However, he often turned around at the very door of the house, ran back, and promised he would teach for two hours the next day, it was simply not possible for him that day. His own personal hardships were not enough to persuade him to teach, only perhaps the thought of his family and in particular of his dear mother.

Later on, after Beethoven had become prominent in Vienna, he developed a similar, if not even stronger, aversion to being asked to play at social occasions, and each time he was asked his good humor disappeared entirely. Many times

---

[1] Now the Furstenberg residence

[2] Like a bad-tempered little donkey—Horace

he came to me, gloomy and upset, complaining that he was forced to play even if the blood burned under his nails. I would try to amuse him and calm him down until gradually a conversation would develop between us. When I had managed this, I would let the conversation drop and seat myself at the writing table. If Beethoven wished to continue talking with me, he had to sit on the chair in front of the piano. Soon he would aimlessly strike a few chords, often still turned away from the instrument, and from these he would gradually develop the most beautiful melodies. Oh, how I wished I understood more about music! Several times I casually placed music paper on a nearby stand, hoping to obtain a manuscript from him; he would cover it with writing, but then he would fold it up and put it in his pocket! All I could do was laugh at myself. — About his playing I was allowed to say nothing, or at best make only a passing remark. He would then leave in a completely different mood and was always glad to come back. His aversion to playing in company, however, remained and was frequently a source of considerable dissension between Beethoven and the best of his friends and patrons.

One day Beethoven was improvising at the von Breuning house (where he was frequently asked to describe in music the character of some mutual acquaintance) and the elder Ries was urged to accompany him on the violin. After some hesitation he consented and that may well have been the first time the two artists had improvised together — a beautiful, very charming manner of playing. Later Ries and his son Ferdinand sometimes treated their audiences at public concerts to this unexpected pleasure.

***

Rather than continue to observe a more or less chronological order as I have done so far, I would like to offer the rest of my remarks on Beethoven as notes to the letters which follow. This method offers a very convenient way for me to recall interesting details and I imagine the reader will also welcome this change. Some of Beethoven's letters to me, therefore, are reprinted below. The first has already been published in the *Allgemeine Theater-Zeitung* in Vienna (No. 37, 25 March 1828). It is, as I remarked then, probably the longest my immortal friend ever wrote. On its first publication I commented: "This serves to prove a great deal of what I have said about Beethoven's education in the von Breuning household. Also in this letter, certainly for the first time, he pours out his heart concerning his deafness, and confides many details of his plans, etc. Taken as a whole, it is a pure expression of his heart, and the reader will know our friend much more closely when he has examined this letter." From now on Beethoven is no longer a stranger. A relationship has been established, based on deeper knowledge of the man and the artist, the foundation of sincere and lasting respect.

The letter was copied from the *Theater-Zeitung* by the *Bonner Wochenblatt* (No. 25, 1829) and finally, to the best of my knowledge, by the *Conversations-Blatt* in the *Ober-Post-*

Amts-Zeitung a few days after it had been read in the Frankfurt museum (at the beginning of 1836). A French translation appeared in the *Journal des débats* on 20 March 1838.

None of the other letters has ever been printed or reproduced before.[17]

***

Vienna, June 29, 1800[1]

My dear, kind Wegeler,

How grateful I am for your remembering me; I have so little deserved it or sought to deserve it from you. Yet you are so very good; you are not deterred by anything, not even my unpardonable negligence. You always remain the loyal, good, honest friend. — You must not believe, no, never, that I could forget you and all the others who were so kind and dear to me. There are moments when I long for your company; indeed I yearn to spend some time there with you. — My fatherland, the lovely countryside where I first saw the light of day, seems in my mind's eye still as beautiful and vivid as when I left you. I know that rejoining you and greeting our Father Rhine once more will be one of the happiest moments in my life. — When that may be I cannot yet say. — But I do want to assure you that when we meet again you will find me a much greater person, not only as an artist but also as a man, better and more fully developed. And if our Fatherland is more prosperous then, my art shall prove itself only for the benefit of the poor.[2] *O blessed moment, how happy I am that I can summon you, that I can create you myself! — You want to know something about my situation; well, at present it is not so bad. Unbelievable though it may seem when I tell you, Lichnowsky*[3] *has always been and remains my warmest friend (of course there have been slight misunderstandings between us, but haven't they only strengthened our friendship? —).*[4] *Last year he set aside a fixed sum of 600 florins on which I can draw until I find a position suitable for me. My compositions bring in a considerable amount, and I might say that I have more commissions than I can possibly honor. I can also count on six or seven publishers for each thing, and even more, should I want them. People no longer bargain with me; I demand and they pay. You can see that this is a rather nice situation. For instance, when I see a friend in need and my own purse at the moment does not allow me to help him immediately, I have only to set myself down and within a short time I can come to his aid. — I also live more economically than I used to; if I remain here for good I should certainly be able to secure one day for an Academy concert every year. I have already given several.*[5] *Only the evil demon, my miserable health, has done me an evil turn: my hearing has become steadily weaker these last three years. This disability is supposed to have been originally caused by the condition of my stomach, which, as you know, has always been wretched*[6] *but has become worse here. I suffer constantly from diarrhea and consequently am extraordinarily debilitated. Frank wanted to restore the* **tone** *of my body with strengthening medicines*[7] *and my hearing with almond oil, but — wouldn't you know it! — nothing changed. My hearing grew steadily worse and my stomach remained in the same state as before; this continued until the autumn of last year, and I was sometimes really in despair. Then a medical* **asinus** *advised cold baths for my condition. A more intelligent doctor ordered the customary lukewarm bath in the Danube. That worked miracles — my stomach improved, although my hearing problem remained, or worsened. This last winter I was truly miserable. I had really dreadful attacks of colic and deteriorated again to my previous state. And thus it remained until about four weeks ago, when I went to Vering*[8], *thinking that my condition might also require the attention of a surgeon. Besides I have always had confidence in him. He managed to stop the violent diarrhea almost*

always had confidence in him. He managed to stop the violent diarrhea almost entirely. He prescribed baths in lukewarm water from the Danube, into which I had to pour a flask of strengthening stuff each time, and gave me no medicine at all until about four days ago when he ordered pills for the stomach and an infusion for the ear. As a result I can only say I feel stronger and better, save my ears which buzz and ring day and night. I spend my life miserably, I must confess; for almost two years I have avoided all society, because I cannot say to people: I am deaf. If I had any other profession, it would not be quite so bad, but in my profession this is a dreadful position. And then there are my enemies, whose number is not small; what would they not say! — I can give you some idea of this peculiar deafness, when I must tell you that in the theater I have to get very close to the orchestra to understand the performers, and that from a distance I do not hear the high notes of the instruments and the singers' voices. It is surprising that there are people who have never noticed this when speaking to me[9]. *Since I have usually been absent-minded, they take it for that. Sometimes too I hardly hear people who speak softly. The sound I can hear, it is true, but not the words. And yet if anyone shouts, I can't bear it. What will become of me Heaven alone knows. Vering tells me that my hearing will certainly get better, even if I don't recover completely. I have so often cursed . . . my existence. Plutarch has taught me resignation. I will, if at all possible, defy my fate, although there will be moments in my life when I shall be the unhappiest of God's creatures. I beg you to say nothing about my condition to anyone, not even to Lorchen*[10]. *I am confiding this to you only as a secret, but I would be grateful if sometime you would correspond with Vering about it. If my condition continues like this, I will visit you next spring. You will rent a house for me in some beautiful part of the country, and for half a year I will lead the life of a peasant. Perhaps that will change things. Resignation! What a miserable refuge, and yet for me it is all that remains.*

You will forgive me, I'm sure, for burdening you with a friend's troubles, even though you yourself already have troubles enough. Steffen Breuning is now here and we are together almost daily. It does me so much good to revive the old feelings of friendship again. He really has become a fine, splendid young man who knows a thing or two and — like all of us, more or less — has his heart in the right place. I now have very nice rooms overlooking the Bastei[11] *which is extremely beneficial to my health. I do believe I will be*

Stephan von Breuning (1774-1827)

able to arrange for Breuning to join me. — I shall send you your Antiochum[12]*, and also a great deal of my music, if you do not think that it will cost you too much. Honestly, your love for art still gives me great pleasure. Just write to me how it may be done and I will send you all my compositions, a fair number by now, of course, and daily increasing. — In return for the portrait of my grandfather which I beg you to send me by mail coach as soon as possible, I am sending you one of his grandson, your ever fond and loving Beethoven. This portrait is being published here by Artaria, who often asked me for it, as did many others, including foreign art dealers. — I shall write to Stoffel*[13] *soon and give him a piece of my mind about his stubbornness. I will shout at* *him about our old friendship until his ears ring, and he will have to promise me faithfully not to hurt you anymore, especially as your present circumstances are troubled enough. —— I shall also write to dear Lorchen. Never have I forgotten any of you dear, kind people, even though I have not written you at all. But writing, as you know, was never my strong point. Even my best friends have not received any letters from me for years. I live only in my music, and one composition is hardly finished before another is already begun. The way I write now, I often do three or four things at once. —— Do write me more often now; I shall make an effort to find time to write you occasionally. Give my regards to everyone, and to our kind Frau Hofräthin[14] and tell her that I still have a "raptus" now and again[15]. As regards the Koches I am not at all surprised about the change in circumstances. Fortune is round as a ball and naturally does not always come to rest with the noblest and the best. —— Concerning Ries, to whom I send my warmest greetings, I will write to you in more detail about his son, although I think that he would probably have better luck in Paris than in Vienna; Vienna is deluged with people and even the best talents are sometimes hard put to make a living. —— In the autumn or winter I shall see what I can do for him, because then everyone rushes back to town. Farewell, my good, faithful Wegeler. Rest assured of the love and friendship of your*

Beethoven

Explanation of the notes to the above letter.

(1) The year is missing; from the subsequent letter, however, it becomes clear that it was in all probability 1800.[18]

(2) Because of the war, Bonn had lost its Elector, court, government offices, university, army, in short all its sources of income. It had never had any factories or commercial traffic.

(3) Carl, Prince Lichnowsky, Count Werdenberg, Dynast of Granson, was a great benefactor, indeed friend, of Beethoven. The Prince even took Beethoven into his house as a guest, where he remained for at least a few years. I found him there toward the end of 1794 and he was still there when I left in the middle of 1796. However, Beethoven almost always kept lodgings in the country at the same time.

The Prince was a great lover and connoisseur of music. He played the piano, and by studying Beethoven's works and playing them more or less well, he tried to prove to Beethoven that he did not need to change anything in his style of composition, even though the difficulties of his works were often pointed out to him. There were performances at the Prince's house every Friday morning. Besides our friend, four salaried musicians — Schuppanzigh, Weiss, Kraft,[19] and one other (Link?) — and usually also an amateur, Zmeskall, took part. Beethoven always accepted the comments of these gentlemen with pleasure. Once when I was there, for instance, Kraft, the famous cellist, pointed out to him that he should mark a passage in the finale of the third trio, *Opus 1*, with *sulla corda G* and that in the second of these trios the finale, which Beethoven had marked 4/4, should be changed to 2/4.

Beethoven's new compositions were first performed here whenever they were suitable. Several great musicians and music lovers usually gathered to hear them. When I lived in Vienna I was usually, though not always, present. It was there that Beethoven first played for old Haydn the three sonatas which he dedicated to him. Here, in 1795, Count

Appony commissioned Beethoven to compose for a fixed fee a quartet, a form in which he had produced nothing so far. The Count announced that he did not wish to reserve the quartet for himself for six months before publication as was customary, nor did he demand a dedication, etc.

I repeatedly reminded Beethoven of this commission and he settled down to work on it twice. However, the first attempt turned into a grand string trio *(Opus 3)*, the second into a string quintet *(Opus 4)*.[1]

It was also here that another Hungarian count (at my encouragement when he told me what he wished to do) once laid before Beethoven a manuscript copy of a difficult Bach composition. Beethoven played the piece at sight and,

---

[1]— Whether and to what extent the following remarks about the composition of quartets apply to Beethoven, I gladly leave to the judgment of the masters in this art. They naturally reminded me of the facts mentioned here.

"The string quartet is the most difficult and in a way also the most thankless form of musical composition both for the composer and for the performer. No work requires such profound, thorough knowledge of freeflowing yet controlled composition, so much delicacy, tenderness, and refinement of taste, and so much depth and profundity of the spirit as does the string quartet. It is, so to speak, the ultimate in all instrumental music. Here, in a small framework and with but four instruments, all the life and depth of music must be delineated in its most delicate outline. For this the orchestra commands a hundred instruments as well as an overall canvas where, because of its very scale, painstaking individual and delicate attention to detail becomes a mistake."

— *Frankfurter Conversations-Blatt* No. 353 — 21 December 1836

according to the owner, just as Bach himself had played it. A Viennese composer, Förster, once brought him a quartet which he had copied out only that very morning. When in the second part of the first movement the cello lost his place, Beethoven stood up and sang the bass accompaniment while continuing to play his own part. When I pointed this out as a mark of extraordinary skill, he answered with a smile: "The bass line had to run that way, unless the author knew nothing about composition." — In response to another remark that he had played a presto passage which he had never seen before so fast that it must have been virtually impossible to have seen the individual notes, he replied, "That is not in the least bit necessary. When you are reading rapidly there may be any number of printing errors which you do not see or pay any attention to, provided the language is familiar to you."

The musicians usually stayed for dinner after the concert. Artists and scholars also gathered here with no distinction made as to social standing. Princess Christiane was the highly educated daughter of Count Franz Joseph von Thun, incidentally a most charitable and honourable gentleman himself though prone to excess romanticism because of his association with Lavater. It was well known that he believed he could heal sickness through the power of his right hand (See *Conversations-Lexicon*, Seventh edition, vol. 11, p. 236).

(4) Beethoven was highly sensitive and consequently easily provoked. However, if one waited for the air to clear after the first explosion, he could listen to remonstrances with a willing ear and a conciliatory heart. As a result he always apologized

for much more than he was guilty of. For instance I possess a little note from him which I received in Vienna where he says, among other things, "What a detestable picture you have shown me of myself! Oh, I realize I do not deserve your friendship, . . . It was not intentional, premeditated malice on my part that made me treat you that way; it was my unpardonable thoughtlessness." . . . He goes on in this manner for three pages and then ends with: "But — no more. I come to you and throw myself into your arms, asking for my lost friend back. Return yourself to me again, to your remorseful, loving friend who will never forget you,

Beethoven[20]

The two letters to Fraulein von Breuning which follow prove the same point.

Subsequently he also broke off his friendship with Stephen von Breuning for some time (— with what friend did he not behave so?).[21] When at last he was convinced by someone else that he had done von Breuning a great injustice, he wrote him in the same contrite fashion. A most sincere reconciliation took place between two and the closest friendship continued uninterrupted until Beethoven's death.

(5) Much has been written about Beethoven's prosperity and poverty. This much I know from personal experience: Beethoven had been brought up in very straitened circumstances and had always been placed under some guardianship, as it were, even if it was only that of his friends. He did not know the value of money and was somewhat less than economical. For instance lunch at the Prince's house was set for

four o'clock. "Consequently," Beethoven said, "I would have to be home by half past three every day, change into something better, see that I was properly shaven, etc. — I can't stand all that!" And so it happened that he frequently went out to eat in taverns. This made matters even worse for him economically because as I've said he understood neither the value of money nor what things should cost.

The Prince, who had a very loud metallic voice, one day instructed his footman that if both he and Beethoven should ring at the same time, Beethoven should be served first. When Beethoven overheard this, he hired a servant of his own that very same day. Similarly, when on a whim he decided that he would like to learn to ride, he hired a horse of his own, even though he had the Prince's entire stable at his disposal.

The financial arrangements for his own works (or as Beethoven aptly put it in a letter to Ferdinand Ries: The honorarium *avec ou sans honneur*) were much more frequently negotiated by his brother Caspar[22] than by himself.

That Beethoven had very little understanding of money matters even as late as 1821 is apparent in one of his letters. I am indebted to the kindness of Polizeirath Guisez in Aachen for providing me with a copy of the original of this letter which is in his possession.

Baden, 27 September 1821

Dear Sir,

Pardon my liberty in troubling you. I have entrusted the bearer of this letter, Herr von —, with a banknote for exchange or sale. Being unfamiliar with everything necessary to such a transaction, I would ask you very kindly to give him your advice and counsel. Several illnesses last winter and summer have set me back somewhat in my financial affairs. I have been here since 7 September and must remain here until the end of October. All this costs a great deal of money and prevents me from earning it in my usual manner. I expect money from abroad, but since the exchange rate is so high at the moment, I feel this is the easiest way to get by for the time being. Later on I shall purchase a new banknote again . . .

Your friend,
Beethoven

(In great haste)

This unsealed letter was in an envelope on which the following was written as a kind of postscript by Beethoven:

You will easily see what kind of business genius I am. Only after the enclosed letter had been written did I discuss the banknote with a friend. It then became clear to me that one need only detach a coupon and the whole matter is taken care of. I am glad therefore that I need not trouble you with it . . . . . .

Yours,
Beethoven.

One might ask: why bother the gentleman with the letter at all, since he apparently did not know him well? The brief request to look after the bearer of the letter, which has been deleted here, would not have required as many lines as did the postscript.

Incidentally, I fully support von Seyfried's opinion on page 27: "Beethoven knew neither ambition (?)[23] nor extravagance, nor did he know the true value of money which he considered merely a means to acquire the absolute bare necessities. Only in his last years did the traces of anxious frugality appear, a trait which did not affect his innate charitable impulses."

Other evidence of his completely disordered financial habits is cited by our friend Ries.

(6) Only on the afternoon two days before the performance of his first concerto (C major) did he write the Rondo. He was suffering at the time from the rather severe colic which plagued him frequently. I helped him with minor remedies as best I could. Four copyists sat in the hallway working from the manuscript sheets he handed over to them one at a time.[24]

Here I may be permitted another digression. During the first rehearsal which took place in Beethoven's room the next day, the piano was a semitone flatter than the wind instruments. Beethoven immediately ordered the winds as well as the other instruments to tune to B flat rather than A and he played his part in C sharp.

(7) Peter Frank, director of medical studies in Pavia, later director of the General Hospital in Vienna, the first classical writer on the enforcement of health laws, etc.[25]

Beethoven probably underlined the word *tone* because he could not harmonize it with the beautiful meaning he knew — or was he laughing about it?

(8) Chief medical staff officer, Imperial Councilor, born in Hungary, father of the general practitioner Joseph von Vering, who was famous in France and Germany. This and the following letter show that Beethoven suffered from various other maladies apart from his deafness and that von Seyfried's

statement (p. 13) — "He (Beethoven) knew nothing of illnesses despite his peculiar way of life." — must be taken with great reservation.[26]

(9) Even Ries did not notice it during the first two years, as we will see.

(10) Eleonore von Breuning, Wegeler's wife.

(11) In Pasquillati's house?[27]

(12) A well-known picture by Füger, Director of the Academy of Painting in Vienna, showing Erasistratus perceiving the love of Antiochus for his stepmother Stratonice.

(13) Christoph von Breuning, *Geheimer Revisionsrath* in Berlin.

(14) The mother of the von Breuning family.

(15) When Beethoven, as recounted earlier, suddenly rushed back to the mother of the Breuning family as she stood watching him rather than going to give his lessons, or when he had another of his strokes of genius, as his whims were called, the kind mother used to say with a shrug: "He has his *raptus* again today." A passage in Goethe's *Correspondence with a Child, Part 2*, p. 200, proves that the word and its meaning remained dear to him. Bettina reports: "Last night I wrote down everything. This morning I read it to him (Beethoven)" and he said, "Did I really say that? — Well, then I had a *raptus* again."

Vienna, 16 November 1801

My dear Wegeler! I thank you for the renewed proof of your solicitude on my behalf, all the more since I deserve it so little. — You want to know how I am, what I need. Much as I dislike discussing such matters at all, I would prefer talking about it with you rather than with anyone else.

For several months now Vering has put plasters on both arms. These consist, as you know, of a certain kind of bark.[1] This is a most disagreeable procedure, since I am always robbed of the free use of my arms for a few days (until the bark has had sufficient effect), not to mention the pain. True, I cannot deny that the buzzing and ringing in my ears is somewhat less than it was, particularly in my left ear where my hearing problems actually started. But my hearing has certainly not improved; indeed I must admit that I think it is perhaps even worse than before. —— My stomach is better. Especially when I take the lukewarm baths for several days running, I am reasonably well for eight or even ten days. I very occasionally take something strengthening for my stomach; I am now starting to use the herbs you prescribed for my stomach. —— Vering will hear nothing about shower baths. In general I am thoroughly dissatisfied with him; he shows too little care and understanding of my illness. If I did not go to see him myself, which costs me a great deal of effort, I would never see him at all. —— What do you think of Schmidt?[2] I do not like changing, but it seems to me that Vering is too much of a practitioner to acquire many new ideas through reading. —— Schmidt seems to be quite a different sort of man in this respect and might possibly not be quite so negligent. —— One hears marvelous things about galvanism; what do you think of it? One doctor told me he had seen a deaf and dumb child recover his hearing (in Berlin) and a man who was deaf for seven years also regained his hearing. —— I have just heard that your Schmidt[3] is experimenting with it.

I live now somewhat more pleasantly again, getting out among people more. You can hardly believe how desolate, how sad my life has been these last two years. My weak hearing haunted me everywhere like a spectre and I fled from people, making myself appear a misanthrope, though I am far from that.

— This change was wrought by a dear, enchanting girl who loves me and whom I love. After two years there are now some blissful moments again; this is the first time that I feel marriage might bring one happiness. Unfortunately she is not of my class[(4)] — and now — I could not, of course, marry. — I must struggle on bravely. If it were not for my hearing, I would have long ago traveled round half the world; I still must do it. — For me there is no greater joy than to practice and present my art. — Do not believe that I would be happy with you. And yet what should make me happier? Your very solicitude would hurt me; every moment I would read pity in your faces and that would make me even more miserable. — That lovely native countryside, what was my lot there? Nothing but the hope for a better life. That would be mine now, but for this affliction! Oh, I would embrace the entire world if I were free from this miserable disease. Yes, I really do feel as if my youth were just beginning; was I not always a sickly person? More than ever before my physical strength has been increasing for some time now, my mental powers as well. Each day I get closer to the goal which I can feel but cannot describe. Only in this way can your Beethoven thrive. — Do not speak of rest! — I know of no rest but sleep and it pains me enough that I must now devote more time to sleep than usual. Were I but half delivered from my affliction, then — I would come to you and, as a completely mature man, renew the old feelings of friendship. You shall see me happy again, as happy as it is my lot to be in this world below; not unhappy. — No, I could not bear that. I shall take Fate by the throat; it shall not subdue me entirely. — Oh, it is so beautiful to live life a thousand times! A quiet life — No, I feel I am no longer meant for that — Do please write me as soon as possible. — Make sure that Steffen decides to take a position somewhere in the Teutonic Order.[(5)] Life here is too strenuous for his health. Besides, he leads such an isolated existence that I cannot see at all how he means to get any further. You know what it is like here; I do not mean to say *that a social life might lessen the strain on him, but one cannot even persuade him to go anywhere. — Some time ago I held a musicale in my home, but our friend Steffen stayed away.*[6] *Advise him to rest more and to quiet down; I have tried everything I could; without rest and patience he can never again be happy or healthy*[7]. *Tell me in your next letter whether or not it matters if I send you a great deal of music. You can of course sell what you do not need and thereby retrieve your postal expenses. — My portrait, too. Fondest and kindest regards to Lorchen — also to Mama — also to Christoph. You do love me a little, don't you? Be assured of this (my love) as well as the friendship of*

Your Beethoven

Notes

(1) The bark of *Daphne mezereum*

(2) Joh. Adam Schmidt, Councilor, staff surgeon, professor of medicine at the Josephinum, eye specialist, author of several classic writings.

(3) Your Schmidt. I lived in the closest, friendliest contact with Schmidt and Hunczovsky until their deaths. The former wrote under his portrait which he sent to me: *Cogitare ut esse tui, idem est. Wegelero suo Schmidt.*[28]

(4) In the biographical notes which Ignaz von Seyfried appended to his *Studies of Beethoven* the following passage occurs on page 13: "Beethoven was never married and strangely enough never involved in a love affair."[28a] The truth, as my brother-in-law Stephan von Breuning, and Ferdinand Ries, and Bernhard Romberg, and I learned to know it, was that Beethoven was never not in love and was usually involved to a high degree. His and Stephan von Breuning's first love was Miss Jeanette d'Honrath from Cologne, Neumarkt

No. 19 (now the house of Mr. Biercher, the master builder) who often spent a few weeks with the von Breuning family in Bonn. She was a beautiful, vivacious blonde, with a good education and an amiable disposition; she enjoyed music enormously and had a pleasant voice. She often teased our friend by singing a song popular at that time:

> Parting from you this very day,
> And unable to prevent it,
> Is much too painful for my heart!

The favored rival was the Austrian recruiting officer in Cologne, Carl Greth, who married Miss d'Honrath and died as Field Marshal General, Commanding Officer of the 23rd Infantry Regiment, Commander of Temeswar, etc., on 15 October 1827.

Then followed the tenderest affection for the beautiful and cultured Fraulein v. W.[29] Bernhard Romberg told me anecdotes about this Werther-like love just three years ago.

These affairs, however, belonged to his adolescence and left no more profound impression than they had aroused in the pretty young ladies.

In Vienna Beethoven was always involved in a love affair, at least as long I lived there, and sometimes made conquests which could have been very difficult indeed, if not impossible, for many an Adonis.

Whether one could have composed *Adelaide* and *Fidelio* and many other things without knowing love in its innermost depths, I leave to the judgment of experts and amateurs. But what can be clearer than Beethoven's statement in this letter as to how essential love was to him.[30]

I should also mention that, as far as I know, each of his loves was of much higher social standing than he.

(5) The von Breuning family had long occupied one of the highest positions in the Teutonic Order. Stephan's great-grandfather von Meirhoven was Chancellor of the Order; his successor to the office was Grandfather von Breuning, then the uncle, and finally the cousin. Stephan himself had been employed by the government in Mergentheim.

(6) The ill-feeling between these friends must have been considerable, since von Breuning was a music-lover, an excellent violinist trained by Ries's father, and himself had played in the Elector's chambers on several occasions.[31]

(7) Beethoven's assessment of his friend's well-being was only too correct. He rarely enjoyed sound health for long, but worked relentlessly all the same. He continued unflaggingly until his death on 4 June 1827, $2\frac{1}{3}$ months after Beethoven's. Von Breuning was in every respect an excellent and widely respected man and his death is especially to be regretted since he was the only one who combined all the talents necessary to become Beethoven's biographer. With brief interruptions he had, after all, lived in intimate contact with him from the time he was ten until his death. As an indication of his high regard for von Breuning, Beethoven had also appointed him one of the two executors of his will.

*
* *

In the meantime the correspondence between us was

not exactly very lively, as the next letter shows. I might have, even then, already mislaid some letters or perhaps given them to autograph collectors. At that time, still surrounded by a large number of people who had known Beethoven, I had no desire to collect his letters; every friend was a source of memories concerning him. And of course these were the accounts from secondary sources I mentioned in the foreword.

Vienna, 2 May 1810

Dear old friend — I can imagine that these lines of mine will occasion some astonishment — and yet, even though you have no written proof, you are still very much in my thoughts. — For a long time there has even been among my manuscripts one intended for you, which you will certainly receive this summer.[1] *For a couple of years now the quieter, calmer life has no longer been mine, and I have been forced to move in society. I have still not decided in favor of it, possibly more against it, but who is not affected by the storms all around him? Yet I would be happy, perhaps one of the happiest of men, if the demon had not settled in my ears. — If I had not read somewhere that man must not voluntarily part with his life as long as he can still perform a good deed, I would long ago have ceased to be — and, indeed, by my own hand. — Oh, life is so beautiful, but for me it is poisoned forever. —*

You will not refuse a friendly request if I ask you to obtain for me my baptismal certificate. — Whatever the expenses involved, maybe Steffen Breuning can pay you directly, since you have an account with him, and I will immediately reimburse Steffen here. — Should you consider it worth while to investigate the matter and if you would be willing to make the trip from

Coblenz to Bonn, then charge all the expenses to me. — There is one particular thing to watch out for, and that is: there was another brother born before me, but who died; he was also called Ludwig, though with the middle name Maria. To determine my age for certain you must first locate that Ludwig, then me; I do know that others have mistaken him for me and I have been said to be older than I am.[2] *Unfortunately, I have lived for some time without knowing myself how old I am. — I used to have a family register, but it has gotten lost, heaven knows how. — So please do not take it amiss if I urge you earnestly to find out all about Ludwig Maria and then the present Ludwig, who was born after him. — The sooner you send the baptismal certificate, the more indebted I shall be to you.*[32] *— I am told that you sing one of my songs in your Masonic Lodges, probably in E major, which I myself do not possess. Send it to me! I promise to recompense you three or four times over in some other way.*[3] *— Think of me kindly, however little I may seem to deserve it. — Embrace and kiss your dear wife, your children, everyone you hold dear — on behalf of your friend*

Beethoven.

(1) My lot in this respect was the same as that of his pupil Ries; the dedication remained in the letters. But are not they of higher value?

(2) A reference to an account by Ries, as it turned out later.

(3) Beethoven is wrong here; it was not a song especially composed by him of which he no longer had a copy; it was instead a different text using Beethoven's music of the *Opferlied* by Matthisson. I had done the same with a very early song composed by him: *Wer is ein freier Mann?* I have taken the liberty of placing these texts in the Appendix,[33] along with the vocal line and the text of an Adagio which was printed with

Beethoven's approval. Beethoven also requested a text to the theme of the variations which begin the grand sonata dedicated to Prince Lichnowsky (Opus 26), but I did not send it to him since I was not satisfied with it, nor with any other version.

Vienna, 29 September 1816

I am seizing this opportunity of reminding you of me through J. Simrock.[1] — *I hope you have received my copper engraving*[2] *and also the Bohemian glass. As soon as I travel through Bohemia once again, you will receive something like it again. All my best wishes. You are a husband, a father; so am I, but without a wife.*[3] *Give my greetings to all your family and our friends.*

Your friend, L. v. Beethoven.

Beethoven's portrait (1814) engraved from a drawing by Letronne

(1) Joseph Simrock, present owner of the business.
(2) *Dessiné — par — Letronne et gravé par — Hoefel. 1814.* At the bottom it says: For my friend Wegeler. Vienna, 27 March

1815. Ludw. van Beethoven. — Our mutual friend, the Director General of the Rhine Customs, Herr Eichhoff, brought it back for me from the Congress.

(3) Beethoven was educating the son of his brother Caspar who had died the previous year.

***

The letter below is written by another hand and only signed by Beethoven.

Vienna, 7 October 1826

My beloved old friend,

I cannot express what pleasure the letter from you and your Lorchen gave me. Of course the answer should have come back as quick as an arrow, but I am generally rather negligent in my correspondence because I think that the best people already know me well anyway. I compose an answer often in my head, but when I come to write it down, I usually throw my pen aside because I am not able to write as I feel. I remember all the love you have always shown me — for instance that time you had my room whitewashed, which was such a pleasant surprise.[1] *— The same with the Breuning family. If we drifted apart, it was simply the nature of circumstances. Each of us had to pursue his destiny and try to attain it. However, the eternally unshakeable principles of goodness have always bound us firmly and closely together. Unfortunately, I cannot write so much to you today as I would wish because I am confined to my bed and must restrict myself to answering a few points of your letter.*

You write that I am somewhere alleged to be a natural son of the late King of Prussia. That has also been said to me before, a long time ago. I have, however, made it a principle never to write anything about myself, nor to answer anything that is written about me. I therefore gladly leave it to you to make known to the world the integrity of my parents, and of my mother in particular. — You write about your son. It goes without saying that if he comes here, he will find a friend and father in me, and where I am in a position to be of any use to him or help him in any way, I will do so with pleasure.

I still have the silhouette of your Lorchen, which shows how much I still treasure all the love and kindness which surrounded me in my youth.(2)

Concerning my diplomas I would write only briefly that I am an honorary member of the Royal Society of Sciences in Sweden, as well as in Amsterdam, and I am also an honorary citizen of Vienna. — Recently a certain Dr. Spieker took my last grand symphony with choruses to Berlin; it is dedicated to the King, and I had to write the dedication in my own hand. I had previously requested permission from the Embassy to dedicate the work to the King, and this was granted. Dr. Spieker requested that I turn over the corrected manuscript to him myself, with emendations in my own hand, for presentation to the King, since it is to be placed in the Royal Library. There was also some mention of the Order of the Red Eagle, Second Class; whether anything will come of this, I do not know. I have never sought such honors, but at this age such an award would be rather welcome for various other reasons.

Incidentally, my motto has always been: Nulla dies sine linea, *and if I do let the Muse sleep, it is only that she may awake all the stronger.*(3) *I still hope to bring forth a few more great works into the world and then to finish my earthly course like an old child somewhere or other among kind people. — You will soon receive some music from the Schott Brothers in Mainz. — The portrait I am enclosing with this letter is indeed an artistic masterpiece, but it is not the latest done of me. — As for honors conferred on me — which, I know, give you pleasure — I must also report that the late King of France sent me a medal with the inscription:* Donné par le Roi à Monsieur Beethoven. *It was accompanied by an exceedingly courteous letter from the* Duc de Châtres, premier gentilhomme du Roi.[4]

My beloved friend! This will have to suffice for today. The memory of the past affects me anyway, and you receive this letter not without many tears. Now a start has been made, and soon you will have another letter from me. And the more often you write, the more pleasure you will give me. Our friendship precludes the need for questions on either side. My warmest regards. Please embrace and kiss your dear Lorchen and your children for me; and think of me then. God be with you all!

As always your true, faithful, and most respectful friend

Beethoven

(1) At the time Beethoven lived in the Peretti house in Wenzelgasse in Bonn.

(2) All the members of the von Breuning family and their closer friends had their silhouettes cut by the painter Neesen in Bonn over a period of two evenings. That is how I came into the possession of Beethoven's silhouette, which is reproduced here. Beethoven was probably then in his sixteenth year.

(3) On 24 July 1804 Beethoven wrote to Ries from Baden: "In all my life I would not have believed that I could be as lazy as I am here. If a burst of activity should follow, then it might really produce something worthwhile."

(4) What has been said about Beethoven's alleged indifference to, or even contempt of, such honors should be judged in this light.

Vienna, 17 February 1827[1]

My worthy old friend,

Fortunately I received at least your second letter from Breuning. I am still too weak to answer it, but you can imagine that everything it contained is welcome to me and what I wanted.[2] *My recovery, if I may call it that, is still progressing very slowly. I presume I must expect a fourth operation, although the doctors have said nothing about that yet. I remain patient and think: Occasionally some good may come of all that's evil.* — *But I was surprised, indeed, when I read in your last letter that you still had received nothing.* — *You will see from this letter that I had already written you by 10 December last year. The same applies to the portrait, as you will see from the date on it when you receive it.*[3] — *"Mrs. Steffen spoke."*[4] — *In short, Steffen asked that these things be sent to you when the opportunity arose, but they were left lying around until now, and really it was difficult enough to request their return before today. You will now receive the portrait by post through Messrs. Schott, who also forwarded the music to you.* — *There is so much I would still like to tell you today, but I am too weak; I can do no more than embrace you and your Lorchen in spirit. In true friendship and affection for you and yours,*

Your old devoted friend
Beethoven

This last letter was also written by another hand but signed by Beethoven.

(1) That is, one month before his death!

(2) If my memory serves me right, I had reminded him of Blumauer in my letter, who, once the water had been tapped, lived in good health for many years (see his letter to Stoll). I had told him of the plan to fetch him from the Bohemian baths, to take a few sidetrips with him along the upper Rhine, then down the Rhine to Coblenz, where he was to regain his strength completely, etc.

(3) Beethoven had written on the portrait above his name: "To my old, honored, beloved friend F. v. Wegeler." There is no date.

(4) The beginning of the second verse of the well-known song: "Zu Steffen sprach im Traume," etc.

*
* *

The following are two letters from Beethoven to Fräulein von Breuning.

Vienna, 2 November 93

My adored Eleonore!
My dearest friend!

Only now, after I have lived for almost a whole year here in the capital, do you receive a letter from me, and yet you were surely always in my liveliest thoughts. I have often conversed in spirit with you and your dear family, although frequently not as calmly as I would have wished. Each time I recall that unfortunate quarrel, my behavior then seems most detestable. But it happened. How much would I give if I could but blot out of my life completely the way I behaved then, which so dishonored me and which was in complete contradiction to my character![1] *There were of course many circumstances which kept us apart, and I believe the whisperings of third parties who passed on all that was said by both sides against the other was the main thing that prevented reconciliation. Each of us believed that we were speaking from absolute conviction, and yet it was no more than fanned-up anger, and we were both deceived. Your kind and noble nature, my dear friend, assures me that you have long since forgiven me. But they say that confession of one's own crime is the most sincere repentance; and that is what I wanted to do. — And now let us draw the curtain over the whole story and just learn from it the lesson that, when friends have quarrelled, it is always better not to use a mediator, but for one friend to go directly to the other.*

Here is a dedication to you from me, and I only wish the work greater and more worthy of you. They have been pestering me here to publish this little work, and I have taken the opportunity to give you, my very dear Eleonore, proof of my esteem and friendship for you and of my undying memory of your home. Accept this trifle and remember that it comes from a friend who admires you very much. Oh, if only it gives you pleasure, my wishes will be fulfilled.[2] *May it be a small remembrance of that time when I spent so many and such blissful hours in your home. Perhaps it will continue to remind you of me until I return one day — which, unfortunately, will not be for a long time. Oh, how happy we shall then be, my dear friend! You will then find your friend a more cheerful person; time and better fortune will have smoothed away the furrows on his brow etched by his earlier adverse lot.*

Should you see B. Koch[3], *please tell her that it is not at all nice of her not to have written me even once. After all, I have written twice to her. I wrote three times to Malchus*[4] *and — no answer. Tell her that if she does not want to write, she should at least urge Malchus to. In concluding my letter I should like to venture one more request: I would like very much once again to possess a waistcoat of rabbit hair knitted by your hand, my dear friend.*[5] *Forgive your friend this presumptuous request. It springs from a great weakness for anything made by your hands. And between the two of us I might also add that there is a touch of vanity at the bottom of it too, namely, to be able to say that I possess something from one of the best, most admirable girls in Bonn. I still have, of course, the first waistcoat, which you had the great kindness to give me in Bonn, but fashion has made it so unfashionable that I can only keep it in my wardrobe as something from you and very dear to me.*

You would give me great pleasure if you would soon delight me with a friendly letter. Should my letters afford you pleasure, then I certainly promise to write faithfully, as much as it is within my power; I welcome every opportunity to show how very much I am

> *Your admiring*
> *true friend*
> *L. v. Beethoven*

P.S. *The V. (variations) will be somewhat difficult to play, in particular the trills in the coda.*[6] *This should not frighten you. It is so arranged that you need not play anything but the trill; leave out the other notes; they are also in the violin part. I would never have written it that way, had I not often noticed that after I had been improvising of an evening, someone somewhere in Vienna would write down many of my own ideas the next day and go about boasting they were his.*[7] *Since I could expect that such things would appear soon, I decided to get ahead of them. There was also another reason: to embarrass the local pianists. Many of them are my mortal enemies, and I wanted to take my revenge on them in this way, because I knew full well that the V. would be placed in front of them here and there, and the gentlemen would then make a very poor showing.*

> *Beethoven*

Excerpt from Beethoven's letter to Eleonore von Breuning, 2 Nov. 1793

(1) Compare what is said in note 4 to the first letter above, namely that Beethoven always begged forgiveness for more than he had committed.

(2) These are the variations on "Se vuol ballare" from Mozart's *Figaro*. In the Dunst edition, 4th section, No. 27.

Later a sonata, or rather sonatina, was also dedicated to her, which is No. 64 in the first section of the Dunst edition.[34]

(3) Barbara Koch from Bonn, later Countess Belderbusch, an intimate friend of Eleonore von Breuning, a lady who, of all the members of the female sex whom I came to know in the course of a rather long and eventful life, came closest to the ideal of a perfect woman. This claim is confirmed by all who were fortunate enough to know her well. She was surrounded not only by younger artists like Beethoven, the two Rombergs, Reicha, the twin brothers Kügelgen, and others, but also by intellectuals of all classes and ages, such as D. Crevelt, who lived in the same house, Professor Velten, who died young, Fishenich, later State Councilor, Professor, later Canon, Thaddäus Dereser, the future Bishop Wrede, Heckel and Floret, both private secretaries to the Elector, the private secretary of the Austrian Ambassador, Malchus, von Keverberg, subsequently Netherlands State Councilor, Hofrath von Bourscheidt, Christoph von Breuning, mentioned above, and many others. — Really, it was a wonderful, and in many ways lively time in Bonn while the personally gifted Elector Max Franz, Maria Theresa's youngest and favorite son, reigned there in peace.

(4) Subsequently Graf von Marienstadt, Minister of Finance in the Kingdom of Westphalia and later in the Kingdom of Würtemberg. Classical writer.

(5) Angora rabbits.

(6) A trill is continued over several bars with changing fingering, while the three other fingers are also occupied at

the same time. The fingering is marked.

(7) Beethoven complained to me about this kind of espionage. He named H. Ab.G.,[35] a most prolific composer of variations, who always took lodgings in his vicinity. This may have been one more reason why Beethoven always looked for lodgings on an open square or on the ramparts.

Elector Maximilian Franz (1756-1801)

Second letter to Fräulein von Breuning.[36]

I was most surprised by the beautiful cravat made by your own hand. Pleased though I was with the actual gift, still, it stirred sad feelings within me. It called up memories of bygone times, and a sense of shame as well, because you have treated me so generously. In truth, I did not think you still considered me worth remembering. Oh, if you could have seen how I felt when I received your gift, you would certainly not think I exaggerate when I tell you that your thoughtfulness made me weep; I was touched by sadness. — However little I may, in your eyes, deserve to be believed, I beg you, my friend (do still let me call you that!), to believe I have suffered greatly and am still suffering from the loss of your friendship. I will never forget you and your dear mother. You were so kind to me that your loss cannot and will not easily be restored; I know what I have lost and what you have been to me, but — were I to detail all this, I would have to return to scenes both unpleasant for you to hear and for me to describe.

As a small token of gratitude for remembering me so kindly, I am taking the liberty of sending you these variations and the rondo for one violin. I am very busy, otherwise I would have transcribed the sonata I promised you so long ago. In my manuscript it is little more than a sketch, and even someone as clever as Paraquin[1] would have difficulty in copying it. You can have the rondo copied and then return the score to me. What I am sending you are the only things I have that might possibly suit your purposes. Besides, since you are now travelling to Kerpen[2], I thought these trifles might perhaps give you some pleasure.

Farewell, my friend — it is impossible for me to call you anything else, however indifferent you may be to me. Do believe that I continue to cherish you and your mother as much as ever. If I can contribute in any other way to your happiness, I beg you not to forget me. This is still the only means left for me to express my gratitude for the friendship I so enjoyed.

Have a happy journey and bring your dear mother back in full health again. Think occasionally of your true and devoted friend

Beethoven

Since this letter is without date and address and, moreover, starts at the very top of the page without the slightest spacing, I presume it is the third page of a letter whose first sheet has been lost. It demonstrates first of all the contention that Beethoven always apologized for more than he had committed and secondly his close relationship with the von Breuning family.

(1) A singer and double bass player in the Elector's orchestra, extremely reliable as an artist and highly respected as a man.

(2) An uncle of the von Breuning family lived in Kerpen. The family and their friends visited him on vacation every year for five or six weeks. Beethoven also spent some very happy weeks there several times and was often asked to play the organ.

**

From the letters of my brother-in-law Stephan von Breuning I would like to quote just one which is concerned with Beethoven's opera *Fidelio*.

Vienna, 2 June 1806

Dear Sister and dear Wegeler,

. . . If I remember correctly, I promised in my last letter to write you about Beethoven's opera. Since this will certainly interest you, I want to fulfill my promise. The music is among the most beautiful and perfect one could hope to hear; the subject is interesting because it concerns the liberation of a prisoner through the loyalty and courage of his wife. But nevertheless, probably nothing has caused so much aggravation for Beethoven as this work, whose worth will come to be fully appreciated only in the future. It was first performed seven days after the occupation by the French troops; in other words, at the most inopportune time. Naturally the theaters were empty, and Beethoven, who had also noticed some imperfections in the treatment of the text, withdrew the opera after three performances. After order was restored, he and I took it up again. I reworked the whole libretto for him, which made the action livelier and faster. He shortened many pieces and it was then performed three times

and was exceedingly well received. Then, however, his enemies at the theater got together and since he had offended several people, particularly during the second performance, they managed to prevent any further performances. Even before this they created many difficulties for him, of which one example may suffice. He could not even ensure that the second performance of the opera was announced under the new title Fidelio, *as it was also called in the French original, and under which it was printed after the changes had been made. Despite various promises the first title* Leonore *appeared on the announcement posters.* [37] *The conspiracy is all the more disagreeable for Beethoven since he was depending on payment derived from a percentage of the receipts. Now that the opera is no longer being performed, he has been set back somewhat in his financial affairs and will recover all the more slowly because he has lost a great deal of his desire and love for his work through the treatment he has received. I may well have given him the greatest pleasure then since completely without his knowledge I had a little poem printed and distributed in the theater in November as well as for the performance at the end of March. I shall copy them both for Wegeler here, for I know from former times that he sets store by such things, and since at one time I composed verses on his investment as* Rector magnificus celeberrimae universitatis Bonnensis, *he can now see by comparison whether my talent for occasion poetry has made any progress. The first little poem was in unrhymed iambics:*

> We hail thee now; thou trav'lest noble paths;
> Applause of connoisseurs dost spur thee on.
> Thy modesty restrained thee much too long.
> Though thou hast but begun, the prize is thine
> And older warriors gladly welcome thee.
> Thy dazzling sound has glory, strength, and depth —
> A golden river flowing rich and full.
> In perfect union skill and grace entwine.
> Thy vital soul taught thee to touch mankind.

> *Our hearts were stirred, elated, pierced in turn*
> *By Leonore's courage, love, and tears;*
> *Now jub'lant echoes vie to praise her faith*
> *And dread anxiety gives way to bliss.*
> *Press on; be brave; pour out triumphant sound.*
> *Awestruck posterity will wonder were*
> *There ever other wonders in the world.*

The second consists of two stanzas and contains a reference to the presence of the French troops at the time of the opera's first performance:

> *We hail thee once again upon this course*
> *Where thou embark'st in anxious fearful times*
> *When dreadful truth ripped down the beauteous veil*
> *Of sweet illusive dreams, and all were cast*
> *In turbulence, as when a fragile ship*
> *Is flung from wave to wave on angry seas;*
> *Art fled in fear from cruel scenes of war,*
> *Nor did we weep from tenderness, but woe.*

> *Thy mighty stride must bring us to our feet,*
> *Thy holy vision raise our eyes on high*
> *Where thou hast seen pure art and feeling wed.*
> *With golden wreaths the muse has crowned thy brow,*
> *While from the laurel grove Apollo smiles*
> *And consecrates, himself, your deity.*
> *Long may you reign! For in your music pure*
> *Is heard eternal beauty's pow'r and grace.*

Copying this has really tired me out completely, so I shall close this letter which is rather long anyway. I want to tell you just one more bit of news — Lichnowsky has sent the opera to the Queen of Prussia and I hope that the performances in Berlin might just show the Viennese the worth of what they have at home.†[37a]

---

† [Translator's note: The English translation of these and other poems in this volume are not meant to be literally exact, but to reflect the spirit and tone of the originals.]

Appendix

Words set to Beethoven's music for Mathisson's *Opferlied* (page 46).

On the Initiation of a Mason

Commence the task! With sacred fire
Exalt in youth the firm desire
    For worthy work concluded;
Give strength to weakness; make weak youth believe
He will the highest goals achieve
    The victor's palm included.

Root from his soul all but the Truth.
Ensure, great Lord, that our brave youth
    With righteous men will wander.
Oh grant he will find favor in your eyes.
Urge him and his companions wise,
    The Mason's life to ponder.

      *
    * *

Words set to the song: *Wer ist ein freier Mann?* (page 46)

Masonic Questions

Chorus: What is the Mason's goal?
A Voice: With noble purpose striving
          On higher plains arriving
          Set free from Chance's role
Chorus: That is the Mason's goal,
          The Mason's splendid goal.

          What marks the Mason's bond?
          Through thought and deed deserving
          Humanity preserving
          On earth and then beyond
          That marks the Mason's bond,
          The Mason's holy bond.

          To whom are Masons drawn?
          Men comforting and giving
          By habit selfless living
          And ruled by heart, not brawn,
          To such are Masons drawn
          They willingly are drawn.

Whom does a Mason scorn?
Those weak and yielding ever
Those heeding conscience never
Abject, like beasts, forlorn
They earn the Mason's scorn
The Mason's deepest scorn.

What weighs on Masons' hearts?
Each tear still daily flowing,
Each life that should be growing
That misery imparts
That weighs on Masons' hearts
On Masons' feeling hearts.

When does his heart beat glad?
When in our grand tradition,
Not asking recognition,
He makes the world less sad.
Then does his heart beat glad
His heart beat warm and glad.

Who praises Masons' deeds?
Our end and our beginning
Who set the planets spinning
And answers all our needs,
He praises Masons' deeds,
The Masons' noble deeds.

In the Dunst edition, 4th section, No. 27.

The Elegy

Voice line and words set to the Adagio from the first of the three sonatas dedicated to J. Haydn (*Opus 2*) with the inscription: Elegy. This arrangement appears in the edition of Beethoven's work erroneously identified as a composition unified from its inception. (See the musical illustration at page 143)

(These three poems were composed in 1797.)

PART TWO

Ferdinand Ries (1784-1838)

Foreword

My friend Wegeler has frequently urged me to collect a few incidents from the life of my immortal teacher and affectionate friend Beethoven and, should I not wish to make other arrangements, to publish the material jointly with some of his reminiscences. So I have been persuaded, not without a certain feeling of anxiety, to set down a few particularly memorable recollections and to place these notes, as well as some of Beethoven's letters, of which I possess a large num-

ber, at the disposal of my friend for him to use as he sees fit.*
The truth of the facts recorded here and the fame of the man
whom they concern should alone lend value to these notes.
My remarks should serve as a continuation and an elaboration
of the preceding notes by my friend Wegeler, and our mutual
endeavors should form, as we intend, a genuine source on
which a complete biography of our immortal friend can be
based. The plainness of style will, I hope, be graciously overlooked, since hitherto I have communicated with the public
only through musical compositions. The same goes for a possible lack of order in presentation. I shall relate the events as
they occur to me; should the reader be of a mind to do so, he
will find it easy enough to put them in order. And so without
further ado I will begin.

Frankfurt/Main, December 1837.

---

* A comment by the author of the article on Ferdinand Ries in the *Conversations-Lexicon der neuesten Zeit* was a strong incentive for me to encourage my friend to collect these reminiscences. The writer remarked: "The accounts which Ries could give about his close relationship with Beethoven are fascinating, particularly since his contact with the Master not only covers the period when the composer rose to the highest peak of his creative power, but also includes that time when Beethoven had the misfortune to lose his hearing. The Beethoven biographies and his literary remains contain all too little about the psychological effect of this tragedy on the great man, and we possess so few detailed reports on his character, on his method and habit of composition, that it would be especially appropriate for Ries to publish his memoirs of that remarkable period."

Wegeler.

From my father I received my first, and most fortunately for my whole career, very thorough instruction on the piano and in music in general. When he decided that it was time for me to continue my studies abroad, since Bonn had suffered greatly because of the war, I was sent to Munich and then to Vienna at the age of fifteen.

My father's friendly relationship with Beethoven as a boy and as a young man justified the expectation that I would be well received by him. A letter of recommendation introduced me. When I presented it to Beethoven on my arrival in Vienna in 1800,[38] he was intensely occupied with the completion of his oratorio *Christ on the Mount of Olives*, since this was due to receive its first performance in a great Academy concert in the Wiener Theater, an event very advantageous to him. He read the letter through and said: "I cannot answer your father now; but write to him that I have not forgotten how my mother died; he will be satisfied with that." Later I learned that my father had been active in supporting him in every way on that occasion, since the family was in great want.

*
* *

Right from the very first few days Beethoven found that he was able to make use of me, and thus I was frequently called as early as five in the morning, as indeed happened on the day the oratorio was to be performed. I found him in bed, writing on single sheets of paper. When I asked him what they were, he answered: "Trombones." — The trombones played from these manuscript pages at the performance.

Had these parts been forgotten in copying? Were they an afterthought? I was still too young at the time to give any thought to questions of such artistic interest. Presumably it was an afterthought, though, since Beethoven would probably have had in his possession the uncopied sheets, as well as those which had been copied. The rehearsal began at eight o'clock in the morning and, beside the oratorio, the following new things were also performed for the first time: Beethoven's Second Symphony in D Major, the Piano Concerto in C Minor and yet another new piece which I do not remember. It was a dreadful rehearsal, and by half past two everyone was exhausted and more or less dissatisfied.

Prince Karl Lichnowsky, who attended the rehearsal from the beginning, had bread and butter, cold meat, and wine brought in great baskets.[39] He most graciously invited everyone to help themselves, which everyone eagerly did, and a general good humor was restored. The Prince then asked that the oratorio be rehearsed once more so that everything might go really well in the evening and that the first work of this type by Beethoven might be presented to the public in a manner worthy of it. The rehearsal was therefore resumed. The concert began at six o'clock, but it was so long that a couple of the pieces were not performed.

There was something most striking about the *Larghetto quasi andante* of the Symphony in D just mentioned; which Beethoven had presented to me in a score in his own hand, out of pure friendship (and which was unfortunately stolen from me by a friend). This *Larghetto* is so beautifully, so purely and happily conceived and the melodic line so natural that one can hardly imagine anything in it was ever changed. The design was indeed from the beginning as it is now. However, in the second violins, almost in the very first lines, a most significant part of the accompaniment was changed in many places and at some points also in the violas. But everything is so carefully crossed out that I could never discover the original idea despite great effort. I also questioned Beethoven about it, who retorted drily: "It is better this way."

In 1802[40] Beethoven composed his Third Symphony (now known as the *Sinfonia eroica*) in Heiligenstadt, a village one and a half hours outside Vienna.

Heiligenstadt c. 1821

When he was composing, Beethoven frequently had a certain subject in mind, even though he often laughed at and inveighed against descriptive music, particularly the frivolous sort. Occasionally Haydn's *Creation* and *The Seasons* came under fire in this respect, though Beethoven did recognize Haydn's greater achievements, especially the many choral works and certain other things for which he properly lavished praise on Haydn. In this symphony Beethoven had thought about Bonaparte during the period when he was still First Consul. At that time Beethoven held him in the highest regard and compared him to the greatest Roman consuls. I myself, as well as many of his close friends, had seen this symphony, already copied in full score, lying on his table. At the very top of the title page stood the word "Buonaparte" and at the very bottom "Luigi van Beethoven," but not a word more. Whether and with what the intervening space was to be filled I do not know. I was the first to tell him the news that Bonaparte had declared himself emperor, whereupon he flew into a rage and shouted: "So he too is nothing more than an ordinary man. Now he also will trample all human rights underfoot, and only pander to his own ambition; he will place himself above everyone else and become a tyrant!" Beethoven went to the table, took hold of the title page at the top, ripped it all the way through, and flung it on the floor. The first page was written anew and only then did the symphony receive the title *Sinfonia eroica*. Prince Lobkowitz later bought this composition from Beethoven for his own use for several years, and it was performed many times at his palace. Here it happened that Beethoven, who was himself conducting, once threw the whole orchestra out of rhythm in the second part of the first *Allegro*, where it runs on so long in half-notes on the off-beat, that they had to start all over again from the beginning.

***

In that same *Allegro* Beethoven plays the horn a mean trick. Several bars before the theme re-enters completely in the second part, Beethoven has the horn suggest it, while the two violins are still holding a chord on the second. To anyone who does not know the score it inevitably gives the impression that the horn player has miscounted and entered too early. During the first rehearsal of this symphony, which went appallingly, the horn player did come in correctly. I was standing next to Beethoven and believing the entry wrong said: "That damned horn player! Can't he count? — It sounds terrible!" I believe I was very close indeed to having my ears boxed. — Beethoven was a long time in forgiving me.

***

That same night Beethoven played his Quintet for piano and wind instruments; the famous oboist Ram from Munich played also and joined Beethoven in the quintet. — In the last Allegro a pause occurs several times before the theme begins again; in one of these pauses Beethoven suddenly started improvising, taking the Rondo as his theme, and entertained himself and his listeners for quite some time, although the other players were not so amused. They were annoyed and Mr. Ram was even angry. It did indeed look rather comic to see these gentlemen, expecting to begin any moment, constantly raising their instruments to their mouths, only to put them down quietly again. At last Beethoven was satisfied and returned to the Rondo. The whole company was enchanted.

***

The funeral march in A flat minor in the sonata dedicated to Prince Lichnowsky (Opus 26) originated in the great praise Paer's funeral march in his opera *Achilles* received from Beethoven's friends.*

***

When the immensely famous Steibelt came from Paris to Vienna, several of Beethoven's friends were worried lest he should cast a shadow on Beethoven's reputation.[42] Steibelt did not visit him; they met for the first time one evening in the house of Count Fries, where Beethoven gave his new trio in B flat major for piano, clarinet, and cello (Opus 11) its initial performance. This work offers the pianist no opportunity to display his virtuosity. Steibelt listened with a certain condescension, paid Beethoven a few compliments, and felt confident of his own superiority. — He played a quintet of his

---

* Beethoven was in a box with a lady who was very dear to him when *La Molinara* was being performed. During the famous "Nel cuor piu non mi sento" the lady said that she had once had variations on this theme, but had lost them. During the night Beethoven wrote VI variations on the aria and sent them to the lady the next morning with the inscription, *Variazioni*, etc. *perdute par la* —— —— *ritrovate par Luigi van Beethoven*. They are so easy that they were presumably intended for the lady to sight-read.

Wglr.[41]

own composition, improvised, and produced a great effect with his *tremulandos*, which were something quite new then. Beethoven could not be persuaded to play again. A week later there was another concert at Count Fries. Steibelt again played a quintet with much success, but more important (one could sense this) he had prepared a brilliant improvisation, choosing the identical theme on which the variations in Beethoven's trio were written. This outraged Beethoven's admirers as well as Beethoven himself. It was now his turn to improvise at the piano. He seated himself in his usual, I might say unmannerly, fashion at the instrument, almost as if he had been pushed. He had picked up the cello part of Steibelt's quintet on his way to the piano, and placing it upside down on the music rack (intentionally?), he hammered out a theme from the first few bars with one finger. — Insulted and irritated as he was, he improvised in such a manner that Steibelt left the room before Beethoven had finished, never wanted to meet him again, and even made it a condition that Beethoven not be invited when his own company was desired.

Beethoven almost always postponed the composition of the majority of his works due by a certain date until the very last moment. He had, for instance, promised Ponto,[43] the famous horn player, that he would compose a sonata (Opus 17) for piano and horn and would play it with him at a concert given by Ponto. The concert with the sonata was announced, but the sonata was not yet started. On the day before the performance Beethoven began the work and had it ready for the concert.

***

Beethoven had originally written the famous Sonata for Violin in A minor[44] (Opus 47), dedicated to Rudolph Kreuzer in Paris, for an English artist named Bridgetower. Matters were not much better in this case, either, even though a large part of the first Allegro was ready in plenty of time. Bridgetower began to put a great deal of pressure on Beethoven, since his concert was already arranged and he wanted to practice his part. One morning Beethoven had me called at half past four and said: "Copy the violin part of the first Allegro for me quickly." — (His usual copyist was already hard at work.) The piano part was only sketched in here and there. The particularly beautiful theme with variations in F major Bridgetower had to play from Beethoven's own manuscript at his concert in the Augarten at eight in the morning, because there was no time to copy it.

On the other hand, the violin and piano parts of the last 6/8 Allegro in A major were very beautifully copied because it had originally belonged to the first sonata with violin (Opus 30) in A major, dedicated to the Emperor Alexander. Beethoven later substituted the present variations, because the other movement was too brilliant for this sonata.

*
**

Beethoven gave a grand Academy concert in the Theater an der Wien, where his C minor and Pastoral Symphonies (the fifth and the sixth) together with his Fantasy for Piano with Orchestra and Chorus were performed for the first time. In the last work the clarinetist inadvertently took a repeat of eight bars just where the last cheerful theme begins its variation. Since there were only a few instruments playing, this mistake was naturally all the more grating to the ear. Beethoven jumped up in anger, turned round, and abused the orchestra members in the coarsest manner so loudly that the whole audience heard it. Finally he shouted: "From the beginning!" The theme began again, everyone made their correct entrances, and it was a splendid success.[45] When the concert was over, however, the artists recalled all too vividly the sarcastic names Beethoven had publicly bestowed upon them and flew into the greatest rage, as though the insult had only just happened. They swore they would never play again if Beethoven were with the orchestra, etc. This lasted until he had composed something new again, when their curiosity overcame their anger.

***

A similar scene is supposed to have occurred on another occasion. This time, however, the orchestra made him feel his injustice more forcefully and insisted in all seriousness that he not conduct. Beethoven then had to stay in an adjoining room during the rehearsal and it took a very long time before the dispute was straightened out again.

***

Of all composers Beethoven thought most highly of Mozart and Handel, followed by S. Bach. If I found him with music in his hand, or if anything lay on his desk, it was sure to be a composition by one of these heroes. Haydn rarely got away without a few digs. Beethoven's grudge here probably went back to earlier days. The following tale may account for one reason. Beethoven's three trios (Opus 1) were to be introduced to the music world in a soirée at Prince Lichnowsky's. Most of Vienna's artists and music lovers had been invited, above all Haydn, whose opinion was anxiously awaited by everyone. The trios were played and caused a tremendous stir. Haydn, too, said many fine things about them but advised Beethoven not to publish the third one in C minor. This astonished Beethoven, since he considered it the best, and in fact to this day it is always found to be most pleasing and has the greatest effect. Haydn's remark of course made an unpleasant impression on Beethoven, leaving him with a feeling that Haydn was envious and jealous and did not wish him well. I must admit that I did not really believe Beethoven when he told me this. I therefore later took the opportunity to ask Haydn himself about it. His answer confirmed Beethoven's story in that he said he had not imagined that this trio would be so quickly and easily understood nor so favorably received by the public.

On the same occasion I asked Haydn why he had never written a violin quintet and received the laconic answer that he had always found four voices sufficient. I was actually told that Haydn had been asked for three quintets but that he could never compose them because he had become so accustomed to the quartet style that he could not find the fifth voice. He had started, but one attempt turned into a quartet, another into a sonata.

Haydn had wished Beethoven to put on the title page of his first works "Pupil of Haydn." Beethoven did not want to do that because, as he said, he had had some instruction from Haydn but had never learned anything from him. (During his first stay in Vienna Beethoven had received some lessons from Mozart, but, he complained, Mozart had never played for him.) Beethoven had also taken lessons in counterpoint from Albrechtsberger and in dramatic music from Salieri. I knew them all well; all three thought very highly of Beethoven but were of one opinion of him as a student. Each said Beethoven was always so stubborn and so bent on having his own way that he had had to learn many things through hard experience which he had refused earlier to accept through instruction. Albrechtsberger and Salieri in particular were of this opinion. The former's pedantic rules and the latter's trivial ones concerning dramatic composition (following the Italian school of the time) could not appeal to Beethoven. Whether the studies published by Ritter von Seyfried furnish irrefutable proof that Beethoven, during his two years of apprenticeship under the eyes of Albrechtsberger, devoted himself to theoretical studies with indefatigable perseverance therefore remains open to doubt.[46]

What follows may serve to corroborate the above. Once during a walk together I mentioned two perfect fifths which resound noticeably and beautifully in one of his first string quartets in C minor. Beethoven did not remember them and claimed that it was wrong to call them fifths. Since he was in the habit of always carrying music paper with him, I asked for a sheet of it and wrote down the passage for him with all four voices. When he saw that I was right, he said: "Well, and who has forbidden them?" Since I did not know how to take that question, he repeated it several times, until, very taken aback, I finally answered: "Well, it is one of the very first basic principles." The question was again repeated, whereupon I said: "Marpurg, Kirnberger, Fuchs, etc., etc., all the theoreticians!" "Well, I allow them!" was his answer.[47]

***

Beethoven had promised the three solo sonatas (Opus 31) to Nägeli in Zurich, while his brother Carl (Caspar), who unfortunately always meddled in the business arrangements, wanted to sell these sonatas to a publisher in Leipzig. There were frequent altercations between the brothers over this, because Beethoven wanted to honor his promise. When the sonatas were ready to be sent off, Beethoven was living in Heiligenstadt. During a walk the brothers quarreled again, eventually coming to blows. On the next day he gave me the sonatas, to send to Zurich immediately, as well as a letter to his brother which was enclosed within another letter to

Stephan von Breuning, meant for his brother's perusal. No one could have preached a more beautiful sermon in a more kind-hearted way than did Beethoven to his brother about his behavior of the day before. First he showed him the true, despicable character of his conduct, then forgave him completely, but also predicted a miserable future for him if he did not radically change his life and behavior. The letter which he sent to Breuning was also exquisitely written.

Those same sonatas occasioned another unusual incident. When the proofs arrived, I found Beethoven busy writing. "Play the sonatas through for me," he said, while he remained sitting at his writing desk. There were an uncommon number of mistakes in the proofs, which made Beethoven very impatient indeed. At the end of the first Allegro in the Sonata in G major, however, Nägeli had even included four measures of his own composition, after the fourth beat of the last firmata:

When I played these, Beethoven jumped up in a rage, ran over, and all but pushed me from the piano, shouting:

"Where the devil does it say that?" — His astonishment and anger can hardly be imagined when he saw it printed that way. I was told to draw up a list of all the errors and to send the sonatas immediately to Simrock in Bonn who was to reprint them and add, *Edition très correcte*.[48] — This designation still stands on the title page. The four measures, however, are still to be found in several of the subsequent editions.

Nikolaus Simrock (1752-1833)

The following notes written to me by Beethoven belong in this context:

Be so kind as to correct the mistakes and send a list of the errors to Simrock immediately with a note telling him he should see to it that these be published soon. — I will send him the sonata and the concerto the day after tomorrow.

Beethoven

I must ask you once again to undertake the disagreeable job of making a clean copy of the Zurich sonatas with the mistakes corrected and to send it to Simrock; you will find the list of mistakes which you prepared in my rooms at the Wieden.

Dear Ries! The marks are badly indicated and also the notes are transposed in several instances — please do pay more attention! — otherwise the work is again all in vain. Ch'a detto l'amato bene?

*
* *

Beethoven arranged an engagement for me as pianist with Count Browne. The Count spent some time in Baden near Vienna, where for a group of enthusiastic Beethoven fans I frequently had to play Beethoven's compositions in the evenings, partly from the music, partly by heart. Here I became convinced that for most people the name alone is enough for them to find everything in a work beautiful and admirable, or mediocre and inferior. One day, tired of playing by heart, I played a march which I made up as I went along without giving it any further thought. An old countess, who was really plaguing Beethoven with her devotion, went into ecstasies over it, since she believed it to be something new by him. I readily went along with this idea, making fun both of her and the other enthusiasts. As ill luck would have it, Beethoven himself came to Baden the next day. He had hardly entered the room at Count Browne's when the old lady began rattling on about the exceptionally magnificent march which bore the stamp of genius. Imagine my embarrassment! Knowing full well that Beethoven could not stand the old countess, I quickly pulled him aside and whispered to him how I had only wanted to make fun of her foolishness. Luckily he took the matter very well, but my embarrassment grew when I had to repeat the march, which this time turned out much worse, with Beethoven standing beside me. He then received from everyone the most extravagant praise for his genius, to which he listened in utter confusion and anger, until at last he dissolved into roaring laughter. Later he said to me: "There you see, my dear Ries! Those are the great connoisseurs who aspire to judge all music so correctly and astutely. Just give them the name of their darling; more than that they do not need."

***

This march, incidentally, did have the positive result that Count Browne immediately commissioned from Beethoven the composition of three marches for four hands, which were dedicated to the Princess Esterhazy (Opus 45).[49]

***

Beethoven composed part of the second march while — this still seems unbelievable to me — giving me a lesson on a sonata which I was to perform in the evening at a small concert in the Count's house. I was also to play the marches there with him.

While we were playing these duets, the young Count P. . . .[50] spoke so loudly and freely with a beautiful lady in the doorway to the next room that, after several attempts to restore silence had proved futile, Beethoven suddenly pulled my hand away from the piano in the middle of the piece, jumped up, and said very loudly: "I will not play for such swine."

All attempts to bring him back to the piano failed; he would not even permit me to play the sonata. This, to everyone's displeasure, put an end to the evening's music.

***

One evening at Count Browne's I was to play a Beethoven sonata (A minor, Opus 23), a work not often heard. Since Beethoven was present and I had never studied this sonata with him, I declared myself willing to play any other sonata, but not that one. Beethoven was then approached and finally said: "Well, you will surely not play it so badly that I cannot listen to it," and so I was obliged to play it. As usual, Beethoven turned pages for me. At a leap in the left hand, where one particular note must be brought out, I missed the note completely, and Beethoven tapped me on the head with one finger. Princess L. . . . .,[51] who sat leaning against the piano facing me, noticed this with a smile. When I had finished, Beethoven said: "Very good. You did not need to study this sonata with me first. The finger-tap was just meant to prove that I was paying attention." Later that evening Beethoven was also obliged to play and chose the D Minor Sonata (Opus 31), which had just been published.[52] The Princess, who probably expected that Beethoven too would make a mistake somewhere, now stood behind his chair while I turned pages. In bars 53 to 54 Beethoven missed the entry, and instead of descending two notes and then two more, he struck each quarter note in the descending passage with his whole hand (three or four notes at once). It sounded as if the piano was being cleaned. The Princess rapped him several times on the head, not at all delicately, saying: "If the pupil receives one tap of the finger for one missed note, then the Master must be punished with a full hand for worse mistakes." Everyone laughed, Beethoven most of all. He started again and performed marvelously. The Adagio in particular was incomparably played.

∗∗

Many of Beethoven's compositions appeared with the notation *Arrangé par l'Auteur même*, but only four of these are genuine: from his famous septet he arranged (1) a violin quintet and (2) a piano trio; from his quintet for piano and four wind instruments he arranged (3) the quartet for piano and string trio; he also arranged (4) the violin concerto (Opus 61) dedicated to Stephen von Breuning as a piano concerto. Many other things were arranged by me, checked through by Beethoven, and then sold by his brother Caspar under Beethoven's name.[52a]

∗∗

When Beethoven gave me lessons, I must say that contrary to his nature he was extraordinarily patient. I could only attribute this, and his almost unfailingly amicable behavior toward me, mainly to his love and affection for my father. Thus he sometimes made me repeat a thing ten times or even more often. In the Variations in F major, dedicated to Princess Odescalchi (Opus 34), I had to repeat the last Adagio variations entirely seventeen times. Still he was not satisfied with the expression in the little cadenza, even though I thought I had played it just as well as he did. I received nearly two full hours of instruction that day. If I made a mistake somewhere in a passage, or struck wrong notes, or missed intervals — which he often wanted strongly emphasized — he rarely said anything. However, if I lacked expression in crescendos, etc. or in the character of a piece, he became angry because, he maintained, the first was accident, while the latter resulted from inadequate knowledge, feeling, or attention. The first happened quite frequently to him, too, even when he played in public.

Beethoven was extremely good natured, but he was also easily irritated or quick to be suspicious. This was caused in part by his deafness but even more so by the behavior of his brothers. Any stranger could freely slander his most loyal friends, since he believed everything all too quickly and unconditionally. He would then neither reproach the victim nor demand an explanation but would immediately show the greatest insolence and contempt in his behavior toward him. Since he was extremely passionate in everything he did, he also looked for the most vulnerable side of the supposed enemy to show his anger. Therefore one frequently did not know how one stood with him until the matter was cleared up, usually by accident. Then, however, he always tried to compensate for the injustice he had done as quickly and effectively as possible. From among many instances I have chosen the following incidents as proof of this.

Beethoven was to become Kapellmeister at the court of the King of Westphalia. The contract, ensuring him six hundred ducats salary besides (unless I am mistaken) a carriage at his disposal, had been drawn up and awaited only his signature. The Archduke Rudolph and the Princes Lobkowitz and Kinsky countered this offer with the guarantee of a salary for life under the sole condition that he remain within the Empire. I knew of the first arrangement but not of the second offer when suddenly Kapellmeister Reichard came to me and said Beethoven would certainly not accept the position in Cassel; wouldn't I, as Beethoven's only pupil, like to go there at a smaller salary. I did not believe him and immediately went to Beethoven to inquire as to the truth of the matter and to ask his advice. For three weeks I was refused admittance; even my letters remained unanswered. Finally I saw Beethoven at a fancy ball. I immediately approached him and acquainted him with the reason for my request, whereupon he said in a cutting tone: "So — you think you could fill a position that was offered to me?" — He then remained cold and shunned me. The next morning I went to him to explain the misunderstanding. His servant told me in a rude tone that his master was not at home, even though I heard him singing and playing in the next room. Realizing that the servant had no intention of announcing me, I decided to go straight in anyway, but the fellow jumped for the door and shoved me away. This infuriated me. I grabbed him by the throat and roughly threw him to the floor. Beethoven, alarmed by the tumult, came running out, found the servant still on the floor and me pale as death. Absolutely distraught by this point, I deluged him with such vehement reproaches that he was speechless with astonishment and just stood there, immobile. When the whole misunderstanding was cleared up, Beethoven said: "I did not know it was like that; I had been told you tried to get the position behind my back." Upon my assurance that I had not responded to the offer at all, he immediately went out of his way with me to make good his injustice. However, it was too late; I did not receive the position, even though it would have been very advantageous for me at the time.[53]

His brothers went to great lengths to keep all close friends away from him. It did not matter how badly they behaved toward him (if one were able to convince Beethoven of this at all); they needed only to shed a few tears and he immediately forgot everything. He then used to say: "He is, after all, my brother," and the friend would be reproached for his solicitude and frankness.* The brothers eventually had their own way in that many friends retreated from him,

---

* Von Seyfried wrote: "Two younger brothers had followed him to Vienna and lifted from his shoulders the oppressive burden of worrying about his economic needs. They really found it necessary to act as guardians for this high priest of art, who was virtually a total stranger to ordinary life." Just how that remark is to be taken must be left to the reader's discretion.

Wglr.

especially when his deafness made it increasingly difficult to converse with him.

***

Beethoven's hearing began to suffer as early as 1802, but the trouble disappeared for a time.* He was so sensitive to the onset of his deafness that one had to be very careful not to make him feel the disability by talking loudly. If he had not understood something, he usually blamed it on his absent-mindedness, which was indeed a strongly developed trait. Much of the time he lived in the country, where I often went to take a lesson from him. Occasionally he would say at eight in the morning after breakfast: "Let us go for a little walk first." So we would go for a walk, often not returning until three or four o'clock, after we had eaten something in one of the villages. On one of these outings Beethoven gave me the first startling proof of his loss of hearing, which Stephan von Breuning had already mentioned to me. I called his attention to a shepherd in the forest who was playing most pleasantly on a flute cut from lilac wood. For half an hour Beethoven could not hear anything at all and became extremely quiet and gloomy, even though I repeatedly assured him that I did not hear anything any longer either (which was, however, not the

---

* His first and second letters to me show that he suffered even earlier than 1800 from this malady, which he confided to me as his friend and as a doctor, but managed to conceal even from his closer associates for several years.

Wglr.

case). — When at times he did seem in good spirits, he often became actually boisterous. However, this happened very rarely.

*
* *

During a similar walk we went so far astray that we didn't get back to Döbling, where Beethoven lived, until nearly eight o'clock. The entire way he had hummed, or sometimes even howled, to himself — up and down, up and down, without singing any definite notes. When I asked what this was, he replied: "A theme for the last Allegro of the sonata has occurred to me" (in F minor, Opus 57). When we entered the room he rushed to the piano without taking off his hat. I took a seat in the corner and he soon forgot all about me. He stormed on for at least an hour with the new finale of this sonata, which is so beautiful. Finally he got up, was surprised to see me still there, and said: "I cannot give you a lesson today. I still have work to do."[54]

*
* *

Among pianists he praised only one to me as an excellent player: John Cramer. All the others found little favor with him. He was most reluctant to play his own things.

Once he seriously planned a grand tour with me, where I was to arrange all the concerts and play his piano concerti and other compositions. He himself wanted only to conduct and to improvise. His improvising was, of course, the most extraordinary thing one could ever hear, especially when he was in a good mood or was irritated. All the artists I ever heard improvise did not come anywhere near the heights reached by Beethoven in this discipline. The wealth of ideas which poured forth, the moods to which he surrendered himself, the variety of interpretation, the complicated challenges which evolved or which he introduced were inexhaustible.

***

One day when we were talking about themes for fugues after the lesson was over, I sat at the piano and he sat down beside me. I played the theme of the first fugue from Graun's *Death of Jesus*. He began to imitate it with his left hand, then added a voice with his right hand, and continued on without the slightest interruption for a good half hour. I still cannot understand how he managed to remain in such an uncomfortable position for so long. His own enthusiasm rendered him insensible to external conditions.

***

When Clementi came to Vienna, Beethoven wanted to call on him at once; his brother, however, put it into his head that Clementi had to visit him first. Clementi, despite the fact that he was much older, would probably have done this anyway, if some gossip about it had not arisen. So it came about that Clementi was in Vienna for a long time without knowing Beethoven other than by sight. We frequently ate lunch at the same table at the Swan, Clementi with his pupil Klengel[55] and

Beethoven with me. Everyone knew who everyone was, but no one spoke with each other or even nodded a greeting. The two pupils had to imitate their masters, because presumably each one risked losing his lessons otherwise. I certainly would have suffered that, since with Beethoven compromises were never possible.

※

Originally there had been a grand Andante in the sonata (C major, Opus 53) dedicated to his first patron, the Count Waldstein. A friend of Beethoven's suggested to him that the sonata was too long, whereupon he was taken to task most severely. Calmer deliberation, however, soon convinced my teacher that the remark had some truth to it. He then published the grand Andante in F Major, in ⅜ time,[56] on its own and later composed the interesting introduction to the rondo in its present form.

※

This Andante, however, left me with a sad memory. When Beethoven first played it for our friend Krumpholz[57] and me, we liked it enormously and pestered him until he repeated it. On the way home I passed the house of Prince Lichnowsky, and I stopped in to tell him about Beethoven's marvelous new composition. I was compelled to play the piece as best I could remember. Since I kept remembering more of it, the Prince entreated me to repeat it. So it happened that he, too, learned part of it.

To surprise Beethoven, the Prince went to him the next day and told him that he too had composed something, which was not at all bad. Despite Beethoven's firm statement that he did not want to hear it, the Prince sat down and to the composer's astonishment played a good portion of his own Andante.

Beethoven was furious and this was the reason why I never heard Beethoven play again. He never wanted to play again in my presence and repeatedly demanded that I leave the room when he played. One day when a small party including Beethoven and myself breakfasted with the Prince after the concert in the Augarten (at eight o'clock in the morning), the suggestion was made that we all drive to Beethoven's lodgings to hear his opera, *Leonore*, which had not yet been performed. After we arrived, Beethoven of course demanded that I should leave, and since even the most urgent entreaties of everyone present were to no avail, I did so with tears in my eyes. The whole party noticed it. Prince Lichnowsky followed me and asked that I wait in the anteroom, because since he himself had been the cause of the situation, he now wanted to settle the affair. My injured self-respect, however, would not permit that. I heard afterward that Lichnowsky had become very incensed with Beethoven for his behavior, since only our love for his works had been to blame for the whole incident and consequently also for his anger. These remonstrations, however, only had the effect that he no longer played for that company.

***

He composed the third of his string quartets, Opus 18 in D Major, first; the one in F Major, which is now the first, was originally the third.

***

Now another anecdote about *Fidelio* or, more correctly, *Leonore*.

Herr Röckel (presently in London) was a tenor at the Wiener Theatre in 1807. The cordiality of his relationship with Beethoven is demonstrated, for example, by the gift of an English dictionary, which is mentioned in a note in my possession. He told me the following anecdote in the spring of 1837. Beethoven's opera, *Leonore*, acknowledged as a failure in 1805, was to be revived in 1807.[58] The main reason for its earlier lack of success was that it premiered just after the French had occupied Vienna. At that time all music lovers and any of the wealthier people who possibly could had fled Vienna, so that there were only French officers in the theater for the most part. Moreover, both the text and the music were extremely drawn out in many places, so that the action proceeded at a very slow pace. Beethoven's friends consequently decided to shorten the opera and called a meeting at Prince Lichnowsky's. This was attended by the Prince, the Princess (who assumed her place at the piano and was known as an excellent pianist), Hofrath von Collin, Stephan von Breuning — these last two had already discussed the cuts to be made — as well as Herr Meyer, the first bass, Herr Röckel, and Beethoven. At the beginning Beethoven defended every measure, but when everyone expressed the opinion that whole numbers would have to be dropped and Herr Meyer claimed that no singer could sing the aria of Pizarro to good effect, Beethoven became rude and angry. Finally he promised to compose a new aria for Pizarro (the present No. 7 in *Fidelio*), and the Prince eventually got him to the point where

Prince Karl (1756-1814) & Princess Maria Lichnowsky (1765-1841)

he conceded that certain things should be left out in the first performance (but only on a trial basis). One could, they maintained, always put them back in again or use them elsewhere. Things couldn't possibly be left to stand as they originally were, since the piece in that condition had failed once already. After long discussion Beethoven gave in — and the cuts were never performed again.

This session lasted from seven o'clock at night until two in the morning, when a cheerful meal brought everything to a pleasant conclusion.

Among the numbers that were cut was a duet in 3/8 for

two sopranos and, I believe, also an aria and a trio in ¾. The first two numbers are, or were, in the possession of Herr Dunst in Frankfurt. The duet is with violin obligato and was performed here in Frankfurt at the benefit concert for the Beethoven memorial. Easy though it appears, it is difficult and strenuous. Whether more was cut in the revision I do not know.

For this performance Röckel was given the part of Florestan. Florestan's aria, No. 11 (at the beginning of the second act), had concluded with the Adagio in ¾ time in the first version. The Allegro in F major was added later by Beethoven for a tenor who refused to perform otherwise. In the first version Florestan had to sustain the high F at the end for four whole measures of the Adagio, while the orchestra slowly faded away. That particular tenor was incapable of this, and so, presumably in the revision, the part of the Adagio which returns to the key of F major or F minor was left out, and it now passes from A flat major, ¾ Adagio, straight into the Allegro in ¾ time in F major. This version of events was told me by Herr Röckel, who also claims to possess the vocal score in Beethoven's own hand.

The following note from Beethoven belongs in this context:[59]

Dear Röckel! Do make sure you do a good job with Milder. Just tell her that you are my emissary, sent in advance to beg her not to sing anywhere else. Tomorrow I will come myself to kiss the hem of her gown. Do not forget

Marconi either, and do not get annoyed with me for troubling you with so many things.

Entirely yours,
Beethoven.

*
* *

I remember only two occasions where Beethoven instructed me to add a few notes to any of his compositions, once in the Rondo of the *Sonate pathetique* (Opus 13) and another time in the theme of the Rondo in his first Concerto in C Major. There he dictated several octaves to make it more brilliant.

Incidentally, he performed this particular Rondo with a very special expressiveness. In general he played his own compositions most capriciously, though he usually kept a very steady rhythm and only occasionally, indeed, very rarely, speeded up the tempo somewhat. At times he restrained the tempo in his crescendo with a ritardando, which had a beautiful and most striking effect.

In playing he would give various passages, sometimes in the right hand and sometimes in the left, a beautiful, truly inimitable expression, but very rarely indeed did he add notes or embellishments.

*
* *

Artistically a most remarkable thing happened with one of his last solo sonatas (in B flat major with the great fugue, Opus 106)[60], which is forty-one pages in print. Beethoven had sent it to me in London to be sold, so that it would be published there at the same time as in Germany. The engraving was finished and I expected daily a letter specifying the day of publication. When the letter arrived, it included the surprising instruction: "Insert at the beginning of the Adagio (which is nine or ten pages long) these two notes as an opening measure."

I must admit that involuntarily I began to wonder if my dear old teacher had really gone daft, a rumor which was going about at the time. To add two notes to such a great work, which had been thoroughly reworked and completed half a year ago!! And yet, I was amazed at the effect of these two notes. Never again could such effective, important notes be added to a completed work, not even if they had been intended from the very beginning. I advise every music lover to try the beginning of this Adagio first without and then with these two notes, which now form the first measure, and there can be no doubt that he will then share my opinion.

Beethoven sent the score of the *Battle of Vittoria* to King George IV of England through the Austrian embassy and dedicated the work to him. He did not hear anything about it for a very long time, except that it had been performed every night at the gala evenings produced by the directors of the Drury Lane Theater, and that it had been greeted with much applause each time. Suddenly I received a letter, sealed in a special envelope and addressed to the King in Beethoven's own hand, enclosed in a letter to me with the instructions to hand it to the King personally. That was impossible, particularly with this king, since only persons of the highest rank, and of these just a select few, were admitted to his presence. Also, the letter itself was rather frightening to look at, since Beethoven had composed it himself and, in his own opinion, had penned it rather beautifully. I turned to Herr von Bauer, the secretary of the Austrian embassy. He answered that the Ambassador could not possibly hand the letter to the King in his position, but that he would try to get it into the King's hands through a private person. This attempt also proved fruitless. Finally I succeeded in arranging the letter's delivery through a page who loved Beethoven's compositions very much. What the letter contained I do not know, but I can state with absolute certainty that no present, not even a word of thanks, ever reached poor Beethoven. He complained greatly about this, and the rebuff probably gave impulse to a droll remark in one of his letters to me: "The King might at least have presented me with a butcher's knife or a turtle." Presumably, Beethoven had heard that the King loved rich and delicate foods, therefore the allusion.

Beethoven had hardly ever travelled. In his younger years, towards the end of the century, he had once been in Pressburg and Pest and once in Berlin. Even though he showed no difference in his behavior toward people in elevated positions and those of low station, he was not insensitive to the attentions of the former. In Berlin he played at the court (of King Friedrich Wilhelm II) several times, where he also composed and played the two sonatas for cello and piano, Opus 5, for Duport (the King's first cellist) and himself. Upon his departure he received a golden casket filled with Louis d'ors. Beethoven related with pride that it was no ordinary box but one suitable for presentation to ambassadors.

In Berlin he associated often with Himmel, whose talent he regarded as agreeable, but not much more — his piano playing was elegant and pleasant but not to be compared even with that of Prince Louis Ferdinand. The Prince he had once paid a great compliment, in his opinion, when he observed that he did not play at all like a king or prince, but like a skillful pianist. Himmel and he fell out in the following manner:

Friedrich Heinrich Himmel (1765-1814)

when they were together one day Himmel asked Beethoven to improvise, which Beethoven did. Afterwards Beethoven insisted that Himmel do the same. Himmel was weak enough to agree. After he had played for quite some time, Beethoven said: "Well, when are you actually going to begin?"

Himmel flattered himself that he had been achieving something marvelous already; he therefore left the piano, and both of them exchanged rude remarks with one another.

Beethoven said to me: "I thought Himmel had just been offering a little prelude." They did make up later, but Himmel could only forgive, never forget. They corresponded for some time until Himmel played a mean trick on Beethoven. Beethoven always wanted to hear the latest news from Berlin. This bored Himmel, who finally wrote that the newest thing there was the invention of a lantern for the blind. Beethoven ran about with this piece of news, and of course everyone wanted to know how this was possible. He therefore immediately wrote to Himmel, charging that it was negligent of him not to have sent him any further details.[61]

The answer he received, which cannot be quoted here, not only put an end to their correspondence forever, but brought ridicule on Beethoven, since he was rash enough to show it here and there.

***

When Prince Louis Ferdinand was in Vienna, an old countess held a small musical evening, to which Beethoven was also naturally invited. When the company went in to dinner, the Prince's table was set only for the high nobility — not for Beethoven. He became very agitated, made a few blunt remarks, took his hat and left.

A few days later Prince Louis gave a luncheon, to which some of that party, including the old countess, were invited. When they sat down at the table, the countess was placed on one side of the Prince and Beethoven on the other. He always mentioned this distinction with pleasure.

Archduke Rudolph (1788-1831)

Beethoven was a stranger to rules of etiquette and all that they imply; he never concerned himself about such things. Thus he often acutely embarrassed the entourage of Archduke Rudolph when he first started to frequent that circle. Attempts were made to coerce Beethoven into behaving with the proper deference. This was, however, unbearable for him. He did promise to mend his ways, but that was as far as it went. One day, finally, when he was again, as he termed it, being "sermonized on court manners," he very angrily pushed his way up to the Archduke and said quite frankly that though he had the greatest possible reverence for his person, a strict observance of all the regulations to which his attention was called every day was beyond him. The Archduke laughed good-naturedly about the incident and gave orders to let Beethoven go his own way in peace; he must be taken as he was.

***

Beethoven needed a great deal of money, even though he enjoyed very little benefit from it, for he lived very modestly. While he was composing *Leonore*, he had free lodgings in the Wiedner Theater for a year, but since the rooms faced the courtyard, he was not satisfied with them. He therefore rented at the same time lodgings in the Rothes Haus, near the Alster Barracks, where Stephan von Breuning also lived. In the summertime he took rooms in Döbling in the country. As a result of a quarrel with Stephan von Breuning (referred to in Beethoven's letter to me dated 24 July 1804, when Breuning had called in the caretaker as a witness for some statement of his), he commissioned me to find lodgings near the walls of the city. I chose a flat on the Mölker-Bastei in the Pasquillati house; it was on the fourth floor and had a very lovely view. Beethoven then had four apartments at the same time.

He moved out of the last several times but always returned again so that, as I afterwards heard, Baron Pasquillati generously remarked when Beethoven moved out: "The rooms are not to be rented; Beethoven will come back."[62]

***

Beethoven attached no importance to his manuscripts.

Once they had been printed, they usually lay about in an adjoining room or scattered on the floor in the middle of his study with other pieces of music. I often put his music in order, but when Beethoven was looking for something, everything was turned upside down again. At any time I could have carried off all those originally autographed compositions that had been engraved already; he would also most likely have given them to me without hesitation had I asked him for them.

Beethoven had given me his beautiful Concerto in C Minor (Opus 37), still in manuscript, so that I might make my first public appearance as his pupil with it. I am also the only student to make such an appearance during Beethoven's lifetime.

Other than myself he acknowledged only the Archduke Rudolph as his pupil. (Cf. Seyfried, op. cit., p. 12.)[63] Beethoven himself conducted, though he only turned the pages, and it may well be that no concerto was ever accompanied more beautifully.[64] We held two major rehearsals. I had asked Beethoven to compose a cadenza for me, which he refused, and instructing me to write one myself, said he would correct it. Beethoven was very satisfied with my composition and made few changes; but there was an extremely brilliant and very difficult passage in it which, though he liked it, seemed too daring to him. He therefore instructed me to compose another in its place. A week before the performance, he wanted to hear the cadenza again. I played it and bungled the passage; again, somewhat irritated, he told me to change it. I did, but the new one did not satisfy me; I therefore diligently practiced the first one as well, without ever becoming absolutely confident of it. At the concert Beethoven sat down quietly when we reached the cadenza. I could not bring myself to choose the easier one. When I boldly began the more difficult one, Beethoven started violently in his chair. Nevertheless, it came off perfectly, and Beethoven was so pleased that he cried "Bravo!" aloud. This electrified the whole audience and immediately assured my standing among the artists. Later, while expressing his satisfaction with my performance, he added: "But you are stubborn all the same! If you had missed that passage, I would never have given you another lesson again."

One day Beethoven visited me, carrying his fourth Concerto in G major (Opus 58)[65] tucked under his arm, and said: "Next Saturday you have to play this in the Kärnther-Thor Theater."[65a] Only five days remained to practice it. Unluckily I remarked to him that the time was too short to learn to play it well; would he not permit me to perform the C minor Concerto. Beethoven was annoyed about this, turned on his heels, and went to young Stein, whom he did not really like much. Stein was also a pianist, and in fact had played longer than I. He was clever enough to accept the suggestion immediately. Since he could not finish learning the Concerto either, he went to Beethoven the day before the performance and asked, as I had, if he could play the other one, in C minor. Beethoven more or less had to give in and agreed to this.

However, whether the fault lay with the theater, with the orchestra, or with the pianist himself, the performance was not successful. Beethoven was most annoyed, especially since he was asked by several people: "Why didn't you let Ries play it, since he had such a success with it?" These remarks gave me a great deal of pleasure. Later Beethoven said to me: "I thought you just did not particularly care to play the G major Concerto."

The piano part of the C minor Concerto was never completely written out in the score. Beethoven wrote it down on separate sheets of paper expressly for me.

***

In his letter of introduction to Beethoven my father opened a small account for me with him in case I should need it. I never had to make use of it; however, when on several occasions Beethoven noticed I was short of money, he sent me some of his own accord and would never accept repayment. He was really fond of me and once gave me very amusing proof of this affection, in his absent-minded way. I had just returned from Silesia after serving for a rather long period on Beethoven's recommendation as a pianist on the estates of Prince Lichnowsky. I entered the room as he was just about to shave and was covered with soap up to the eyes (his frightfully strong beard grew that high). He jumped up, embraced me cordially, and behold! he had transferred every last bit of lather from his left cheek onto my right.* How we laughed! Beethoven must also have heard privately some stories concerning me, because he knew several of my youthful indiscretions, about which, however, he only teased me. He behaved in a truly paternal fashion toward me on many occasions, including when he was displeased. Once, in 1802, he wrote to me, quite annoyed, about an awkward situation Carl Beethoven had gotten me entangled in: "Don't bother coming to Heiligenstadt, since I have no time to waste." Count Browne was at that time leading a rather high life, and since he took great pleasure in my company, I too often went carousing and neglected my studies in the process.

Beethoven very much enjoyed looking at women; lovely, youthful faces particularly pleased him. If we passed a girl who could boast her share of charms, he would turn around, look at her sharply again through his glass, then laugh or grin when he realized I was watching him. He was very frequently in love, but usually only for a short time. When once I teased him about his conquest of a certain beautiful lady, he

---

* *Nil novi sub sole.* "When he (Lord Lovat) was brought to London, Hogarth visited him, finding him in the hands of a barber. Lord Lovat's happiness at seeing his old friend was extraordinary; he jumped up, embraced Hogarth, and conferred upon him, of course, a large portion of the lather on his face." Lichtenberg's *Erklarung*, etc. ninth edition, p. 155.

Wglr.

confessed that she had captivated him more intensely and longer than any other — seven whole months.

One evening I went to Baden to continue my lessons with him. There I found a handsome young woman sitting beside him on the sofa. Feeling that I had come at an inopportune moment, I wanted to leave immediately, but Beethoven detained me and said: "Sit down and play for a while!"

He and the lady remained seated behind me. I had already been playing for a long time when Beethoven suddenly called out: "Ries! Play something romantic!" Soon after: "Something melancholy!" Then: "Something passionate!" and so on.

From what I could hear I deduced that he had evidently offended the lady somehow and now was trying to make up for it by amusing her. Finally he jumped up and shouted: "Why, those are all things I have written!" I had been playing nothing but movements from his own works all the time, connecting them with small transition passages, which seemed to please him. The lady soon left and, to my great surprise, Beethoven did not know who she was. I learned that shortly before I arrived, she had come to the house, wishing to meet Beethoven. We immediately followed her to find out her lodgings and through them later her social position. We could still see her from a distance (it was a moonlit night), but suddenly she disappeared. We then walked another hour and a half or more in the lovely neighboring valley, talking about all kinds of things. In parting, however, Beethoven said: "I must find out who she is and you must help me." A long time afterward I met her in Vienna and discovered that she was the mistress of a foreign prince. I told Beethoven the news but never heard anything more about her, neither from him nor from anyone else.

Beethoven never visited me more frequently than when I lived in a house owned by a tailor with three very beautiful but absolutely irreproachable daughters. He refers to them at the conclusion of his letter of 24 July 1804 when he says: "Do not do too much tailoring, remember me to the fairest of the fair, send me half a dozen sewing needles!"

***

Even after he reached Vienna, Beethoven had taken violin lessons with Krumpholz,[66] and when I was first there, we sometimes played his piano and violin sonatas together.* That was, however, truly dreadful music-making, because in the throes of his enthusiasm he did not hear when he attacked a passage with the wrong fingering.

Beethoven was most awkward and bungling in his behavior; his clumsy movements lacked all grace. He rarely picked up anything without dropping or breaking it. Thus he frequently knocked his inkwell into the piano, which stood beside his writing desk. No piece of furniture was safe from him, least of all anything valuable. Everything was knocked

---

* That is, when his hearing was already impaired!

Wglr.

over, soiled, or destroyed. How he ever managed to shave himself at all remains difficult to understand, even considering the frequent cuts on his cheeks. — He never learned to dance in time with the music.

Beethoven's String Quintet (Opus 29) in C Major had been sold to a publisher in Leipzig but was stolen in Vienna and suddenly published by A. & Company.[66a] Since it had been copied in one night, there were innumerable mistakes in it, and even whole bars were missing. Beethoven here behaved in such an exemplary fashion that one would be hard put to find anywhere another example of such manners. He asked A. to send the fifty copies which had already been printed to me at home for correction, while at the same time telling me to make the corrections in ink so roughly on the poor paper and to cross out several lines in such a way that it would be impossible to use a copy or to sell it. Most of the crossing out occurred in the Scherzo. I followed these instructions faithfully, and A. had to melt down the plates to avoid a law suit.

Beethoven was very forgetful in many things. Once he had received a handsome riding horse as a present from Count Browne in return for the dedication of the Variations in A Major, No. 5 on a Russian Song.[67] He rode it a few times but soon forgot all about it and, what was worse, forgot all about feeding it too. His servant soon noticed this and started hiring out the horse to his own advantage. So as not to attract Beethoven's attention, he put off handing him a bill for the fodder for a very long time. At last, however, to Beethoven's great astonishment, a rather large feed bill was presented to him, which suddenly recalled to him the existence of his horse, as well as his negligence in regard to it.

***

During the short bombardment of Vienna by the French in 1809, Beethoven was very frightened. He spent most of the time in the cellar at his brother Caspar's house, where he covered his head with pillows to shut out the sound of the cannons.*

***

Beethoven was at times extremely short-tempered. One day we were eating lunch in the Swan Inn, and the waiter brought him the wrong dish. No sooner had Beethoven remarked about it — and received an answer from the waiter which was not exactly deferential — then he took the dish (it was a so-called Lungenbratel, a roast with plenty of gravy) and flung it at the waiter's head. The poor man carried a large number of plates piled with food on his arm (the Viennese

---

* Might not the thundering of the cannons have been painful for his suffering ears?

Wglr.

waiters are extremely clever at that), so he was quite helpless. The gravy ran down his face. He and Beethoven shouted and abused each other, while all the other guests laughed loudly. Finally, Beethoven broke off and burst out laughing also at the sight of the waiter with gravy dripping down his face, licking it up with his tongue, trying to curse but forced to resume licking up the gravy and pulling the most ludicrous faces. A picture worthy of a Hogarth.

***

Beethoven had hardly any understanding of money, which often gave rise to the most unpleasant scenes. Suspicious by nature, he often thought himself cheated when this was not the case. Easily excited, he would call people cheats to their faces and, where waiters were concerned, often had to make up for it with a tip. Finally his peculiarities and absent-mindedness became known in the inns he most frequently visited, and anything was tolerated, even if he left without paying.

***

Beethoven remembered his early youth and his friends from Bonn with much pleasure, although in reality those had been difficult times for him. Of his mother especially he spoke with love and affection, and often called her a fine, truly kindhearted woman. He did not like talking about his father, who was mainly at fault for the family misfortunes, but any harsh word about him let fall by a third person made

Beethoven angry. In general he was a thoroughly kind person at heart, only his temper and irritability often got him into trouble. And no matter what insult or injustice had been done him by anyone, Beethoven would have forgiven him on the spot, had he met him when crushed by misfortune.

Johann van Beethoven (1740-1792), the Composer's Father

Beethoven's way of conducting business was rather peculiar. For instance, he had commissioned me to sell his piano sonatas (Opus 110 and 111) and 33 Variations on a Waltz (Opus 120), which he expected to send to me soon in London. I had already come to an agreement with Messrs. Clementi & Company for the sonatas and with Boosey, the music publishers, concerning a fee for the Variations, and I was waiting for the works to arrive. Finally they came. To my surprise I saw that in very large, even letters drawn by his own hand on the title page, Beethoven had dedicated the Variations to my wife.[67a] This dedication, however, is found only on that single copy which is still in my possession.

Beethoven had put off sending these manuscripts for so long that in the meantime he had completely forgotten what he had instructed me to do. When I took the Variations to Boosey, we discovered not only that they were the same as a set already published in Vienna — with a dedication to Madame Brentano — but also one sonata had already been printed in Paris!

Beethoven had enclosed a few insignificant pieces (the second collection of Bagatelles), some of which had better not been published at all; I sold them immediately for twenty-five guineas, wrote Beethoven how everything worked out, and received a response charging me with negligence. He apologized for the double dedication. To my amazement he pointedly mentioned: "I would never think of accepting a present or a gratuity for it!" One could hardly find a more sudden about-face or a more obvious self-contradiction!

I have grave doubts about manuscripts found in Beethoven's estate. I shall only recognize *oeuvres posthumes* as genuine after I have seen a copy in his own handwriting or otherwise properly verified.

My reasons are as follows:

First: when I was with him from 1800 to November, 1805, and later in 1809, when I returned to Vienna, there were no manuscripts on hand because Beethoven was always behind with commissioned compositions right up to the day of his death.

Secondly: all his little compositions and many things he never wanted published, since he did not consider them worthy of his name, were secretly brought out by his brothers. Thus songs he had composed in Bonn years before his departure for Vienna became known only after he had achieved a high reputation. Even small compositions which he had written in autograph albums were purloined and engraved in this way.

Thirdly: as almost all the letters I received from him in England dealt with money difficulties, why wouldn't he have sent me the manuscripts straightaway, had he had any?

Once, after a lot of effort, I had prevailed upon the Philharmonic Society in London to order three overtures from him, which were to remain the Society's property. He sent me three, none of which we could perform in view of Beethoven's great reputation, since at these concerts everyone eagerly looked forward to, indeed demanded, something out of the ordinary from him. He had all three published a few years later, but the Society did not find it worth its while to complain. Among them was the *Overture to the Ruins of Athens*, which I consider unworthy of him.

If Beethoven had had anything better in manuscript, he would certainly have sent it to this Society; that is quite clear from all his letters. Since he also frequently claimed he could live by his pen alone, I find it virtually impossible to acknowledge as authentic the three piano quartets published by

Artaria after his death.

Beethoven could not have patched together from old themes a gigantic work like the three sonatas (Opus 2), dedicated to Haydn, which immediately created such a sensation throughout the world. Nor would he have employed these themes years later in empty, badly written quartets. After all, his mind kept pouring forth new ideas incessantly until his death.

Without wanting to hurt the feelings of any composers living or dead, I must reaffirm my belief that no one else showed such a wealth and variety of ideas, nor such originality, as Beethoven did in his works. Even though Beethoven as a friend and teacher stood and remains above all others for me, I have never been known as the sort who reveres only one or at the most two musical idols and judges everything not by them as mediocre, if not downright bad, in advance. Such one-sidedness never will be a failing of mine.

Now follow several letters of Beethoven's, all of them written entirely in his own hand except one, which is but partially in autograph. Most of them offer corroboration of what I have written above, providing more details or elaborating on what I have said.

A small portion of the following consists of notes, usually without specific dates, through which he gave instructions to me as his pupil. Then come more recent letters up to the year 1825. During my travels through North

Germany, Russia, Sweden, and Denmark I very rarely heard from him directly, but after I settled in England, he wrote quite often. Everything else should be clear from the letters themselves.

Some of Beethoven's letters containing very definite and most uncomplimentary opinions about certain people I have withheld, at least for the time being, if not, I hope, forever. Why should anyone either through harassment or negotiation put pressure on me to do otherwise?*

*
**

(Undated, no address)

(Vienna, probably 1801)

Here, dear Ries! Take these four parts which I have corrected and check the other copies against them right away. . . . Here is the letter to Count Browne; it says that he must give you the 50# (ducats) in advance because you have to buy some things for yourself. He cannot take offense at that since as soon as you have finished, you will be leaving for Baden with him next Monday. All the same, I must reproach you for not having turned to me much sooner. Do you not consider me your true friend? Why do you hide your need from me? None of my friends should go in want as long as I have anything; I would have

---

* I, too, possibly even more sensitive about this since the death of my friend than he was during his own life, would like to add a warning to this effect, since I am fully familiar with the subject as well as the content of the letters in question.

Wglr.

already sent you a small sum if I were not relying on Browne to help you; if that fails, turn immediately to your friend.

Beethoven

(1803)

You probably already know that I am here. Go to Stein and find out whether he cannot deliver an instrument to me — to be paid for. I am worried about having mine carted out here. Come this evening around seven o'clock. My lodgings are in Oberdöbling No. 4[1]*, on the left side of the road where one goes down the hill to Heiligenstadt.*

(1) A village near Vienna.

(Probably 1804)

Dear Ries! Please do have the kindness to copy this Andante[1]*, even if only poorly. I must send it off tomorrow and — since heaven alone knows what might happen to it — I would like to have a copy. But I must have it back around one o'clock tomorrow. The reason why I am bothering you with this chore is that one copyist is occupied with other important things and the other is ill.*

(1) It was, unless my memory fails me, the Grand Andante with Variations from the Kreuzer Sonata for piano and violin, Opus 47.**68**

Baden, 14 July 1804

If you, dear Ries, could manage to find me better lodgings, I would really appreciate it. —— I would very much like to have rooms on a large, quiet square or on the ramparts.* —— I will make every effort to be at the rehearsal on Wednesday. I am not pleased that it is at Schuppanzigh's. He ought to be grateful if all my insults have caused him to lose a little weight. **68a** *Stay well, dear Ries! We are having bad weather, and I am not safe from people here. I must run away to be alone.*

(Vienna, probably the beginning of July, 1804)

Dear Ries, Since Breuning did not hesitate in his behavior to represent my character to you and to the caretaker in such a light that I appear as a miserable, wretched, mean man, I choose you to convey my answer to Breuning directly. But tell him I am replying to only one point, indeed the first point, of his letter, which I do only to justify my character to you! —— Tell him, therefore, that I had not considered reproaching him for the delay in giving notice and that, if it was really Breuning's fault, any harmonious relationship in this world is far too valuable and dear to me to hurt one of my friends for the sake of a few hundred or even more. You yourself know that I had accused you in jest that it was your fault that the notice had come too late. I know for sure you will remember that; I had forgotten the whole affair. Then my brother started up again, while we were sitting at the table, and said that he thought Breuning was at fault there. I denied it on the spot and said that it was your fault. Surely it was clear enough that I did not blame Breuning. Breuning, however, jumped up in a frenzy and said that he wanted to call in the

---

* A verification of my opinion in Note 6 to the letter of 2 November 1793.   Wglr.

caretaker. I was not used to such behavior from any of the the people with whom I am friends, and it made me lose my self-control. I also jumped up, knocked my chair over, went away and did not come back. This behavior then induced Breuning to present me in such a nice light in front of you and the caretaker. He also wrote me a letter, which, incidentally, I have answered only with silence. —— I have nothing more to say to Breuning. His way of thinking and acting toward me proves that no friendly relationship should ever have existed between us and certainly shall not exist in future. I wanted to acquaint you with this, because you have misunderstood and belittled everything I said and did in this instance. I know that, had you known the affair as it really was, you certainly would not have done so and am satisfied with that.[1]

Now I ask you, dear Ries, immediately upon receipt of this letter to visit my brother the pharmacist and tell him that I shall be leaving Baden in a few days and that he should rent the lodgings for me in Döbling immediately now that you have notified him. I almost came today; I loathe it here and am tired of it all. For heaven's sake urge him to rent it immediately, for I want to stay there straight away. Don't tell Breuning anything, and show him nothing of the letter written on the other side. I want to demonstrate in every way possible that I am not as small-minded as he and that I have written him only after sending this letter to you, even though the decision to sever our friendship is and remains fast.

Your friend,

Beethoven.

(1) The real basis for the quarrel mentioned in this letter is probably connected with an eviction notice on one of Beethoven's lodgings postponed or ignored by Breuning (because at that time Beethoven lived in the Theater auf der Wieden). Breuning, a hothead like Beethoven, was especially

enraged about his friend's behavior, as Beethoven himself describes it here, because it happened in the presence of Beethoven's brother. — The following letter is a sequel to this. A few months later the two met by accident, and a complete reconciliation took place. Beethoven's hostile intentions, however vehemently proclaimed in the two letters, were completely forgotten. Beethoven's letters prove this at several points, as do those written by Breuning. Beethoven dedicated a work (Opus 61) to him and named him as joint guardian of his nephew. Incidentally, the two letters with their sequels, particularly the next letter, are too fine a documentation of Beethoven's character not to be included here.

Baden, 24 July 1804

...You will surely have been surprised over the affair with Breuning. Do believe me, my dear friend, when I say that my explosion was only the culmination of several previous unpleasant incidents with him. I have the gift of concealing and containing my sensitivity to a great many things. But if I am ever irritated at a time when I am more susceptible to aggravation, then I lose my temper more violently than any other man. Breuning certainly has some excellent qualities, but he thinks himself free of all fault, and usually suffers to the greatest degree from precisely those weaknesses he divines in other people. He has a meanness of spirit which I have despised since childhood. My good sense had almost discerned the way things would go with Breuning, since our way of thinking, acting, and feeling differs too much, but I had thought that these difficulties might also be overcome. — Experience has proved me wrong. And now I have had enough of friendship! I have found only two friends in this

world with whom things have not gone sour at some point. And what men! One is dead; the other is still living.[69] *Even though for almost six years now neither of us has heard anything of the other, nevertheless, I know I occupy the first place in his heart, as he does in mine. True friendship must be grounded in the rarest similarity of heart and soul in men. I wish only that you should read the letter which I wrote to Breuning and his letter to me. No, never again will he command the place in my heart that he once had. Anyone who can attribute to his friends such a low level of thinking and could respond himself by behaving in such a manner — he is not worth my friendship. — Don't forget to see to my lodgings. Stay well. Don't do too much tailoring, remember me to the fairest of the fair, and send me half a dozen sewing needles. — I would never in my life have believed that I could be so lazy as I am here. If I should suddenly be seized with energy, something worthwhile might possibly result.*

Vale.

Beethoven

*(Undated. Written a few days before
the French invasion, 1805)*

Please excuse me, your serene highness, if the bearer of this message should alarm you. Poor Ries, my pupil, is being forced to take up arms in this miserable war[1]*, and is, furthermore, obliged to depart from us in a few days because he is an alien. — He has nothing to his name, absolutely nothing, and must make a long journey. The opportunity to appear at an Academy concert is now entirely lost to him because of these circumstances. He must take refuge in charity. I recommend him to you. I know you will excuse my boldness. Only extreme need could force a high-minded man to plead in such a way for another's generosity.*

> *With complete confidence I send the poor fellow to you, hoping you will lighten somewhat the burden of his difficulties. He must appeal to everyone who knows him.*
>
> > *With deepest respect,*
> > *L. van Beethoven*[2]

(Address) *Pour Madame la Princesse Liechtenstein*, etc.

(1) Being born on the left bank of the Rhine, I was conscripted under a French law.

(2) The letter was never delivered (to Beethoven's great anger), but I preserved the original written on a small, unevenly cut quarter sheet as a testimony of Beethoven's friendship and love for me.

***

(1809)

Your friends, my dear fellow, have certainly given you some bad advice. But I know them — they are the same people from whom you gathered the pleasant news you sent me from Paris[1], *who inquired after my age, which you were so good as to supply,*[2] *who have lowered you in my estimation several times already and now forever.*

Farewell,[3]

> *Beethoven.*

(1) I no longer recollect the immediate reason for this note. I had written from Paris that musical taste there was poor and few of Beethoven's works were known or performed there.

(2) Some friends of Beethoven wanted to know his birthday for certain. I spent a great deal of effort looking for his baptismal certificate when I was in Bonn in 1806. Finally I

found it and sent it to Vienna. He never liked to speak of his age.

(3)   He soon got over his irritation, and the former friendship was resumed.

Wednesday, 22 November. Vienna, 1815

Dear Ries, I write in haste to tell you that I have sent the piano score of the Symphony in A to Messrs. Thomas Coutts & Co. by post today. Since the Court is not here, couriers are rare or nonexistent, and in any case this is the safest way. The symphony should come out about March; I will determine the day. It took too long this time for me to set a more specific deadline. —— The trio in the violin sonata may take longer; both will be in London within a few weeks. —— I urgently entreat you, dear Ries, to take an interest in this matter yourself to be sure that I receive the money. It costs so much before everything is finished and I need it.

I have suffered a reduction of 600 florin annually in my salary; at the time of the B.N. (bank notes) it was nothing at all. —— Then came the redemption (certificates), and because of them I lost these 600 florin. After several years of aggravation and virtual disappearance of the stipend we have now reached the point where the R.C. (redemption certificates) are worse than the B.N. (bank notes) ever were. I pay 1,000 florins rent. Imagine the misery this paper money causes. My poor unfortunate brother (Carl) has just died; he had a wretched wife; I might add that he had consumption for several years, and what money I gave to make life easier for him I figure must come to some 10,000 florin V.C. (Viennese currency). This is of course nothing to an Englishman, but, for a poor German —— or worse yet, an Austrian —— it amounts to a great deal. The poor man changed terribly in his last years, and I must say I pitied him with my whole heart. Now I am very happy to be able to say that as far as his support was concerned I have no need to reproach myself. — Tell Mr. Birchall that he should recompense Mr. Salomon[70] and yourself for the postal expenses incurred by your letters to me and mine to you; he can deduct it from the sum due me; I want those who act on my behalf to suffer as little as possible.

Wellington's Victory in the Battle of Vittoria[(1)] must long since have arrived at Th. Coutts & Co. Mr. Birchall need not pay the fee until he has received all the compositions. Do hurry and let me know the date when Mr. Birchall can bring out the piano reduction. Other than that I urge you heartily to keep looking after my affairs generally; I am at your service however you might need me. With my most cordial wishes, dear Ries!

Your friend

Beethoven

(1) This is the same as the title on the piano reduction. (Beethoven's own note.)

Excerpt from Beethoven's letter to Ferdinand Ries, 20 January 1816

***

Vienna, 20 January 1816

My dear Ries, The Symphony is dedicated to the Empress of Russia. The piano score of the Symphony in A must not be published before the month of June; the publisher here cannot manage it earlier. Convey this immediately to Mr. Birchall, dear, kind Ries. — The sonata with violin, which will be sent by next post, can also be published in London in the month of May. The trio, however, later.[71] *(It will also arrive with the next post.) I will set the date for it later.*

And now, my heartfelt thanks, dear Ries, for all the kindness you show me, and particularly for the corrections. May heaven bless you and may you enjoy ever greater success, in which I take the most cordial interest. My greetings to your wife!

As always. Your true friend

Ludwig van Beethoven.

28 February 1816

. . .I have not been well for some time; my brother's death affected my spirits and my work.

Salomon's death grieves me greatly because he was a noble man whom I remember even from my childhood. You have become the executor of his will, and I at the same time guardian of my poor late brother's child. You will hardly have had as much aggravation as I have had over that death; yet I have the sweet consolation of having saved a poor innocent child from the hands of an unworthy mother.

Fare well, my dear Ries! If I can serve you in any way, do consider me only as your very true friend

Beethoven.

Vienna, 8 March 1816

My answer comes somewhat late. —— However, I was ill and have had a lot to do. . . . Not a penny of the 10 ducats in gold has arrived so far, and I am beginning to believe that the English, too, are only generous when abroad; the same applies to the Prince Regent, from whom I did not even receive the copying costs for my Battle *which I sent him, indeed not even a word or note of thanks. . . . My salary amounts to 3400 florins in paper money. I pay 1100 rent; my servant and his wife 900 fl.; you can work out what remains. At the same time I have to support my little nephew completely; until now he has been*

Karl van Beethoven, Beethoven's nephew (1806-1858)

at boarding school; this costs up to 1100 fl. and is still not satisfactory, so that I have to establish a regular household before I can bring him to live with me. How much one has to earn merely to be able to live here; and yet there is no end to it, because — because — because — you know what I mean.

A few commissions from the Philharmonic Society, apart from an Academy concert, would also be highly welcome.

Incidentally, my dear pupil Ries ought to sit down and dedicate something good to me, whereupon the Master would respond by repaying him in kind.[1] *How shall I send you my portrait? . . . My best wishes to your wife; alas, I have none; I found only one, whom I will most likely never possess,*[72] *but that does not make me a woman hater.*[2]

Your true friend
Beethoven

(1) Cf. Note 1 to Beethoven's letter of 2 May 1810.
(2) Cf. Note 4 to Beethoven's letter of 16 November 1801.

*

Vienna, 3 April 1816

. . .Neate must presumably be in London now; I have given him several compositions of mine and he promised to put them to good use, to my advantage. . .Archduke Rudolph also plays your works with me, my dear Ries! of which il sogno *has pleased me particularly well. Fare well, salutations to your dear wife, and to all the beautiful English ladies whom it might please — my personal greetings.*

Your true friend
Beethoven.

*

Vienna, 11 June 1816[73]

My dear Ries,

I am sorry that you have to spend postage money again for my sake; though I may be happy to help and serve all men, it hurts me to have to take advantage of others. Of the 10 ducats still nothing has appeared,[74] *and so one must come to the conclusion that in England, as here, there are windbags and people who do not keep their word. — I do not blame you for this at all.*[75]*. . .Since I also have not heard a single syllable from Neate,*[76] *I would ask you simply to inquire*[77] *whether he has succeeded in selling the Concerto*[78] *in F Minor.*[79] *I am almost ashamed to speak of all the other works which I gave him. I mean ashamed of myself*[80] *for handing them over to him so trustingly,*[81] *so completely without any conditions other than those which I invented in the name of friendship and conscience.*[82]

Someone gave me a translation of a notice in the Morning Chronicle[83] *about the performance of the Symphony (probably the one in A).*[84] *Undoubtedly this and all the other works taken along by Neate*[85] *will turn out like the* Battle (of Vittoria)[86], *that is, I'll most likely enjoy no benefit from them*[87] *other than reading about performances in the newspapers.*[88, 89]

*
* *

Vienna, 9 July 1817

Dear friend,

The offers made in your kind letter of 9 June are most flattering. This letter is meant to show you how much I appreciate them. Were it not for my unfortunate handicap, which entails considerable care and expense, especially while traveling and in a foreign country, I would unquestionably accept the proposal of the Philharmonic Society.

Put yourself, however, in my position. Consider how many more obstacles I must fight than any other artist, and then judge whether my demands are excessive. Here they are, and I would ask you to convey them to the directors of this particular Society.

1) I will be in London by the first half of the month of January, 1818, at the latest.

2) The two grand symphonies, entirely new, shall be ready then, and they will be and will remain the whole property of the Society.

3) The Society is to give me 300 guineas for them, plus 100 guineas toward traveling expenses, which will, however, come to much more, as I absolutely must take a companion with me.

4) Since I shall commence work on the composition of these grand symphonies immediately, the Society shall (upon acceptance of my conditions) transmit the sum of 150 guineas to me here, so that I can equip myself with a carriage and other necessities for the journey without delay.

5) The condition that I not appear with another orchestra in public, that I not conduct, and that the Society have precedence in other, similar instances, I accept. These restrictions could have gone without saying, from my sense of honor alone.

6) I may rely on the Society's assistance in preparing and promoting one or more benefit concerts for myself, depending on circumstances. Both the special friendship of several directors of your honorable Reunion, *as well as the musician's kind interest in my works, vouches for such cooperation and encourages me all the more to fulfill the general expectations.*

7) I further ask to receive the agreement or confirmation of the above in the English language, signed by three directors in the name of the Society.

You can well imagine how much I look forward to meeting the worthy Sir George Smart and to seeing you and Mr. Neate again. If only I could fly there myself in place of this letter!

Your loyal admirer and friend

L. v. Beethoven

(P.S. in Beethoven's own hand)

Dear Ries, I embrace you warmly; I intentionally engaged another hand for the above so that you could read all of it better and present it to the Society. I am convinced of your good will toward me and hope that the Philharmonic Society will agree to my proposal. You need have no doubt that I will do everything within my power to fulfill properly the commission with which such a select society of artists has greatly honored me. —— How large is your orchestra? How many violins, etc., etc.? Single or double winds? Is the hall large, resonant?

⁎
⁎⁎

Vienna, 5 March 1818

My dear Ries,

Much though I wished to, I could not come to London this year; please inform the Philharmonic Society that weak health prevented me. I do hope, however, to be fully recovered this spring. Then later in the year I will be able to respond to the Society's offer and shall fulfill all the conditions thereof. Tell Neate I wish he would not play, in public at least, any of the various works which he took from me until I come. However things may have worked out with him, I have reason to complain about his behavior.

Botter[90] visited me a few times; he seems to be a good man and has a talent for composition. — I hope and pray that your fortune improves daily. Alas, I cannot say the same about myself. . . . I can never see anyone in need but what I must give him something. Therefore you can well imagine that this whole business is even more insufferable. Please do write me soon. If at all possible, I shall leave here even earlier to avoid being utterly ruined, and then I shall arrive in London during this winter at the latest. I know you will stand by your unhappy friend. If only it had been within my power and I had not been fettered by circumstances here as always, I would certainly have done far more for you. — Farewell; give my regards to Neate, Smart, Cramer — even though I hear that he is a creature at odds with you and me. In the meantime I have learned the art of dealing with things like that a little better, and in London we will produce a pleasant harmony in spite of everything.

I greet and embrace you with all my heart.

Your friend

Ludwig van Beethoven.

My cordial greetings to your dear, beautiful (so I hear) wife.

My dear Ries,   *Vienna, 30 April 1819*

Only now can I answer your last letter of 18 December. Your sympathy comforts me. For the moment it is impossible to come to London, entangled as I am in various difficulties. But God will speed me to London next winter surely; then I will also bring the new symphonies. I am expecting the text for a new oratorio I am writing for the Musik-Verein *here, which ought to serve us in London too. Do for me what you can, because I need the help. Commissions from the Philharmonic Society would have been very welcome to me. The reports which Neate has sent me in the meanwhile about the virtual failure of the three overtures have upset me; each one was, in its own way, not only well received here, but those in E Flat and C Major even made a great impression.*[91] *The fate of these compositions with the Philharmonic Society is incomprehensible to me. You will by now have received the Quintet arrangement*[92] *and the Sonata. Make sure that both works, but especially the Quintet, are printed immediately. It may take a little longer with the Sonata, but I would wish for it to appear at least within two or, at the most, three months. The earlier letter you mention I did not receive. Therefore I did not hesitate to hawk the two works here as well — but for the rights in Germany only. Meanwhile, it will be another three months before the Sonata is published here;*[93] *just hurry with the Quintet. As soon as you send me the money, I will send a statement certifying the publisher as owner of these works for England, Scotland, Ireland, France, etc.*

You will receive the tempi *according to Mälzel's metronome with the next post. De Smidt, Prince Paul Esterhazy's courier, has taken the Quintet and the Sonata with him. At the earliest opportunity you will also receive my portrait, since I hear you really want one.*

Farewell. Remain fond of me, your friend   *Beethoven*

Beautiful greetings to your beautiful wife!!!
From me!!!!!

Vienna, 16 April 1819

Herewith, dear Ries, the tempi for the Sonata.
1st Allegro, but only Allegro, the *assai* must go.

Mälzel's metronome ♩ = 138.

2nd piece scherzoso.[94] *M. metronome* ♩ = 80.

3rd piece *M. metronome* ♪ = 92.

Note here that the first bar still must be inserted as follows:

1st bar

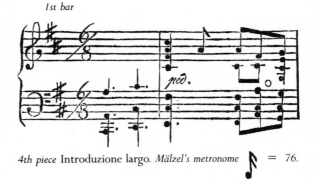

4th piece Introduzione largo. *Mälzel's metronome* ♪ = 76.

5th piece, ¾ time.

and the last, *Mälzel's metronome* ♩ = 144.

Forgive the confusion; if you knew my situation, you would not be surprised. You would instead be amazed at what I can still achieve. The Quintet finally cannot be delayed any longer and will appear very soon. The Sonata, however, must not appear, until I finally receive your answer and the payment I yearn for. De Smit is the name of the courier through whom you received the Quintet as well as the Sonata. — I beg you for an early answer. More next time! Hurriedly

Your

Beethoven

***

19 April 1819

Dear friend,

Please do forgive me the inconvenience I am causing you. It is incomprehensible to me how so many mistakes could creep into the copy of the Sonata — the incorrect copying probably stems from the fact that I can no longer keep my own copyist. Circumstances have brought that about, and may God preserve me until . . . finds himself in a better position! This may take yet another full year. — It is truly dreadful how this business has transpired and what has become of my salary. No one can say what will happen until the year is over. Should the Sonata (Opus 106) not be suitable for London, I could send another one; or you might also leave out the Largo *and start straightaway with the fugue in the last section; or use the first section, then the Adagio, and for the third movement the* Scherzo *and the Largo and* $A^{llo.}$ *risoluto. — I leave all that to you; do as you think best.*[1] *. . . The Sonata was written under oppressive circumstances. It is hard to write almost for the sake of bread alone, and it has now come to that.*

Let us keep in touch about my coming to London. It would certainly be the only way to extricate myself from this miserable anxiety-producing situation. Here I shall never regain my health nor achieve what I might under better circumstances.

(1) What a wealth of suggestions! What range! Was this because he foresaw the difficulty of selling it?

*
* *

Vienna, 25 May 1819

. . . In the meantime I have been burdened with more worries than ever before in my life, all as a result of excessive generosity to other people.

Compose diligently! My dear little Archduke Rudolf and I play your things too, and he says that the former pupil is a credit to the master. — Farewell for now. Since I hear that your wife is beautiful, here is a kiss for her — in thought only, though I hope to have that pleasure next winter. — Do not forget the Quintet and the Sonata and the money, I might say the honorarium, avec ou sans honneur.

I hope to hear the very best from you, not in allegro fashion but Veloce Prestissimo. *The bearer of this letter is a clever Englishman. Most of his countrymen are bright fellows with whom I would like to spend some time in their own country.*

Prestissimo — Responsio, il suo amico e Maestro

Beethoven

Vienna, 10 November 1819

Dear Ries,

I am writing to tell you that the Sonata has already been published. But only about fourteen days ago. Because it is almost six months since both works — the Quintet and the Sonata — were sent to you. I will send you engraved copies of both the Quintet and Sonata by courier within a few days. Then you can correct everything in both works.

Since I did not receive a letter from you about the receipt of the two works, I thought nothing was coming of it. — After all, I have already been shipwrecked once this year, thanks to Neate. I wish now that you would just find some way to obtain the 50 ducats, since I was counting on them and really need a lot of money. I've written enough for today. I'd just like to add that I have nearly completed a new grand mass. Let me know what you could do with it in L. (London). But soon, very soon, and send the money for the two works soon as well. — Next time I will write you in more detail. In haste! Your good true friend

Beethoven

*
* *

Vienna, 6 April 1822

Dear, most excellent Ries,

Again for over half a year now I have not been well and could never answer your letter. I received the 26 pounds sterling in good order, and thank you cordially for it. I have received nothing of your symphony dedicated to me. . . . My greatest work is a grand mass[95] *which I wrote recently, etc., etc. Time is too short today, therefore only the essentials. . . .What would the Philharmonic Society offer me for a symphony?*

133

I am still thinking of coming to London, if only my health would permit, possibly next spring?! — You would find me truly appreciative of my dear pupil, who has now become a great master. Who knows what other good things in the cause of art might result from our joint efforts? I am, as always, completely devoted to my muses, and in this alone do I find the happiness of my life. I also act and work for others as best I can. . . . You have two children; I have one (my brother's son). But you are married, so your two do not cost you as much as the one costs me.

Now farewell. Kiss your beautiful wife, until I can perform this ceremonial act in person.

Cordially your friend

Beethoven

P.S. Make sure that I receive your dedication, so that I can respond in kind, which I shall do immediately upon receiving it. [96]

Vienna, 20 December 1822

My dear Ries!

Since I have been overburdened with work, I am only now able to answer your letter of 15 November — I am delighted to accept the commission to write a new symphony for the Philharmonic Society. Even if the honorarium from the English cannot be compared with fees paid by other nations, I would even compose free of charge for the foremost artists in Europe if I were not still the poor Beethoven. If only I were in London, what wouldn't I write for the Philharmonic Society! Because Beethoven can compose, thank God, though he may not be able to do anything else in this world. If only God would restore my health, which has at least improved, then I could fulfill all the offers from all over Europe, indeed even from North America, [97] *and might yet get my life in order.*

(Excerpt from a letter the beginning of which is missing)

... *Pursue everything as soon as possible for your poor friend. I also await your plans for my journey.*[1] *This really has become annoying; it aggravates me more than ever; if it does not work out, it will be just too much! A* crimen laesae! ... *Apparently you want me to dedicate something to you soon. I shall most gladly comply, more gladly than for any of those great lords,* entre nous. *The devil knows what can happen at their hands. I will dedicate the new symphony to you (the Ninth, with chorus).* — *I hope I shall finally receive a dedication from you.* ... *"B" should open the letter to the King (George IV) which he took with him; he will then see what I wrote his Highness about the* Battle of Vittoria. *The next letter to him (the King)*[2] *contains the same, but with no mention of the mass.* — *Our amiable friend B.* [97a] *might see if he cannot at least extract a butcher's knife or a turtle in return. Obviously the engraved score of the* Battle *should also be given to the King.* — *This letter will cost you a great deal of money*[3], *subtract it from what you are sending me. How sorry I am to inconvenience you!*

God be with you. Cordial greetings to your wife until I am there myself. Watch out! — *you believe me old, I am a young old man. As always*

Yours

(1) The plan of how Beethoven was to arrange his journey to England.

(2) The letter, affixed with two seals, was in Beethoven's own hand, as was the address on the envelope. It was enclosed in a letter to me, which had been inserted in another envelope. Since the first address to me evidently seemed illegible even to him, he made a third envelope around it without taking off the second.

(3) 17 shillings = 10 ⅕ florin.

Vienna, 25 April 1823

Dear Ries!

The Cardinal's (Archduke Rudolph) sojourn here for four weeks, when I had to give him two and a half or even three hours of lessons each day, robbed me of a great deal of time. After such lessons one is hardly able to think the next day, much less compose. —

My permanently depressing situation, however, demands that I write just now whatever brings me enough money for my immediate needs. What sad disclosures you are hearing!! Now, because of the many troubles I have been enduring, I am not well; even my eyes are sore! However, do not worry. You will receive the symphony shortly; really only this miserable situation is at fault. — Within a few weeks you will also receive 33 new variations on a theme (Waltz, Opus 120), dedicated to your wife.

Excerpt from Beethoven's letter to Ferdinand Ries, 25 April 1823

Bauer (First Secretary of the Imperial Embassy) has the score of the Battle of Vittoria, which . . . was dedicated to the former Prince Regent and for which I am still owed the copying costs . . . I only ask you, dear friend! to send me as soon as possible whatever you can get for it. — *After all, we both know those gentlemen, the publishers . . . As regards your tender other half, you will always find a kind of opposition in me, that is, an opposition against you and a proposition for your wife.*

As always,

Your friend

**

Hetzendorf, 16 July 1823

My dear Ries,

How greatly pleased I was to receive your letter the day before yesterday. . . . By now the Variations probably have arrived. I could not write the dedication to your wife myself, since I do not know her name. Do it therefore yourself, on behalf of your and your wife's friend. Surprise your wife with it; the fair sex loves that. — *Between you and me, the best thing possible is a combination of the surprising and the beautiful! . . . As far as the* allegri di bravura *is concerned, I must have a look at yours. To be quite frank, I am no friend of that kind of thing since it unnaturally promotes mechanical playing. At least those I am familiar with do. I do not know yours yet, so I will enquire with . . ., with whom I would suggest you deal with caution. Could I not arrange various things for you here? These publishing presses, who are always trying to press you into service, pirate your works and you get nothing for it; it should be possible to devise some other arrangement. . . . I shall send you a few choruses; also, if necessary, I will compose a few new ones. This is what I really prefer. . . .*

Thank you for the payment for the Bagatelles[98]. *I am quite satisfied.* — *Do not give anything to the King of England.* — *Take whatever you can get for the Variations; I will be satisfied with anything. But I must stipulate that for the dedication to your wife no reward whatsoever will be accepted, other than a kiss, which I am to receive in London. You sometimes write about guineas and I receive only pounds sterling.*[1] *I hear, however, there is a difference. Do not be angry with a* pauvre musicien autrichien *about it. My situation really is still difficult. — I am also writing a new string quartet. Could one perhaps offer it to the musical or unmusical Jews in London? —* en vrai juif? —

Embracing you cordially,

your old friend
Beethoven

(1) For a 25 guinea fee Beethoven had been sent 26 pounds, 5 shillings. Accounts are transferred only in pounds.

**

5 September 1823

My dear, admirable Ries,

I still have no further news about the symphony. However, you can certainly expect . . . that it will soon be in London. Were I not so poor that I must live from my pen, I would not accept anything from the Philharmonic Society. As it is, I have to wait until the fee for the symphony has been transferred to me here. However, as a proof of my love and trust in the Society I have already sent off the new . . . overture[99] *to them. I leave it to the Society to decide what they will be able to arrange for the overture.*

My fine brother (Johann), who keeps a carriage, wanted to be pulled along by me as well and so offered this particular overture to a publisher — Boosey in London — without asking me. Say merely that, concerning the overture, my brother was mistaken. — He bought it from me in a profiteering spirit, I would say. O frater!

I have received nothing of your symphony dedicated to me. If I had not always regarded the dedication as a sort of challenge demanding an answer in kind, I would have dedicated a work to you long ago. But I have always felt that I had to see your work first. How glad I would be to demonstrate my gratitude to you in some way. I am, after all, deeply indebted to you for the loyalty and consideration you have so often shown me. If my upcoming treatment at the spa improves my health at all, I shall kiss your wife in London in 1824.

Ever yours,
Beethoven

Nikolaus Johann van Beethoven, Beethoven's youngest brother (1776-1848)

Vienna, 9 April 1825

My very dear Ries,

Quickly, just the most essential matters! In the score of the symphony I sent you (it was the Ninth, with chorus) there is, as far as I remember, in the first oboe, specifically bar 242

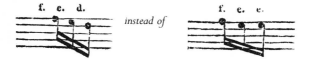

139

I have looked over all the instrumental parts (except the brass, which I have checked only here and there), and I believe they are more or less correct. I would gladly have sent you my own score[1]*, but I still have an academy concert ahead of me (and the manuscript is the only score I have). If only my health allows it, that is, because I must get out into the country soon: only in the country can I thrive at this time of year. — You will soon receive the* Opferlied[100], *copied for the second time. Mark it immediately as having been corrected by me, so that it won't be used along with the one you have already. This will give you a sample of the miserable copyists I have had since Schlemmer's death. One can hardly rely on a single note. — Since you have received all the copied parts for the finale of the Symphony, I have sent you the vocal score as well. You can easily have them written into the score from the parts before the chorus enters, and where the singing begins it will be quite easy, if you are careful, to have the instrumental parts added above the vocal line in the score. It was not possible to have all this written out at the same time, and since it was done in a hurry, you would have been faced with nothing but errors from this copyist. I have sent you an Overture in C in 6/8 time,*[101] *which has not yet been published; the engraved parts you will receive by the next post. A Kyrie and a Gloria, two excellent pieces (from the* Messe solemnelle *in D major) and an Italian vocal duet are also on their way to you. You will also receive a grand march with chorus, well suited for a gala concert.*[2] *A grand overture,*[102] *still unknown abroad, would also be available, but I think you will have enough with what I have sent. . . .*

I hope you are well, there in the Rhineland so dear to me.[3] *I wish you and your wife every happiness in life. To your father all my best wishes from your friend,*

Beethoven

(1) This passage concerns the intended performance of this

symphony — which Beethoven, however, did not send — at the music festival in Aachen. The local committee had also written to him directly, but had received only promises. At last I wrote and asked him, since I knew him and his scores so well, to send me the original score, and I would figure everything out. At the same time, knowing his constant need of money, I promised him a reward. They later handed me 40 Louis d'ors for him.

(2) Probably in the *Ruins of Athens*.

(3) When I left England, I moved to Godesberg, near Bonn, one of the loveliest regions on the Rhine. I had invited Beethoven to visit and urged him to live with me, there in his native country, for a while.

***

Epilogue

With this recollection of his pleasant stay in the beautiful town of Godesberg my late friend Ries finished his notes on Beethoven. His wish to welcome the Master there was not fulfilled. Beethoven died on 26 March 1827, in Vienna. The year previously Ries had moved from Godesberg and settled in Frankfurt am Main, which he did not leave again except for journeys to Italy, England, and France. Here he lived, tirelessly engaged in furthering his art, appreciated near and far, loved and honored by his friends. It is not without significance that memories of his great teacher were constantly at the center of everything he thought and did. Finally, having been asked so often, he decided to write them down. Why did the good humor in which they were started and for the most part completed have to give way so soon to illness and death for him, to melancholy grief for me!

Coblenz, May 1838

Wegeler

Ferdinand Ries

Musical score of Beethoven's "Die Klage" (see page 62)

SUPPLEMENT

*Upon the Occasion of the Dedication of
A Monument to Ludwig van Beethoven
In Bonn, His Native Town*

Foreword

Everything has its own time, during which it flourishes, then passes away; but whatever is noble, significant, and fertile in spirit endures beyond its time. It is now over seventy-four years since Beethoven was born in Bonn and over half a century since his mighty spirit began to spread its wings. Throughout the vicissitudes of his life, despite the disturbing effects of physical and mental distress, his artistic fame grew enormously. The world has now recognized the profundity of his spirit and bows before the power of his music. The statue, masterfully sculpted in bronze and placed in his beloved native city by artists and art lovers, will bear witness to this universal love and reverence for generations to come.

I am now the oldest living friend of the great Master, except for the venerable Great-grandfather Ries, who was born on 10 November 1755. I was in contact with Beethoven from his twelfth year until his death, and I feel unable to remain altogether silent on this happy occasion. It is now seven years since my friend Ferdinand Ries — who left this world all too soon — and I published, in Coblenz in 1838, a small collection of memories of Beethoven entitled *Biographical Notes on Ludwig van Beethoven*. I must say that my unforgettable friend Ries and I were mainly motivated by the idea of providing a respectful memorial to our departed friend. At the same time we wished to rectify errors and false ideas about the great man. We intended thereby to fulfill a duty of friendship to him.

This intention was fully appreciated. The little work was praised by many; it was also severely criticized; — much was learned from it; it also was translated into foreign languages, more or less completely and adequately. On reading these reviews, and also when further surveying our work and my collections, various remarks and additions came to mind. Perhaps they are not unworthy of seeing the light, like a bas relief on the memorial set up previously, as a supplement to the original.

Old age possesses its memories, youth its hopes.

On Beethoven's memorial, however, memory and hope should reach out and join hands.

Coblenz, August, 1845, at the close of his eightieth year.

Wegeler.

Supplement

Beethoven's name

The name is not the man, and yet it can become important for the man.

Ludwig van Beethoven was occasionally taken for a nobleman, because the Dutch *van* was equated with the German von. This happened for three years in Vienna. A lawsuit of Beethoven's was litigated for that long a period before the Landrecht court and was remanded to the city magistrate only after the error was discovered.[103]

(*Kölnische Zeitung*, 6 March 1844)

The house where Beethoven was born

Closer investigation, instigated by the teacher Dr. Hennes, into which house should be considered the birthplace of the great master, has shown at last that it is in fact, as I had said, the Graus house at Bonngasse 515, which now belongs to Dr. Schildt. An extensive discussion of this is to be found in the *Kölnische Zeitung*, in particular in the supplement to Issue no. 240 for 30 August 1838. The teacher Kneisel, who has a great reputation in Bonn, sets forth there clearly and convincingly the reasons for and against the idea that the Graus house is Beethoven's birthplace and proves the assumption correct. Subsequently Beethoven's parents lived in the house owned by a baker named Fischer at Rheingasse No. 934, which has often been erroneously taken for the house where he was born.

Beethoven's birthplace

The Fischer house

BEETHOVEN'S FAMILY

Beethoven's younger brother, Nicolaus Johanes, born in 1776, is still alive in Vienna, a rich man in comfortable circumstances. The foundations for this affluence had already been laid during Ludwig's lifetime. The son of his brother Carl (1774-1815) — Beethoven's nephew, ward, and subject of great concern — is said to be married and employed in Vienna at present.[104]

BEETHOVEN'S EDUCATION AND KNOWLEDGE

Beethoven's father, who was so strict with his son, was rather too lenient with himself. In particular he was somewhat given to drink and was especially irascible in that condition. Young Ludwig did his exercises frequently in tears and was kept at them by the iron hand of his father. Apart from music he learned only reading, writing, arithmetic, and a little Latin at a public school. One admirer of his genius has turned him into a scholar who understood Latin, Italian, and French, as well as his native tongue. Beethoven was even alleged to have studied the philosophy of Kant.* The truth is that Beethoven never went to a gymnasium, understood only a few Latin idioms, and spoke French with difficulty. When private lectures on Kant were held in Vienna — arranged by Adam Schmidt, Wilhelm Schmidt, Hunczovsky, Göpfert, the Royal Physician, and several others — Beethoven did not want to attend even one of them,[105] not even when I tried to persuade him. He probably felt within himself a categorical imperative different from that of the great man from Königsberg. His knowledge was creation.

HIS DEAFNESS

All too early Beethoven's worst misfortune began, his deplorable hardness of hearing, which his beautiful letter to me on 29 June 1800 from Vienna bemoans so heartrendingly (p. 28). What Beethoven says there about his unhappy condition is further attested to in a letter on 13 October 1804 from our mutual friend Stephan von Breuning to me. To excuse his long silence Stephan von Breuning says: "The only friend here who goes back to my childhood is frequently and often the major reason that I am forced to neglect my absent

---

* Supplement to the *Kölner Zeitung* of 22 March 1835.

friends. You would not believe, dear Wegeler, what an indescribable and, I should say, truly dreadful impact the loss of his hearing has had on him. Imagine the feeling of being unhappy — and with such a vehement nature as his. Add to this his shyness, distrust (often of his best friends) and general indecisiveness! For the most part, except for the occasional moments when his original affection expresses itself freely, association with him is a real strain, and one can never be quite off one's guard. From May until the beginning of this month we lived in the same house, and early on I invited him to stay in my flat. He had just barely moved in when he became violently ill. The illness threatened to become really dangerous but finally passed into a lingering, intermittent fever. The worry and the nursing were a considerable strain on me. Now he is quite well again. He lives on the ramparts, and I live in a house newly built by Prince Esterhazy in front of the Alster Barracks.* Since January I have kept my own household, including a sixty-six year old cook, so he eats lunch with me daily. "

HIS FINANCIAL AFFAIRS

Only in his last years did Beethoven appear to have been living in reduced circumstances, even though he complained of this much earlier. That this was not always the case is

---

* The street built within the last few years and bearing Beethoven's name runs straight behind this house, and the one known as the Schwarz-Spanier, where Beethoven died.

proved by the following lines, which he enclosed with a letter to me from Lenz von Breuning in May, 1797:

"God bless you, my dear friend! I owe you a letter, which you shall have very soon, as well as my latest compositions. I am well and, I might say, always improving. If you think anyone would be pleased, give them my regards. Farewell, and do not forget your L. v. Beethoven."

His comfortable financial circumstances lasted until 1800, as he told me himself happily in great detail (p. 28).

Unfortunately this had already changed for the worse by 1806. Stephan v. Breuning wrote to me in October that year: "Beethoven is at present with Prince Lichnowsky in Silesia and will not come back before the end of this month. His circumstances now are not the best, as his opera* has been only rarely performed, through the intrigues of his enemies, and has therefore not brought him anything. He is in a very melancholy mood for the most part, and to judge from his letters the stay in the country has not cheered him up."

His illnesses

Illness Beethoven unfortunately knew all too well, as is clearly demonstrated above (p. 39), contrary to Ignatz von Seyfried's contention. The seeds of his disorders, his hearing problems, and the dropsy which finally killed him, already lay within my friend's ailing body in 1796 (p. 28). The frequent interruptions of any regular regime were bound to aggravate this basic infirmity.

---

* *Fidelio*

However, Beethoven was also susceptible to other illnesses which could not be ascribed to the same source. Stephan von Breuning wrote to me in March, 1808: "Beethoven almost lost a finger because of a parasitic infection. However, he is reasonably well again. He was thus spared a great misfortune which, on top of his hardness of hearing, would have put an end to any good spirits completely, which are rare enough as it is."[106]

He himself complains (p. 136): "Now, I am not well from the many grievances I have suffered, and even my eyes are sore."[107]

And of course Breuning's letter quoted earlier speaks of the severe illness which had overcome Beethoven in von Breuning's flat.

By and large Beethoven lived very moderately, and as far as I know, not one of his friends and acquaintances ever saw him intoxicated. Dr. Waurauch's[108] statement that Dr. Malfatti had prescribed iced punch for him when he suffered from dropsy because as an old friend he knew Beethoven's strong taste for alcoholic beverages, is therefore completely unfounded. (*Frankfurter Conversationsblatt* No. 192, 1842.) Causes for dropsy were unfortunately abundant. — The accusation — which Livius had long ago leveled against musicians in Rome, when he called them *vini avidum ferme genus*,* and with which I frequently teased my friend — should be treated with considerable reservation.

---

* Livius, *Historiarum* lib. IX. Cap. 30

His marriage intentions

It does indeed seem that at one time in his life Beethoven entertained the thought of marriage, having often been involved in love affairs, as these *Notes* point out (p. 105).[109] Several readers have noticed, as I had, the urgency with which Beethoven in his letter on 10 May 1810 asked me to obtain his baptismal certificate. He insisted on underwriting all my expenses, even the travel costs from Coblenz to Bonn, and gave detailed instructions on what to look out for to be certain the certificate I obtained was the correct one (p. 45).[110]

I found the solution to the puzzle in a letter from my brother-in-law, Stephan von Breuning, written to me three months later. He says: "Beethoven tells me at least once every week that he wants to write to you. However, I believe his marriage plans have fallen through, and therefore he no longer feels such an urge to thank you for obtaining his baptismal certificate."

Thus Beethoven had still not given up the thought of marriage in his thirty-ninth year.

A printer's error in the *Notes*

"Graf von Marienstadt" (p.55)[111] should read: "Graf von Marienrode". Therefore, the famous author of the classic work on economics (1830, 2 vols.), Karl August Freiherr von Malchus, who died at the age of seventy on 24 October 1830, was also one of Beethoven's closer acquaintances.

Beethoven's opinion of composers

Beethoven's opinion of Mozart certainly reflects on both of them. Ries asserts that of all composers Beethoven thought most highly of Mozart and Handel, "and then Seb. Bach. If I found him with music in his hand or if anything lay open on his desk, it was sure to be a composition by one of these heroes" (p. 73).[112] Now L. Rellstab (*Weltgegenden*, vol. 3) tells us that during a visit where opera libretti for Beethoven were discussed, Beethoven made the following statement: "I am not too particular about the genre if the material attracts me. But I must be able to approach it with love and tenderness. I could not compose operas like *Don Giovanni* and *Figaro*. I have an aversion to that."[113] Rellstab recognized this, certainly with justice, as a confession of the great Master about the basic difference between himself and Mozart. Beethoven prefers the transcendental; Mozart is greatest when he reaches into the full, sensuous, natural life, penetrating right to the follies and passions of the human heart.

This much can be accepted. However, what should one say to a statement by the Marquise of Abrantes? First she raises Beethoven above Raphael and Michaelangelo, Dante, and Shakespeare. Then she continues (*Memoires*, p. 29):[114]

> Beethoven did not love Mozart. I cannot forgive him for that. It is a fault! That is, for me, something which reveals a lack of taste. — His reason for condemning *Don Giovanni* is sheer buffoonery. He claimed that Mozart should not have prostituted his talent (that's his word) on such a scandalous subject.

So then, the serious Beethoven, the greatest admirer of

Mozart, stooped to "buffoonery", indeed to "buffoonery" at his idol's expense? — He spoke of prostitution, our noble, highly proper master? — Are the requirements of propriety set higher in France or in Germany? — Who does in fact reveal a "lack of taste" — Beethoven or Madame Abrantes?

Concern for his pupil Ries

Ries recounts an incident (p. 84)[115] which demonstrates Beethoven's rash manner and inflexible behavior even toward friends when he thought them guilty of something. In this case Ries lost the most promising prospect of a position in Cassel which Beethoven himself had refused.

But, to Beethoven's credit it must be said that both he and his friend Breuning went to much trouble at that time to improve Ries's situation. "However, it is most difficult," as Breuning wrote to me, "to comply with both the father's wishes (Ries's father) and our own, and to combine our concern for the son's livelihood with the possibility of advancing him in his art. Beethoven is also now often distracted by his present situation, which obliges him to frequent the circles of the great families. — From my own experience I might add Vienna is no longer the place it was. The war has had an immense influence; the great families are curtailing their expenses in those things where they formerly were wont to be generous. And, of course, this is where one first gets a start, etc." Thus it was very probably more a question of the times than of a lack of zeal on Beethoven's part if his excellent pupil (p. 101)[116] did not readily attain the independence he desired.

HIS TRAVELS

"Beethoven hardly traveled at all," according to Ries (p. 96).[117] In our times, when the ease of travel and consequently also the desire for it grows every year, this would be somewhat incomprehensible. But even artists traveled much less thirty, forty years ago than they do nowadays. However, even Beethoven's life was not entirely devoid of travel. In January, 1796, the two elder brothers von Breuning, Christoph and Stephan, met him at Nuremberg on the way back to Vienna. Where he had been is not stated. Possibly Berlin.* Since none of the three had identity cards for Vienna, they were detained in Linz. I intervened for them in Vienna, and they were soon allowed to continue their travels. Later Beethoven also considered a journey to Italy. The war may have prevented it. Who knows what glorious and fruitful, even life-long, impression the lovely South might have left on the artist's soul? After all, his favorite models, Handel and Mozart, had also, to their renown, traveled in Italy at an early age. It was certainly unfortunate that Beethoven did not carry out this plan. Later on, of course, his miserable deafness prevented him more and more from moving beyond his usual circle. The fact, however, that travel did have a beneficial effect on him is clear from passages in Stephan von Breuning's letters to his mother, which I shall quote here:

"From Nuremberg Beethoven traveled always in our

---

* Cf. p. 96[118], where it indicates that he had been visiting King Friedrich Wilhelm II.

company. Naturally, three natives of Bonn together aroused the attention of the police, who were sure they had discovered something. I do not think that one could find a less dangerous man than Beethoven. " (Letter from Stephan von Breuning to his mother, January, 1796.) Beethoven also never came into contact with the police, even though he is supposed to have drawn the attention of those authorities to himself with sharp criticism of administrative regulations and with his democratic ideas. (Cf. the story of the *Sinfonia eroica*, which first bore the title *Bonaparte*; p. 68.)[119]

The same Stephan von Breuning wrote his brother and myself on 23 November 1796 from Mergentheim: "I do not know whether Lenz (the youngest of the brothers von Breuning) has written anything about Beethoven; if he has not, here is the news. I saw him while still in Vienna, and in my opinion, which Lenz also confirms, the journey (or perhaps the outpouring of friendship upon his return!) has made him more stable, or actually a better judge of men, and has convinced him of the rarity and value of good friends. A hundred times, my dear Wegeler, he has wished you were back with us, and he regrets nothing more than that he did not follow many of your suggestions."

Stephan von Breuning also wrote his mother: "I have written Wegeler that since the beginning of this year (1811), I have been keeping my own household with a sixty-six year old cook. Beethoven now eats with me. When he is not here, as was the case during the summer and will probably be again soon since he is supposed to travel to Italy, I eat alone."

This journey was not undertaken.

Immediately after the mutual journey mentioned above, the youngest of the Breunings wrote to me in January, 1796: "Beethoven is back again. He played in Romberg's Academy concert. He is still the old Beethoven, and I am glad that he and the Rombergs still get on so well. Once he almost had a falling out with them; I played the mediator, more or less successfully. In general he now sets great store by me."

General Remarks

Here follows an amusing letter from Beethoven to Stephan von Breuning, which may also serve to throw proper light on the relationship between the two:

(Probably 1820)

You, my esteemed friend, are overburdened, and so am I. And I am still not quite well. — I would have invited you to a meal before now. However, I am going around with a number of people whose most inspiring muse is the cook, and since inspired works are not to be found in their own spirit cellars, they seek them out in other kitchens and other cellars. [120] *— You would not be particularly well served by their company. However, this will soon change. Do not practice Czerny's studies for the pianoforte for the time being; I shall get detailed information about other exercises within the next few days.*

Here is the fashion journal I promised your wife, and something for your children. I can always pass on copies of this journal to you. And if there is anything else you could possibly wish from me, you have only to let me know.

In love and respect,

Your friend

Beethoven

I hope we will be seeing each other soon.*

∗∗∗

Gerhard von Breuning, Stephan von Breuning's only son, wrote to me just this year:

One of Beethoven's most cherished wishes was to possess Handel's complete works. When these arrived during his last illness as a present from England, I had to stack them all up on his bed against the wall — I think about fifty volumes — where he kept them almost all day long, always leafing through them, beside himself with joy, and endlessly praising that great master.[121]

I was using Pleyel's pianoforte studies; but with this, as with all the others, he was not satisfied.** *He said to me once when I was sitting at his bedside: "I felt like writing a set of pianoforte exercises myself once, but never found the time for it; I would have written something quite unconventional though." He then promised my father that he would provide a manual of exercises for me. Some time later he sent me Clementi's, which was not available at the time, with the following note:*

"Dear Werther,

At long last I can carry out what I carried on about. Here are

---

* The original of this letter is in the possession of von Breuning's widow in Vienna.

** See the letter cited above on p. 157.

the Clementi exercises promised for Gerhard. If he practices them according to my instruction, they certainly will produce good results.

I do hope to see you very soon and embrace you fondly.

Yours,

Beethoven "

These are but bits and pieces, remembered from a friendship in years gone by and solemnly offered here at the foot of this memorial, which is intended to proclaim Beethoven's artistic fame to the world. And yet, small as they are, these postscripts to what has been related do not seem insignificant to me, because everything is valuable to me that summons up before my inner self the presence of my friend and his noble qualities.

The sculptor has represented for us in bronze the glorious form of the master of sound, towering above a base of the most solid granite. I have dwelt more on the man, the friend. His works are admired by the world, and in their radiant character we can glimpse the soul from which they emerged.

Well, nothing can or need be added to the works, which will perpetuate Beethoven's fame forever. But I can still limn a few more features on the portrait of the man, the loyal friend.

Here, then, are a letter from Stephan von Breuning and a page from an autograph album.

Of course, these are only isolated outpourings of a great soul. But who can remain unmoved when Beethoven as a twenty-six year old young man confesses to his friend Lenz von Breuning that the union of truth and beauty is the highest ideal for a wise and sensitive man![122] — And he remained true to this belief throughout his life, through happiness and misfortune. His works openly proclaim to the world what he confided to his intimate friend. His own words could serve as inscription on the spiritual memorial a friend's hand has endeavored to erect here:

Truth and Beauty Joined

A letter from Beethoven to St. v. Breuning sent with his portrait:

Miniature portrait of Beethoven on ivory by Christian Hornemann, 1803

(Undated*)

Behind this painting, my dear, good Stephan, may all that has happened between us for some time be hidden forever. — I know I have torn open your heart. My own emotion, which surely you must have noticed, has

---

* The original of this letter is in the possession of von Breuning's widow in Vienna.

punished me severely enough. It was not malice that motivated me against you; no, or I would no longer be worthy of your friendship. Passion worked on both our sides — yet, distrust of you erupted within me. Men came between us who will never be worthy of you or me. — My portrait has long been intended for you; you know that I had always intended it for someone. To whom could I give it with such a warm heart as to you, faithful, good, noble Steffen! — Forgive me if I have hurt you; I myself have suffered just as much. When I did not see you around me for such a long time, I realized all the more sharply how dear you are to my heart and ever will be. — Surely you will rush into my arms again with the same trust as before!

With this letter all the quarrels, which, as Beethoven's letters to Ries clearly demonstrate, were frequent between the friends, seem to have come to an end. "Steffen," wrote Beethoven, "has really become a good, wonderful boy, who knows a few things and has his heart in the right place" (p. 30).

And yet the two were separated so often! Steffen writes on 10 January 1809: "I have not seen Beethoven for over three months. He has during this time written me in a rather friendly way but has not visited me again; I do not know the reason why."

Beethoven wrote into the autograph album belonging to Lenz von Breuning:

> *Truth exists for the wise;*
> *Beauty for a feeling heart.*
> *Both are meant for each other.*[123]

Dear, worthy Breuning,

Never will I forget the time which I have spent with you, both in Bonn* and here. Preserve your friendship for me, in return you will always find me the same.

Vienna, 1797

on the 1st of October

Your true friend,
L. v. Beethoven. "

(The album is in my possession.)

From these outpourings we can see that Beethoven was always in the most intimate relationship with the von Breuning family. Stephan von Breuning indeed especially deserved this friendship. He had a most noble nature, which can be seen very clearly in his behavior toward his dearly beloved mother,** his sister, brothers, and other relatives. These qualities have been handed on to his children. Stephan's son, Dr. Gerhard von Breuning, is well known in Vienna and beyond as a doctor and surgeon. Since a talent for music runs in the family, one of the daughters has tried her hand at

---

* Lenz von Breuning, as the youngest of the three brothers, was closest to Beethoven in age.

** Frau von Breuning (born 1750, died at Coblenz in 1838), to whom Beethoven also owed a great deal in his youth. (p. 15).[124]

composition as an amateur, not without success. In this family, then, Beethoven's memory lives on.

***

Finally, here is also a little offering for the artist, a romance composed by Beethoven and hitherto unpublished. It was found in Stephan von Breuning's family papers and sent to me with the following title to use as I thought fit:

Upon Learning of Lydia's Infidelity
A Song
set to music
by
Ludwig van Beethoven
(The poem is taken from the French)

The original text, as far as I know, is taken from the opera *Le Secret* by Solie.[125] Beethoven, however, set the melody to a German text, finishing it in May of 1806, according to the handwriting and to changes in Stephan von Breuning's manuscript.* The composition of the ballad, then, falls still within our Master's earlier period, although several knowledgeable judges of Beethoven's work place it on the same level as the best of his work in this style at a later period.

This is the poem:

---

* At the same time I happened to translate this same ballad along with several other songs from the same opera for Simrock, my fondly remembered friend, who then published them.

Ballad

Feelings Upon Lydia's Infidelity

So the last gleam of hope has now declined.
She broke her vows; fickle is her mind.
And now I lose, to infinite unrest,
All consciousness that I was once so blessed.

What did I say! Still her charms enslave me.
No power can turn, no resolution save me
Yet when I reach the brink of wild despair
Sweet memories stay with me there.

Ah, gentle hope to me once more returning
Do thou revive the flame within me burning
Though love's sufferings indeed are great
Still he who loves would never change his fate.

And thou who does my love so true disdain
Thy image will within my heart remain.
I could not hate thee, no but love thee more.
Forget thee? Not till this life of woe be o'er.

***

The small worth of these additional remarks I would not deny, but the thought that everything here is an offering to Beethoven's admirers led me on. Thus came into being this little portion of some future biography, which only a true musician will be able to write. Beethoven's entire soul lives on in his works; all his joys and sorrows have been consigned to his art. His music is his real biography, the true and imperishable story of what he strove for and what he accomplished, written for all people and all times.

Finally, I feel justified in claiming also for these additional remarks the same credibility, other sources to the contrary, which was generally accorded the *Notes* themselves when they first appeared.

And now may these modest words of fond remembrance mingle with the noise of ceremonial rejoicing, which is not only to be indulged but also to be cheerfully taken up.

***

Beethoven's study in the Schwarzspanierhaus, 1827

NOTES

The following notes are a complete translation of the annotations by Dr. Alfred C. Kalischer to his edition of the ***Biographische Notizen über Ludwig van Beethoven*** by Wegeler and Ries, second edition, published 1906 in Berlin. The numbers, which appear in the text as boldface superscripts, are in the sequence and location given by Kalischer in his edition. The very few notes which are not translated refer either to minor variants in the German text or to page numbers in Kalischer's editions. Otherwise the translation is true to the meticulous accuracy and idiosyncracies of the original.

1. Despite this quite unequivocal, incontestable claim of Franz Wegeler, the creditable Beethoven biographer A.W. Thayer has attempted to prove that Beethoven's relationship with Wegeler and the whole von Breuning circle did not begin until after the pilgrimage to Mozart, this is, after July 1787 — an effort which of course was bound to fail. Already in 1890 Wegeler's grandson, the independently wealthy Koblenzer Carl Wegeler, successfully rebuffed Thayer's puzzling assertion in the *Kölnische Zeitung* (no. 143, May 24, 1890). Printing the complete version of a letter presented only in fragmentary form in this Beethoven book, he rests his case with complete justification on the following sentence of this letter of Beethoven to Wegeler: "Oh, Wegeler, my only consolation is that you have known me since my childhood." This letter will be printed in its complete form in the appropriate place (note 20, below). Even Dr. H. Deiters, the knowledgeable and perceptive editor of Thayer's Beethoven biography, finds himself forced to concede in the 2nd edition of 1901 (I, 206): "Also all the additional reports gain meaning and fall into place only if one assumes that Beethoven had found acceptance in the von Breuning household as early as his boyhood days. The editor [Dr. Deiters] thus believes that in this matter he must differ from Thayer's opinion."

I myself can and must at this point introduce in favor of F. Wegeler's incontestable assertion yet a few additional pieces of evidence which I found in a little-known publication. In the year 1839 F.G. Wegeler celebrated the fiftieth anniversary of his doctorate. For this celebration a *festschrift* was compiled which consisted of a sizeable number of contributions in both poetry and prose, and which was printed in its entirety. There is a biographical sketch, "Franz Gerhard Wegeler," probably written by his son, Dr. Julius Wegeler. Here one finds (page 8):

> At this point one must also remember the noteworthy relationship in which Wegeler had stood since 1782 to his great countryman, the composer Ludwig van Beethoven. This relationship had been formed between the two youths in Bonn and then cemented through common experience, sufferings, and joys in their mutual development as well as through their association with the von Breuning family, with whom they were intimately connected. Up until September 1787, when Wegeler went to Vienna, this association continued uninterrupted and then was resumed with equal warmth after his return to Bonn in 1789 and up to the time of Beethoven's departure for Vienna at the end of 1791. In Vienna the friends came together again with undiminished feelings and a day rarely went by on which they did not see one another.

The fact that Wegeler had been friends with Beethoven since boyhood days is powerfully celebrated in this *Wegleriana* in the fifth and sixth strophes of a festive poem in classical form:

> Him today I joyfully greet
> With the powerful greeting of the ancients,
> Greet him with splendid Hellenic rhythms,
> Bring him homage with the echoing sounds
> Of the festival ode.
>
> Proudly must the chorus sing
> Of the understanding friend of art
> Who once listened to Beethoven's lofty muse,
> Almost while still in the cradle,
> And early prophesied the future splendor.

It should be mentioned that Wegeler, born on August 22, 1765, was more than five years older than Beethoven.

2. Here Wegeler writes that the day of baptism is "usually the actual day of birth or certainly the day after." This is an unhappy error. If this were true all those who assume that December 17 is the composer's birthday, as Wegeler himself later does, would be correct. This Rhenish custom is in fact a curious matter, as I have already demonstrated (compare the editor's article "When Was Beethoven Born?" in the Sunday Supplement to the *Vossische Zeitung* of January 11, 1891). On inquiring into this Bonn custom I have heard the most varied contradictions. From a Catholic friend born in Aachen, a portrait painter by trade, I learned that in accordance with Catholic custom in all the districts and regions known to him, the newborn child is usually baptized at the latest on the third day after its birth. That arises from the Catholic belief that the child, if it were to die unbaptized, would not be able to partake of heaven. My informant knows no example from his area of the baptism of a newborn child on the first day.

3. Dr. Deiters, the editor of A.W. Thayer's biography of Beethoven, points out in the 2nd edition (1901, p. 107) that Wegeler's presentation of the facts is not precise. The maiden name of the mother — Johann van Beethoven's wife — which Wegeler states as Westorffs, is not given in the *Ehrenbreitstein Parish Book* of December 1746; Westorf is the name of the godmother. Dr. Deiters cites the entry in the Parish Book as follows: "19 nata et 20ma renata (= baptized) est Maria Magdalena Keverichs, D^{ni} Henrici Keverich coqui primarii Em^{smi} et Mariae Catharinae cjugum Legitima filia, eam de sacro fonte Levantibus D^{na} Maria Magdalena Westorfs de Confluentia [Coblenz], et D^{no} Mauritio Wisdorff [sic] itidem de Confluentia."

According to this it is not true that the day of the baptism coincides with the day of birth.

4. As Dr. Deiters has established, this should read: "Henrici Keferich et annae clarae [not Mariae] Westorffs filiam legitimam." "If the Bonn Parish Book calls her a born Westorffs, then it seems that at the issuance of the daughter's baptismal certificate for the purpose of her wedding in Bonn, the same erroneous reading of the

birth certificate prevailed as in the communication of the information to Wegeler." Cf.note 3.

5. To be precise this should read "baptized" on April 2, 1769.

6. This is not clear. Even in the Thayer-Deiters version — "the wife of the locksmith Jean Courtin" — the matter does not become clear.

7. It is incomprehensible that Wegeler can here so positively assert that Ludwig van Beethoven was born on December 17, since the document that follows merely says "baptizatus est Ludovicus": Ludwig was baptized. I have already advanced the argument above that it cannot be claimed with plausibility that Catholic children in the Rhineland were baptized one day after birth. Up to three days may elapse after the birth. The peremptory passage in the *Conversation Books* (no. 63, leaf 6b) of December 1823 remains all the more in force. There Beethoven's nephew writes: "Today is December 15 and that is when you were born, so far as I could see; however, I could not vouch for whether it might be the 15th or the 17th, since one cannot rely on the baptismal certificate, and anyway I only read about this once, in *Janus*, when I was still at your place. I already thought about it yesterday and at some length, but the unpleasantness that occurred yesterday made it inopportune to mention it." If one compares this with the fact that even Dr. Hennes in his article "Beethoven's Birthplace" (printed in Thayer-Deiters' first and second editions) states: "Beethoven was born in Bonn in the year 1770, in the middle of December, probably on December 15," where two days before the baptism are assumed, one can thus assert in bringing this discussion to a close: the 15th of December is far more probable as Beethoven's birthday than the 16th or even the 17th of December.

8. [Missing in Kalischer's edition of 1906.]

9. General material and historical information on the legendary descent of Beethoven from King Frederick William II of Prussia has been published in my article "Beethoven in Berlin" in the November 1886 issue of the journal *Nord und Süd* (Breslau). Although the Englishman who would have Beethoven be a natural son of Frederick the Great has well earned Wegeler's ridicule because of

his ignorance of chronology, he however deserves some credit for being the first and only one to name Frederick II as Beethoven's natural father. For only very recently did I find in Beethoven's *Conversation Books* a noteworthy confirmation of the fact that in the composer's most intimate circle the belief was cherished that he was a natural son of Frederick the Great. In a *Conversation Book* from 1820 (no. 22) in just the place in which Karl Bernhard sings Napoleon's praises to Beethoven, this author and editor writes, among other things (leaf 39a):

> Napoleon was a man of arts and sciences.
> In the Encyclopedia [Conversations-Lexicon] it says that you are a natural son of the great Frederick.

And immediately thereafter Hofrat Peters writes:

> H. v. Janitschek is of the opinion that you love Frederick the Great so much because he is supposed to be your father.
> Such errors must however be corrected. You don't need to borrow anything from Frederick.
> (Bernhard 39b). Someone should put an article in the *Allgemeine Zeitung*.

10. The operetta *L'Amore Artigiano*, with music by Florian Leopold Gassmann († 1774 in Vienna) was performed in Bonn in the arrangement by C.G. Neefe.

11. Tobias Friedr. Pfeiffer was certainly succeeded by the court organist Egidius von der Eeden as the young Beethoven's teacher for piano and organ.

12. What is here said about Chr. Gottlob Neefe, however, is not to the point. Beethoven, as he himself acknowledged, owed much to this extraordinary teacher, particularly in composition instruction. Also, it is sufficiently well known how clearly Neefe recognized Beethoven's genius in *statu nascenti*.

12a. Wegeler here cites the first edition of von Seyfried's book. In the second edition, January 1853, edited by H.H. Pierson (Mannsfeld), these words are followed by: "and conferred on him the status of court organist, along with several years leave and free passage to Vienna" (Appendix, p. 4).

13. The friendship and respect which Beethoven felt for the

Rombergs in general and for the cellist Bernhard Romberg in particular, even when he was already at the zenith of his fame as a composer, becomes clear from Beethoven's first and only letter to Romberg, which the editor published several years ago. See *New Beethoven Letters,* Berlin 1902, page 58f. Bernhard Romberg died long after Beethoven, in Hamburg in 1841. Pleyel, also mentioned here, is Haydn's pupil Ignaz Pleyel († 1831). The time of the anecdote is the year 1791. Cf. A.W. Thayer's *Beethoven* I, 242; 2nd ed., revised by H. Deiter, Berlin 1901.

14. This cantata has in the meantime appeared in print. It is the Funeral Cantata on the Death of Emperor Joseph II, which had its first performance in Vienna in November 1884. Ed. Hanslick first brought the discovery of the cantata to the attention of the music world. This cantata, as well as the other, written on the occasion of the accession of Leopold II to the imperial throne, has been printed in the supplementary volume to the large edition of Beethoven's works by Breitkopf and Härtel.

15. A piano arrangement by Dulcken of the music to a Ritterballet was published in 1872 by Rieter Biedermann in Leipzig. The music appeared in its original form in the Breitkopf & Härtel *Complete Works* in series 25, no. 286.

16. These are, of course, by no means all the compositions dating from the period of Beethoven's life in Bonn. Many others can be added to these, most of which have by now been printed. In this connection one should study the noteworthy chapter in Thayer's *Beethoven,* "What Did Beethoven Compose in Bonn?" (I, 231-241) and especially the creditable enlargement of this chapter by H. Deiters (2nd ed., pp. 272-313).

16a. The words are from Horace's *Satires* (Liber I, Sat. IX, verse 20). In complete form:

Demitto auriculas, ut iniquae mentis asellus,
Cum gravius dorso subiit onus.

17. This can only be understood, however, regarding Beethoven's letters to Wegeler and Ries; other letters from the master had already found their way to public attention. One has only to be reminded of Beethoven's interesting letter to Abbé Stadler on Mozart, which the first small Beethoven biography by Schlosser

(1828) presents in facsimile. One thinks further of the not insignificant number of Beethoven's letters to the music dealer Hofmeister & Kühnel in Leipzig (later C.F. Peters), which were published in 1837 — that is, before the appearance of the *Biographical Notes* — in the *Neue Zeitschrift für Musik* under the editorship of R. Schumann.

18. This extremely important letter belongs to the following year (1801), however, as Thayer has already convincingly demonstrated (*Beethovens Leben* II, 156). — Thus I wrote in the first edition of this annotated reprint. After renewed examination, however, I have arrived at a new opinion. In this matter I now stand with Wegeler, Nohl and Schindler, all of whom place this letter in the year 1800. Compare the notes to this letter (no. 36) in [Kalischer's] *Beethoven's Complete Letters*, I, 51.

19. That the artists performing in the string quartet at Prince Carl Lichnowsky's are listed differently by different Beethoven biographers is connected with the circumstance that these quartet performances later alternated with those at Count (later Prince) A. Razumovsky's so that the artists from the Lichnowsky Quartet became confused with those from the Razumovsky Quartet. The confusion was not lessened by the fact that some of these artists also participated in the house orchestra of Prince von Lobkowitz. Schindler names the members of the Lichnowsky Quartet, mentioned here by Wegeler, as J. Schuppanzigh, the violist Franz Weiss, and the two cellists Anton Krafft and his son Nicholas Krafft (I, 35), whereas A.W. Thayer names Louis Sina as the second violin (I, 275). — The cellist Linke, whom Wegeler problematically introduces, does not in any case belong here. As members of the Razumovsky "Model Quartet" Schindler lists (I, 38): Schuppanzigh first violin, Sina second violin, Weiss viola, and Linke cello. — Despite this peremptory presentation Thayer's own more recent investigations yielded a different result. The name of Sina, the man who was later so zealously active in the introduction of Beethoven's music in Paris (Sina, † 1857 at Boulogne sur mer) is completely absent in Thayer. He explains that in addition to Count Razumovsky, who usually played the second violin himself, Maysede probably stepped in from time to time (III, 48). Anyway, this holds good *non liquet*. In any case, we have here solid testimony that the later Schuppanzigh

String Quartet consisted of the other artists of the Razumovsky Quartet, with the exception of Sina. The part of the second violin was not infrequently played by Beethoven's youthful friend Karl Holz. Before the premier of the first of the late quartets (E-flat, Op. 127), Beethoven directed the following order to the performers:

> *My dear friends, each person is herewith given his own part and is bound by oath, that is to say he pledges on his honor, to perform at his best, to distinguish himself, and to strive for mutual excellence. This paper is to be signed by each one of you who is to take part in this endeavor.*
>
> Beethoven
>
> *Schuppanzigh, m.p.*
> *Weiss, m.p.*
> *Linke, m.p.*
> *The accursed cello of the great master.*
> *Holz, m.p.*
> *Last, however, only in signing.*

(Cf. Schindler II, 113). Karl Holz later also played the first violin in August 1825 in the A Minor Quartet .

20. This letter has attained a larger significance in Beethoven's biography because a passage is found there that offers unmistakeable evidence for the fact that Beethoven's friendship with Wegeler reaches back long, long before 1787, so that Thayer's assertion that Beethoven did not become acquainted with Wegeler and the whole von Breuning circle until after his trip to Vienna, is also deeply shaken by this letter. Wegeler's grandson, Herr Karl Wegeler, with the help of this letter (which he published in the *Coblenzer Zeitung* of May 20, 1890) victoriously fought his case against Thayer (compare the article in the *Kölnische Zeitung*, no. 143, second morning edition of May 24, 1890). This very important letter runs as follows:

My dearest, my best friend! What a detestable picture have you shown me of myself! Oh, I realize I do not deserve your friendship, you are so noble, so thoughtful, and this is the first time that I may not stand near you, I have fallen far beneath you. Oh, for weeks I have caused distress to my best and noblest friend. You think that I have lost some of my goodness of heart. But thank Heaven! — it was no intentional, premeditated evil on my part which caused me to behave so. It was my unforgiveable levity which did not permit me to see the matter in its true light. O how ashamed I am both on your behalf and on mine! I almost no longer trust myself to plead with you again for your friendship. Oh, Wegeler, my only consolation is that you have known me almost since my childhood! And nevertheless — oh, let me say it myself — I was really always good and made the effort to be upright and honorable in my dealings. How else could you have loved me? Can I then in so short a time have changed so terribly and so to my disadvantage? Impossible! Can my feelings for the great and the good suddenly have been extinguished? No, Wegeler, dearest and best friend. But venture once more to throw yourself completely into the arms of your B. Build upon the good qualities which you otherwise have found in him. I promise you that the pure temple of holy friendship you will thereupon erect will endure firmly, eternally. No mischance, no storm will be able to shake its foundation — firm — eternal — our friendship — forgiveness — forgetfulness — revival of our dying, sinking friendship. Oh, Wegeler, do not turn away this hand of reconciliation, put yours into mine. Oh, God — but nothing more. I myself am coming to you and will throw myself into your arms and will plead for the lost friend and you will return to me, your penitent one who loves you and will never forget you.

Beethoven.

I have just now received your letter because I have just now returned home.

Even Dr. H. Deiters has been convinced by Karl Wegeler's arguments. He declares, among other things, in the second edition of Thayer's *Beethoven* (I, 206) in the notes: "The editor thus believes that in this matter he must differ from Thayer's opinion."

21. For example, with Hofrat Peters, with the lawyer Dr. Joh. Bapt. Bach, with Zmeskal von Domanovecz, among others.

22. This is Kaspar Karl von Beethoven, the father of Beethoven's famous nephew, also named Karl.

23. However, in the second edition of this book by von Seyfried, *Ludwig van Beethoven Studies* (1853), it is stated (Appendix, p. 26): "Beethoven knew neither stinginess (!) nor extravagance."

24. That was on the occasion of Beethoven's first public appearance as a pianist and composer, in March 1795. The Piano Concerto in C Major, which he then performed in the Burg Theater, did not appear in print until 1801 as Opus 15. A. Schindler is mistaken in his observation that this concerto, dedicated to the Princess von

Odescalchi, had its first performance in the Kärnthner-thor Theater in the Spring of 1801. This place in Wegeler's notes supports the chronology given in Nottebohm and Thayer.

25. The world famous Dr. Peter Frank was born on March 14, 1745 in Rothalben in the Palatinate and died in Vienna on April 24, 1821. In 1795 he came to Vienna, the second time in 1808. He was the founder of the field of public health.

26. Seyfried, however, contradicts himself. For prior to this pronouncement, he has it quite otherwise (appendix, page 9): "Gradually other bodily ills also arose, forcing the once thoroughly healthy, robust man to seek medical assistance."

27. Correctly, Baron Joh. Bapt. — von Pasqualati (= Pasquillati, Pascolati), in whose house on the Mölker-Bastei Beethoven lodged so frequently, that it gave rise to the Baron's characteristic remark: "The rooms are not to be rented; Beethoven will come back."

28. To this famous physician, Prof. Dr. J.A. Schmidt, Beethoven in 1802 gratefully dedicated the great Septet (Op. 20), which he himself had arranged as a trio for piano, clarinet (violin), and cello. He attended the composer during his grave illness following the break with his beloved Giulietta Guicciardi. The upshot of this was the "Heiligenstadt Testament."

29. This is Fraülein von Westernholt, about whom more can be found in the editor's "Beethoven's Women Friends" in the *Neue Berliner Musikzeitung* of 1892.

30. In this place there will, of course, be no further examination of Beethoven's love for his Giulietta, to whom the Sonata Fantasia in C Sharp Minor was dedicated. Only one particular passage from this letter should be singled out for special recognition: "This transformation has been brought about by a dear enchanting girl who loves me and whom I love." In no other place did Beethoven write of any woman "who loves me and whom I love."

31. That Beethoven appreciated von Breuning's musical talent, particularly as a violinist, is evident from his dedication of his only Violin Concerto in D (Op. 61) to his friend Stephan v. Breuning.

And his own arrangement of the Violin Concerto for piano was dedicated to Frau von Breuning.

32. Beethoven wished to have the baptismal certificate, because in 1810 he seriously considered marrying. Many suppositions have been offered as to which woman was the object of his marriage plans. I myself believe it was Therese von Malfatti. The details appear in my article "The Malfatti Sisters" (from "Beethoven's Women Friends") in the Sunday Supplement to the *Vossische Zeitung* of February 5 and 12, 1905.

33. These deeply felt song texts in favor of Freemasonry, which Wegeler presents in the appendix to this work, were in any case of his own composition. The Matthisson "Opferlied" (in E) begins with the words: "The flame blazes, wild light shines through the gloomy oak grove." The composition (in C), "Der freie Mann," for choir and solo voices, was written by G.C. Pfeffel and begins: "Who is a free man? Only he whose lawgiver is his own free will and not the tyrant's whim."

34. The variations mentioned here are the Twelve Variations for Piano and Violin in F on the well-known theme "Se vuol ballare" from Mozart's *Figaro*, in the Breitkopf and Härtel edition, series 12, no. 12. The variations were published in 1793 by Artaria under the title assigned by Nottebohm: *XII Variations pour le Clavecin ou Piano-Forte avec un Violon obligé Composées et Dediées à Mademoiselle Eleonore de Breuning par Mr. Beethoven. Oeuvre I*. It is thus especially noteworthy that Beethoven's first compositions to be published in Vienna were not the three Trios (Opus 1) but rather these Variations in F (Oeuvre I). Later this composition was designated as No. 1; the publisher was changed from Artaria to F. Mollo. Compare Wegeler's account above (p. 15) of the "Variations No. 1" dedicated to the woman who was later to become his wife. The second composition dedicated to Eleonore von Breuning, Easy Sonata in C (Fragment) (Breitkopf & Härtel, series 16, no. 36) was first published by Fr. Ph. Dunst in Frankfurt am Main under the title given by Nottebohm: *Sonate pour le Pianoforte Composée et dediée à Mlle. Eleonore de Breuning par L. van Beethoven*. The second movement of this little sonata, or rather, sonatina, is known to have been added to by Ferd. Ries; a third

movement is completely missing. Nottebohm's note in the *Thematic Index:* "Eleonore von Breuning is thought to have received the original manuscript from Beethoven in 1796" (cf. Cäcilia XIII, 284; Wegeler's *Notes*, p. 57) is in any case mistaken. Fraülein von Breuning must have already received the sonatina in Bonn, where the little work must have been composed. This small question is connected with Beethoven's second letter to this lady friend of his youth and will be briefly considered along with it.

35. This is Abbé Gelenik, who lived from 1758 to 1825 and was called the Variations-Smith.

36. Concerning this letter I direct the reader to my articles "Beethoven's Women Friends," published in the *Neue Berliner Zeitung* in 1892, in the second of which Beethoven's relationship with Eleonore von Breuning is considered. This letter is discussed in the issue of June 16, 1892. The key sentence reads: "Everyone unreluctantly accepts that this letter was actually written to Fraülein von Breuning from Vienna; however, everyone will be astonished to learn that I now in all seriousness claim that this undated letter of Beethoven to Eleonore von Breuning was written not in Vienna but rather sent to his pupil and friend from Bonn." The evidence is presented there. Independently of this, Dr. H. Deiters ten years later arrived at the same conclusion, as one reads in the first volume of the second edition of the Thayer-Deiters Beethoven biography. However, I will now add the assertion, already hinted at above, that the sonata mentioned in this letter (Easy Piano Sonata in C) was not, as is claimed, composed around 1796 in Vienna but rather much earlier, in about 1790 or 1791, in Bonn.

37. Stephan von Breuning must have confounded the words "Fidelio" and "Eleonore"; one should examine Otto Jahn's article "Leonore or Fidelio?" in his *Collected Essays on Music* — for which, moreover, A.W. Thayer provided the impetus; see his *Beethoven* II, 301. The theater program appended in Thayer's book actually calls the opera "Fidelio." Also compare the excellent introductory essay by Dr. Erich Prieger to his piano arrangement *Leonore, Opera in Three Acts by Ludwig van Beethoven* (Leipzig, Breitkopf & Härtel, 1905).

37a. Nevertheless, Fidelio did not receive its first performance until nine years later, in Berlin, in October 1815.

38. The date of Ferdinand Ries' arrival in Vienna and at Beethoven's is clearly placed too early. The works named, *Christ on the Mount of Olives* and especially the Second Symphony in D provide the strongest evidence. Autumn 1801 may in fact be considered the time at which F. Ries became Beethoven's pupil. Compare Thayer's detailed exposition in his Beethoven biography (II,63). The Second Symphony was composed in the year 1802 and performed for the first time in April 1803. The oratorio *Christ on the Mount of Olives* was not composed in 1800 — as one might conclude from this and as others, even Nottebohm, assert — but rather in 1801 and revised in 1802; it was first performed in April 1803 in the Theater an der Wien. In general, the chronology in Ries' first account seems quite confused. One should not be too surprised by this, since Ries, as he himself assures us, related things as they had impressed themselves upon his memory: "It should be easy for the reader, if it is of importance to him, to bring some order to this." — Let us, therefore, do our part.

39. This large rehearsal with obligato refreshments at Prince Karl von Lichnowsky's, to whom the Second Symphony is dedicated, must have taken place shortly before April 5, 1803 or on the day of the performance itself (April 5).

40. Actually in 1803.

41. It is not easy to understand how Wegeler can link this delightful anecdote with the composer Paër. *Die schöne Müllerin* (La Molinara) is, of course, an opera by Paisiello. Probably Wegeler confused the composers Paër and Paisiello with one another.

42. Daniel Steibelt was born in Berlin in 1755 (by some accounts in 1765) the son of a piano manufacturer. Beginning at about the age of 30 he created a universal sensation as a piano virtuoso as well as a composer, particularly of chamber works. In Paris he achieved great success with his opera *Romeo and Juliet* with libretto by Viscount de Ségur. Later he became music director in Petersburg, where he died in September 1823. His character both as a person and artist was subjected to diverse bitter criticism. Today only a little is still known

about this rival of Beethoven, who was once highly celebrated in almost all creative fields. Occasionally some of Steibelt's études pop up in étude collections.

43. Giovanni Punto (not Ponto) is the Italianized name of the thoroughly German musician Johann Wenzel Stich, who was born in Czaslau, Bohemia as a serf of Count von Thun. 1775, the year of birth given by Schindler, is probably much too late, since he certainly created a sensation in Paris as early as 1778, where he played a silver horn. He died in Prague in 1803. He had fled from his master with several other musicians. As the story goes, the Count was principally concerned about Stich and had him pursued. The Count is said to have given his bloodhounds the noble commission, in the event that they were unable to gain control over the hornplayer, at least to smash in his front teeth so that he could no longer play the horn. Stich, however, changed his name to Punto and under this appelation attained a high reputation as a virtuoso. When he came to Vienna toward the end of the century, he became friendly with Beethoven, who was particularly fascinated by his life story. For Punto's concerts in 1800 Beethoven wrote the Horn Sonata in F (also to be performed by cello and piano). In manuscript form it may have been dedicated to Stich. However, when the sonata was published in 1801, Baroness von Braun received the dedication. Punto's fellow countrymen honored him in extraordinary fashion. At his funeral the Mozart Requiem was performed. Indeed his grave was adorned with the following Latin distichon:

Omne tulit punctum Punto, cui Musa Bohemia
Ut plausit vivo, sic moriente gemit.

(Such praise was won by Punto, who is mourned after
In death by the Bohemian muse that exulted him in life.)

He wrote numerous works for his instrument.

44. The sonata can probably better be designated as the Sonata in A major, as indeed it usually is.

45. This scene took place during the memorable concert of December 22, 1808. Different witnesses give different accounts, of

which A.W. Thayer has already presented a few versions. A complete summary is to be found in my articles "The Prussian Court Conductor J.F. Reichardt and Beethoven" in the illustrated Berlin weekly *Der Bär* nos. 14-16, of January 7, 14, and 21, 1888. Ferdinand Ries is obviously painting here with overly garish strokes, as can be recognized from the accounts of Reichardt, Czerny, Dolezalek, Moscheles and others presented in these articles.

46. In our time this probably goes without saying. Whoever has had the chance to peruse G. Nottebohm's both scholarly and painstaking book, *Beethoven's Studies: Volume One: Beethoven's Studies with J. Haydn, Albrechtsberger and Salieri* (Leipzig and Winterthur, 1873), must be completely convinced of the truth of von Seyfried's statement.

47. This, however, is in no way evidence against von Seyfried's statement. For one finds parallel fifths not only elsewhere in Beethoven, but in Mozart as well. It would probably not be at all difficult to find so-called "perfect fifths" (that is, major fifths) even in Haydn and in every great master. And yet, no one will contest the assertion that all of our truly great masters have pursued the most assiduous theoretical studies.

48. The first two sonatas of Opus 31, in G and in D minor, were composed in 1802 and first appeared without opus numbers at the beginning of 1803 in the fifth part of the collection *Repertoire des clavecinistes,* published by H.G. Nägeli in Zurich. Soon thereafter they were in fact published by N. Simrock in Bonn under the title *Deux sonates, pour le Pianoforte composées par Louis van Beethoven. Oeuvre 31 . . . Édition très Correcte.* Then, however, they were published again by J. Cappi in Vienna as Opus 29: *Deux Sonates pour le Clavecin ou Piano-Forte, etc.* The third sonata, in E flat, did not appear until 1804, without opus number, in the 11th part of Nägeli's *Repertoire* and in 1805 was included in the Vienna edition. From this it becomes clear that Beethoven's great anger with the tyrannical Nägeli had subsided. Cappi in Vienna published all three sonatas in 1805 under the new title: *Trois Sonates pour le Clavecin ou Piano-Forte composées par Louis van Beethoven. Oeuvre 29* (!) (in Nottebohm's *Thematic Index* p. 35, 2nd ed.) What still does not cease to astonish is the fact that such a pro-

found, extensive three-part set of sonatas was published without a dedication.

49. They were published in March 1804 by the Bureau d'Art et d'Industrie.

50. Who this young count may have been cannot be traced with certainty. L. Nohl interprets the P. . . . as Count "Palffy," thereby evoking a polemic from Thayer (Thayer II, 201). But Thayer's argument that the number of dots does not fit with "Palffy" with two "f's" is not very sound, for Ries may very well have written this name with one "f" and have thought of it that way.

51. This may very well signify Princess (Josephine von) Liechtenstein, Landgravine of Fürstenberg, to whom the Sonata in E-flat major, Opus 27, No. 1 (March 1802) is dedicated.

52. In the beginning of 1803.

52a. This judgment is, however, very much in need of qualification, as can be inferred particularly from a Beethoven letter dated July 13, 1802 to Breitkopf & Härtel. See the editor's *Beethoven's Complete Letters,* no. 53, and the accompanying notes.

53. These matters belong in the period of Ferdinand Ries's second stay in Vienna. The story of the Kappellmeister position in Cassel belongs in the year 1809.

54. This may have been in 1804. The sonata itself was not published until February 1807.

55. This is August Alexander Klengel, who later became so famous as a contrapuntist; born 1784 in Dresden, died there 1852. His chief work, *Canons et Fugues dan tous les tons majeurs et mineurs pour le Piano*, was published after Klengel's death in 1854 by Moritz Hauptmann through Breitkopf & Härtel. — "At the Swan" refers to the tavern Beethoven so loved to patronize, known variously as *zum Schwan, Schwann, von der Schwann, im Schwann* and *zur Schwann.* It plays a considerable role in his letters to von Zmeskall.

56. This deservedly beloved composition, known under the name "Andante favori," was composed in 1804 and published in May 1806 with the title *Andante pour le Pianoforte compose par Louis van Beethoven.* See also the explanatory note to letter no. 83 (to

Ferdinand Ries) in *Beethoven's Complete Letters.*

57. This is Wenzel Krumpholz, the famous violinist, under whose guidance Beethoven further improved his skills on the violin. On May 2, 1817 he died suddenly on the Promenade. On the day immediately following Beethoven memorialized this most devoted friend with the "Song of the Monks" from Schiller's *Wilhelm Tell* ("Suddenly death comes to man," etc.), written in Aloys Fuchs' family album.

58. Not 1807 but rather 1806.

59. This in turn belongs at the end of 1808, when Beethoven gave the December 22 academy concert already mentioned above, during which the drastic incident described by Ries, Spohr, Reichardt and others took place. Anna Milder, however, did not sing the aria intended for her at that time, for as a consequence of a quarrel between Beethoven and Hauptmann, her future husband, she was not allowed to participate in the concert and was replaced by Schuppanzigh's sister-in-law, Josephine Killitschky, later the wife of Counselor-at-Law Schulze. — Nanette Marconi, the future Frau Schönberger, whom Beethoven also considered on this occasion, was an outstanding contralto from Mannheim.

60. The great Sonata for the Hammerklavier in B was composed in part in 1818 in Mödling. It was published by Artaria in Vienna, in September 1819.

61. A more detailed and accurate account of this is presented in the editor's article "Beethoven in Berlin" (*Nord und Süd*, November 1886).

62. Compare with note 27.

63. In von Seyfried's *Ludwig van Beethoven's Studies,* second edition, appendix, page 10.

64. The C Minor Concerto (Opus 37), dedicated to Prince Louis Ferdinand of Prussia, had already been composed in 1800, before Ferdinand Ries was even Beethoven's student in Vienna. The C Minor Concerto was performed in Beethoven's academy concert of April 5, 1803, at a time when Prince Louis Ferdinand was most probably in Vienna. It was not published until November 1804. It

does not appear, however, that Ries meant precisely this academy concert, because, according to the account of this concert, Beethoven himself was the pianist in the C Minor Concerto.

65. The Concerto in G was ready for the printer in 1807. It was published by the Industriekomtoir in August 1808, with a dedication to Archduke Rudolf.

65a. The theater is called the Kärnthner-Thor Theater.

66. Compare with note 57.

66a. A. & Company = Artaria and Company. Compare Beethoven's stinging letter to Breitkopf & Härtel on the Quintet, which was published for the first time in the editor's *Beethoven's Complete Letters*, (no. 60, of November 13, 1802, with explanatory notes).

67. These were certainly the Twelve Variations in A on the Russian Dance from the Ballet "The Maid of the Forest" by P. Wranitzky, dedicated to Countess von Browne.

67a. Beethoven found himself in an irresolute state of mind when it appeared necessary to find someone to whom to dedicate the Diabelli Variations. At this time Beethoven felt himself beholden not only to Frau Antonia Brentano but also to his friend Ries, then living in London. He vacillated back and forth until the scale tilted more and more in favor of his Frankfurt friend. In Schindler's collection of Beethoven's posthumous papers in the Royal Library in Berlin there is a particularly small scrap of paper on which is written in Beethoven's hand in a lapidary style: "The dedication of the two sonatas in A Sharp and C Minor is to Frau Brentano, born Edle von Birkenstock. — Ries — nothing." Certainly new material for resentment of Ries had accumulated and to him was dedicated neither one of these sonatas nor even the variations; nor to his wife. One of the consequences of this irresolution may have been the fact that Opus 110 and Opus 111 received no dedication at all. For even the dedication of Opus 111 to the Archduke Rudolf did not derive from Beethoven himself but rather from the publisher.

68. It was not the Andante con Variazioni but rather the so-called Andante Favori in F that was composed in 1804 and published in 1806.

68a. Because of his stoutness Schuppanzigh received the nickname "Mylord Falstaff" from Beethoven. In two canons Beethoven's sense of humor glorifies Schuppanzigh's corpulence.

69. The dead friend was probably Lenz (Lorenz) von Breuning, who had already died in 1798 in Bonn; the other, perhaps Pastor Karl Amenda of Kurland or even Franz Wegeler.

70. The name Salomon can probably only bring to mind the violinist Johann Peter Salomon, who was born about 1745 in Bonn. When in later years he chose London as his place of residence, he did a great deal for German music. In 1790 he prevailed upon Haydn to travel to England. He was also frequently in communication with Beethoven. It is now well known that in August 1815 he suffered a calamitous fall from a horse, dying from the effects of that fall in November of the same year. The above letter from Beethoven to Ries was, as we see, written on November 22, 1815. Thus one can only assume that either Beethoven on this day knew nothing of Salomon's death or that Salomon did not die until after November 22, 1815. Compare this with Beethoven's letter of February 28, 1816, in which maestro Salomon's death is painfully lamented.

71. The two-hands piano arrangement of the A major Symphony, dedicated to Empress Elisabeth Alexiewna, is not by Beethoven himself but merely corrected by him. The violin sonata here named is Opus 96 in G, which was completed at the end of 1812. It was published by S.A. Steiner & Company in Vienna in July 1816 along with the Trio in B flat, Opus 97. The latter had already been composed in March 1811, was not given to the engraver until June 1816, and was published on July 16, 1816. With the trio Beethoven made his last public appearances as a pianist: on April 11, 1814, and some weeks later in one of Schuppanzigh's morning concerts in the Prater.

72. A most curious utterance: "I found only one, whom I will most likely never possess." Who may this woman have been? In all likelihood it was Amalie Sebald, later Frau Justizrat Krause, with whom Beethoven had been quite taken for some years. In this same year he expressed himself in a similar vein to the Giannatasio del Rio family when they visited him in Baden in the company of his

nephew. Beethoven said of himself that he had loved unluckily. Five years ago he had become acquainted with a woman, and it would have been the greatest joy of his life to have bound himself more closely to her. It wasn't to be thought of, a near impossibility, a chimera; however it was now just as on the first day. He had still not found this harmony. However, it could not be explained; he still could not put it out of his mind. And all that had been inflicted upon him by the dark-eyed Amalie Sebald.

73. This letter of Beethoven to F. Ries is presented here in only fragmentary form. I had the opportunity to become acquainted with this letter, which is in the possession of an offspring of the Ries family, namely the Berlin composer and music publisher Franz Ries, and was granted permission to copy it with a view to using it. The original manuscript of this letter must also have been available to Herr A.W. Thayer, who published it in its complete form in his Beethoven biography (III,395f), without, of course, mentioning anything about his copy. Here it is intended only to indicate the variant and missing passages in the footnotes. First: Vienna with "v", as often with Beethoven. Orthographic matters will otherwise be overlooked.

74. [Indication of a variant in the German text.]

75. Here the following portion is missing: "In any case I must ask you to apply to H. Birshall for the 10 ducats and to make sure that you yourself get them. I guarantee you on my honor that I have already paid 21 florins for expenses and that is without taking into account the copyist's charges and several postal fees in bank notes. The money was not even sent to me in ducats, although you yourself had written that it was to be sent to me in Dutch ducats. — Are there also in England such conscientious persons who put no value on keeping their word?" — [On the second page are the following words, written on the side]: "NB did you dedicate your Concerto in E flat to Archduke Rudolf? Why did you not write to him yourself about it?" — "With regard to the Trio the publisher here in Vienna has suggested to me that this be published in London on the last day of August. Therefore, I beg you to be so kind as to speak to Herr B. about it. Herr B. can make preparations for the piano arrangement

of the Symphony in A, for I will let you or Herr B. know as soon as the publisher here names the day."

76. Here the following words are omitted: "since his arrival in L."

77. [Indication of a variant in the German text.]

78. In the original: "the quartet."

79. The following words are omitted: "because I also wish to publish it here soon, and what I have to expect with regard to the Cello Sonatas?" The quartet in question here is the Quartet in F Minor, Opus 95, written in 1810 and published in December 1816; the cello sonatas mentioned here are those in C and D (Opus 102), composed in 1815 and published in 1817 in Bonn.

80-82. [Indication of a variant in the German text.]

83. In the original: "Morning cronigle."

84. [Indication of a variant in the German text.]

85. In the original: "by N."

86. The material in parentheses is missing in the original text.

87-88. [Indication of a variant in the German text.]

89. This concluding passage is omitted by Ries: "The piano arrangement of the Symphony in A was copied rapidly, and after a more thorough review I have had the copies changed in some places, of which I will inform you. My best to your wife. In haste, your true friend, Beethoven."

This letter takes up four quarto sheets, three of which are completely covered with writing. There is no envelope. The address on the fourth page — not in Beethoven's hand — reads: "Mr. Ferd. Ries, in care of Mr. B.A. Goldschmid, London" — without street name or number.

90. This is Cypriani Potter, pianist and composer, born in London in 1792; he became director of the Royal Academy of Music there in 1825 and died in London in September 1872. In his travels he had become acquainted with Beethoven in Vienna.

91. The three overtures are: a) to A. von Kotzebue's "Ruins of Athens," which was performed for the first time in February 1812

and published in February 1823; b) the Overture in C, Op. 115, dedicated to the Princess von Radziwill, also called the Name Day [Zur Namensfeier] Overture; it was given its first performance on December 25, 1815. And this overture, which Beethoven at its publication in 1825 characterized — uniquely — as "composed for large orchestra," found no favor with the English, who were otherwise such Beethoven enthusiasts! c) The Overture in E-flat to von Kotzebue's prologue "Hungary's First Benefactor, King Stephan" (Op. 117); it received its first performance in February 1812 along with the *Ruins of Athens* Overture. This overture, as beautiful as it is original, had been given to the publisher in 1815 but was not published until July 1826; it was characterized by Beethoven as a "Grand Overture."

92. The quintet for string instruments (C minor) named here was arranged by Beethoven himself from his C Minor Trio (Opus 1) and received the opus number 104. As is well known, Beethoven wrote at the head of the manuscript: "Trio arranged as a four-voiced quintet by Mr. Goodwill, brought into the daylight by the radiance of five voices, and raised from great misery to considerable respect by Mr. Benevolence. Vienna, August 14, 1817. NB: The original three part quintet score is presented to the lower gods as a solemn burnt offering." Performed on December 10, 1818 in Vienna, the Quintet in C Minor was published in February 1819. See also the editor's *New Beethoven Letters*, page 31.

93. The great Sonata in B-flat (Op. 106), the subject of this and the following letter, was actually published in September 1819.

94. The second movement of this gigantic sonata later received the marking *Scherzo assai vivace*. The metronome reading is $\d. = 80$, not $\d = 80$. The introduction to the great concluding fugue now reads *Largo*. However, one should not overlook the statement in the following letter to Ries: "The sonata was written under oppressive circumstances." What a triumph of pure genius!

95. The great *Missa Solemnis* in D (Opus 123) had already been started in 1818 but was not completely finished until March 1823.

96. Beethoven may now indeed have seriously considered hon-

oring with a dedication the friend and pupil who had been of such assistance to him.

97. It was the city of Boston which at about this time gave instructions to commission from Beethoven an oratorio for its music society — this, however, produced no positive results.

97a. Beethoven is most likely once more speaking of the charming Potter, whom he had already earlier called Botter; therefore, to be correct this should read P. and not B. — Despite these terrible times his divine sense of humor did not desert him. This is how one must grasp the numerous jokes about Ries' beautiful wife. — The Ninth Symphony was, however, later directed to another address, that of Frederick William III, King of Prussia.

98. These are the eleven (or twelve) new Bagatelles for Piano, published as Opus 119.

99. It is the Overture in C (Opus 124), *The Consecration of the House,* dedicated to the Princess von Galitzin. It was performed at the opening of the Josephstadt Theater on October 3, 1822 and was published in 1825 by B. Schott's Sons in Mainz.

100. It is the *Opferlied,* Opus 121b (text by Matthisson — "The flame blazes" —), which according to Schindler's information (II,152), Beethoven composed for a solo voice with choir and orchestra for Prof. Ehler's benefit concert in Pressburg in 1822. This is an arrangement of a much earlier song composed on the same text. The earlier song "The flame blazes" in E was certainly composed before 1800.

101. The Overture in C already mentioned (Op. 115, *Zur Namensfeier*) also known as the "Hunt Overture." This "poetic" overture was published in 1825.

102. The impressive overture *The Consecration of the House*, Opus 124, also in C.

103. During the years of his lengthy guardianship litigation, Beethoven had ample opportunity to concern himself with his "van." One of the *Conversation Books* from 1819 (notebook no. 30, leaf 42a) contains the following remarks in Beethoven's hand concerning the word "van": "Van indicates nobility and patrician status

only (42b) when it stands in the middle between two proper names [i.e. family names], e.g. Bantink van Dieperheim (Halt van Strenloo [or Streelang]). One would get the best information on this insignificant significance from the Dutch."

104. The nephew, Karl van Beethoven, died as a respected citizen in 1858.

105. However, it is an established fact that Beethoven certainly concerned himself with extracts from Kant's philosophy. Particularly Kantian moral philosophy, with its categorical imperative, seemed to have been written as if from his inmost thoughts. Bearing witness to this are passages in the *Conversation Books*, for example in number 22 from Spring 1820, where Beethoven himself recorded in lapidarian hand (leaf 17a): "The moral law within us, and the starry sky above us. — Kant!!!"

106. Mention of Beethoven's finger infection is also made in a recently released letter from this time in which the composer writes: "My colic is better, but yesterday my poor finger had to undergo a serious nail operation. When I wrote to you yesterday it looked very threatening. Today it is quite limp with pain." (*New Beethoven Letters*, p. 38).

107. In 1823, just as he was composing the Ninth Symphony, Beethoven suffered from a serious eye disease, traces of which lingered until the beginning of 1824. See the editor's detailed treatise "Beethoven's Eyes and Eyelids" in the first Beethoven issue of *Musik* (1903).

108. The name of this ill-famed man was Dr. *Wawruch*.

109-112. [References are to page numbers in Kalischer's edition.]

113. These words of Rellstat resulted from his visit to Beethoven in 1825. All of this as well as everything else concerned with Mozart and Beethoven is exhaustively detailed in the editor's article "Beethoven's Connections with Mozart," in the first Mozart issue of *Musik* (October 1904).

114. The editor should not like to leave unmentioned that he has scrutinized many volumes of the memoirs of the Marquise von

Abrantes (the wife of General von Junot) without finding this statement or anything else from her pen dealing with Beethoven. Dr. Wegeler must have owned a very special edition. I would be very grateful to anyone who could assist me in locating these words in the original.

115-119. [References are to page numbers in Kalischer's edition.]

120. This passage with its terse manner of expression offers further evidence of Beethoven's acquaintance with Kant's works; it appears to me to be reminiscent of Kant's *Anthropology*.

121. This is confirmed by A. Schindler. It was J.A. Stumpf, the German-English Beethoven enthusiast, who in 1826 presented Beethoven with the complete works of Handel in 40 folio volumes. The composer's pleasure at this was infinitely great (see Schindler, II, 139, note).

122. See note 123.

123. These verses are by Schiller — a fact which I discovered quite by chance only a few years ago. They are from *Don Carlos*, the speech of Marquis Posa to the Queen in Act 4, scene 21.

124. [The reference is to the page number in Kalischer's edition.]

125. The name of the librettist is Soulié. However, the song had already been published as a supplement to the *Leipziger Allgemeine Musikalische Zeitung* of November 22, 1809 with the title *Als die Geliebte sich trennen wollte*. For further information on this see Nottebohm, *Thematic Index*, p. 179 (1868).

INDEX

Entries preceded by the letter "k" refer to the number of the note by A. C. Kalischer. These notes are indicated by bold superscripts in the text, and begin after the text, on page 168. Index entries in bold type refer to illustrations.

Albrechtsberger, Johann Georg 75
Apponyi, Count Anton Georg 33
Artaria and Co. (publisher) 30, 107, 113, k34, k60, k66a

Bach, Johann Sebastian 33, 73, 153
Baden 79, 116, 117, k72
Baums, Gertrude (godmother) 9, 11
Beethoven, Caspar Anton Carl van (brother) 12, 76-77, 82, 104, 112, 121, 123, k22
Beethoven, Johann van (father) 7, 13-14, 109, **110**, 148, k3
Beethoven, Karl van (nephew) 123, **124**, 147, k7, k22, k72, k104
Beethoven, Ludwig van (grandfather) 7, 12-14
Beethoven, Ludwig van
    Birth and family history 7-12, 46, 146-147, k2, k3, k4-k8, k103
    Characteristics
        absentmindedness of 86-87, 107-108
        appearance of 14
        clumsiness of 106
        generosity of 28, 114, 121-122, 132
        temperament of 34-35, 52, 67, 71, 73, 76, 80, 83-85, 88-90, 102, 108, 116-119, k20, k22, k45, k48, k59
    Diplomas and honors 49, 97
    Education, general 14, 148
    Education, music
        organ playing 17-18, k11, k12, k12a
        teachers 14, 17, 75, 106, k10, k11, k12
    Finances 28, 36-38, 59, 84, 100, 109, 110, 121-126, 131-132, 133-135, 137, k75
    Health
        deafness 26, 28-29, 40, 45, 86-87
        illnesses 40-41, 51, 132, k26, k28, k106, k107

Beethoven, Ludwig van (continued)
- Music
    - composing 17, 22, 25, 54, 65-67, 71-72, 76, 80, 87-88, 92, 95, k16, k34, k107
    - conducting 66, 73, 102, k45
    - improvisation 21, 25, 68, 71, 87-88, 97
    - performing
        - chamber musician, as 21, 32, 69
        - organist, as 18, 20, k12a
        - pianist, as 23, 25, 32, 34, 54, 70-71, 80-82, 94, k24
        - violinist, as 106, k31, k57
- Portraits *frontispiece*, *viii*, 30, *47*, 50, 51, 124, 129, *160*-161
- Relationships
    - von Breunings 14-16, k1, k31
    - enemies and rivals 56, 59, 70-71, 83, 88-89, 97-98, k42, k55
    - guardian, as 123-124, 134, k103
    - parents, to 13-14, 24, 109, 147
    - women, with 41, 43, 52, 104-105, 124, k28, k30, k32, k34, k36, k72
- Residences 11, *12*, 30, 66, 79, 100, 116-117, *147*, k27
- Teacher, as 14, 24, 80, 82, 83
- Travels, 23-24, 96,155-156

Beethoven, Ludwig Maria van (older brother) 9, 46
Beethoven, Maria Magdalena Kewerich (mother) 7-9, 13, 49, 109, k3
Beethoven, Nicolaus Johannes 12, 83, 85, 116, 138, *139*
Belderbusch, Countess Barbara (née Koch) 55
Bernhard, Karl k9
Berlin 60-94, k9, k37a, k42, k61
Bonaparte, Napoleon 57, 68
Bonn 4-5, 31, k1, k2, k4, k7, k10, k16, k34, k36, k48, k69, k70
Boosey and Co. (publishers) 110, 138
Brentano, Frau Antonie von k67a
Breuning, Christoph von 15, 30, 39, 155
Breuning, Eleonore Brigitte von xii, 15, 31, 48, 50, 52-54, 56-57, k34, k36
Breuning, family von *15*, *16*, 24-25, 30-31, 48, 50, 53, 56, 57, k1, k20
Breuning, Frau Helene von 15-16, 24, 31, 39, 42, 56, k31
Breuning, Gerhard von 158, 162
Breuning, Lenz von 15, 156, k69
Breuning, Stephan von xiv, 5, 15-16, *30*, 42-45, 50-51, 77, 86, 91, 100, 116-119, 155-156, 159-163, k31, k37

Bridgetower, George Augustus 72
Browne-Camus, Count Johann Georg von 79-80, 104
Browne-Camus, Countess Anna Margarette von k67

Clementi and Co. (publishers) 110
Clementi, Muzio 88-89
Cologne 7, 9, 18
Coutts and Co. (bankers) 121, 122
Cramer, Johann Baptist 87, 128
Czerny, Carl 157, k45

Döbling (residence) 86-87, 100, 117
Dunst (publisher) 22, 54, 93, k34
Duport, Pierre (cellist) 97

Eeden, Heinrich van den [van der Eden] 17, k11
Esterhazy, Prince Paul 129
Esterhazy, Princess Marie 80

Förster, Emmanuel Aloys 34
Friedrich Wilhelm II, King of Prussia 10, 48, 96, k9
Friedrich Wilhelm III, King of Prussia k97a
Fries, Count Moritz von 71

Gelinek, Abbé Joseph 57, k35
George IV, King of England 95-96, 136
Giucciardi, Giulietta k28, k30
Goethe, Johann Wolfgang von 39

Handel, Georg Friedrich 73, 153, 155, 158, k121,
Hatzfeld, Countess Maria von 22
Haydn, Franz Joseph 22, 32, 67, 74-75, 113, k13, k46, k47, k70
Heiligenstadt 66, **67**, k28
Heller, Ferdinand 20
Himmel, Friedrich Heinrich **97**-98
Hogarth, William 104, 109
d'Honrath, Jeanette 42

Kant, Immanuel 148, k105, k120
Kärnthner-Thor Theater 102, k65a
Kewerich, Heinrich (maternal grandfather) 9, k3, k4
Kewerich, Maria Westorffs (maternal grandmother) 8, k3, k4
Kinsky, Prince Ferdinand Johann Nepomuk 84

Kirnberger, Johann Phillip 64
Klengel, August Alexander 88, k55
Kraft, Anton 32, k19
Kraft, Nikolaus 32, k19
Krumpholz, Wenzel 89, k57

Lichnowsky, Prince Karl 27, 32-36, 60, 66, 70, 74, 89-91, **92**, 103, k19, k39
Lichnowsky, Princess Maria Christiane 34, 91, **92**
Lichnowsky Quartet 32, k19
Liechtenstein, Princess Josephine Sophia von 81-82, 120, k51
Linke, Joseph 32, k19
Lobkowitz, Prince Franz Joseph von 67, 84, k19
London Philharmonic Society 124, 126-129, 138
Louis Ferdinand, Prince of Prussia 98-99, k64

Malchus, , Karl August von 53, 55, 152
Malfatti, Therese von k32
Matthisson, Friedrich von (poet) 46, 61, k33, k100
Maximilian Franz, Archduke (Elector of Cologne) 17-18, **56**
Maximilian Friedrich (Elector of Cologne) 13
Mayer [Meyer], Sebastian 92
Mölker-Bastei (residence) 30, 39, 100, k27
Mozart, Wolfgang Amadeus 54, 73, 75, 153, 155, k1, k17, k34, k43, k47, k113
Music
    Arrangements 68, k15, k28, k37, k71, k75, k89, k92, k100
    Manuscripts 100-101, 111-112
    Posthumous publications 111-112
    Works *see separate Index of Musical Compositions*

Nageli, Hans Georg (publisher) 76-77
Neate, Charles 125-126, 127-129, 133
Neefe, Christian Gottlob 17, *18*, 14n, k10, k12

Oberdöbling (residence) 115
Odescalchi, Princess Barbara (neé Keglevics) 82, k24

Paer, Ferdinando 70, k41
Paisiello, Giovanni k41
Palffy, Count Ferdinand 80, k50
Paraquin, Johann Baptist 57
Pasqualati [Pasquillati] house (residence) 100, k27
Pfeiffer, Tobias 17, k11

Pleyel, Ignaz Joseph 21, k13
Potter [Botter], Cipriani 128, k90, k97a
Publishers 76-79, 82, 107, 113, 129, 138, k48
Punto [Ponto], Giovanni *see* Stich, Johann Wenzel

Ram, Friedrich 69
Razumovsky, Count Andreas k19
Reichardt, Johann Friedrich 84, k45, k59
Ries, Ferdinand *Introduction*, 1-3, 26, 31, **63**-64, 79-90, 101-106, *142*, k17, k34, k38, k53, k56, k64, k67a, k97a
Ries, Franz (father of author) xix, 13, 21, 23, 26, 31, 44, 65, 82, 103
Röckel, Joseph August 93
Romberg, Andreas 23, 55
Romberg, Bernhard 21, 23, 55, k13
Rudolph, Johann Joseph Rainier, Archduke 84, **99**, 101, 125, 132, 136

Salieri, Antonio 75, k46
Salomon, Johann Peter 122, 123, k70
Schindler, Anton xiv-xix, *xv*, k18, k19, k24, k43, k67a, k100, k121
Schmidt, Dr. Johann Adam 40, k28
Schuppanzigh, Ignaz 32, 116, k19, k68a
Seyfried, Ignaz Ritter von 14, 17, 37, 38, 42, 75, 85, k12a, k23, k26, k46, k47, k63
Simrock, Nikolaus 23, 77, **78**, 163, k48
Smart, Sir George 127-128
Steibelt, Daniel 70-71, k42
Sterkel, Abbé 23
Stich, Johann Wenzel [also Ponto or Punto] 71, k43

Theater-an-der-Wien 65, 72, 78, 100, 117

Vering, Dr. Gerhard Ritter von 28, 44
Vogler, Abbé Georg Joseph 17

Waldstein, Count Ferdinand Ernst von *19*, 20-21, 22, 89
Wegeler, Dr. Franz Gerhard *Introduction*, *1*, 4-6, 16, 25, 27, 48, 51, 59, 63, k1, k4, k20, k33, k69
Weiss, Franz 32, k19, k43

Zmeskall von Domanovecz, Nikolaus 32, k55

Index of Musical Compositions

Entries preceded by the letter "k" refer to the number of the note by A. C. Kalischer. These notes are indicated by bold superscripts in the text, and begin after the text, on page 168.

Orchestral Music

Symphonies
No. 2 in D (Opus 36) 66-67, k38, k39
No. 3 in E-flat (Opus 55) 67
No. 5 in C minor (Opus 67) 72
No. 6 in F (Opus 68) 72
No. 7 in A (Opus 92) 121-122, 126, k71, k75, k89
No. 9 in D minor (Opus 125) 135, 136, 139, k97a, k107
Wellington's Victory (Battle of Vittoria) (Opus 91) 95-96, 124, 126, 135, 136

Piano Concertos
No. 1 in C (Opus 15) 38, 94, k24
No. 3 in C minor (Opus 37) 66, 101-103, k64
No. 4 in G (Opus 58) 102-103, k65
No. 5 in E-flat (Opus 73) 121-122, k71, k75

Other Concertos
Violin (Opus 61) 82, 118, k31

Overtures & Miscellaneous
Consecration of the House (Opus 124) 138, 139, k99, k102
King Stephan (Opus 117) 129, k91
Ritterballet (WoO 1) 22, k15
Ruins of Athens (Opus 113) 112, 129, 139-140, k91
Zur Namensfeier (Opus 115) 129, 140, k91, k101

Chamber Music

Sonatas
Piano and Cello in C and D minor (Opus 102) 125, k79
Piano and Horn (Opus 17) 71, k43
Piano and Violin in A minor (Opus 23) 81-82, k44
Piano and Violin in A (Kreuzer, Opus 47) 72, 115, k68
Piano and Violin in G (Opus 96) 121-122, k71
Variations for Piano and Violin on Mozart's "Se vuol ballare" ("Variations No. 1", WoO 40) 15, 54, 57, k34

Trios

Piano, Clarinet and Cello in B-flat (Opus 11) 70
Piano, Violin and Cello (Opus 1, Nos. 1-3) 32, 74
Piano, Violin and Cello in B-flat (Archduke, Opus 97) 121-122, k71, k75
String Trio (Opus 3) 33

String Quartets

Opus 18, No.1 (F major) 91
Opus 18, No. 3 (D) 91
Opus 18, No.4 (C minor) 76-77
Opus 95 (F minor) 125, k78, k79
Opus 104 (C minor [arrangement of Opus 1 No.3]) 129, k92
Opus 127 (E-flat) k19
Opus 132 (A minor) k19

Other

Quintet for Piano and Wind Instruments (Opus 16) 69
Septet (Opus 20) 82, k28
String Quintet in C (Opus 29) 107, k66a
String Quintet in E-flat (Opus 4 [revision of Opus 103]) 26

Keyboard Music

Sonatas

Opus 2, Nos. 1-3 (F minor, A, and C) 112-113
Opus 13 (C minor, Pathetique) 94
Opus 26 (A-flat) 70
Opus 27, No.1 (E-flat) k30, k51
Opus 27, No.2 (C-sharp minor, Moonlight) k30, k51
Opus 31, Nos. 1-3 (G, D minor and E-flat) 76, 81, k48
Opus 53 (C, Waldstein) 89-90
Opus 57 (F minor, Appassionata) 87
Opus 106 (B, Hammerclavier) 94-95, 129-131, k60, k93
Opus 110 (A-flat) 110
Opus 111 (C minor) 110
WoO 47 (E-flat, F minor, and D, Electoral) 22
WoO 51 (C) 55, k34, k36

Miscelllaneous

Andante in F (WoO 57) 89, k56, k68
Eleven Bagatelles for Piano (Opus 119) 137, k98
Marches for Four Hands (Opus 45) 80
Variations in F major (Opus 34) 82-83
Variations on "Nel cuor più non mi sento" (WoO 70) 70

Variations on "Vieni amore" (WoO 65) 22-23
Variations on a Russian Dance (WoO 71) 107, k67
Variations on a Waltz by Diabelli (Opus 120) 110, 136-137, k67a

Vocal Music

Cantata on the Death of Emperor Joseph II (WoO 87) 22, k14
Cantata on the Elevation of Leopold II (WoO 88) 22, k14
Christ on the Mount of Olives (Opus 85) 65, k38
Fantasia for Piano, Chorus and Orchestra in C minor ("Choral Fantasy", Opus 80) 72-73
Fidelio (Opus 72) 58-60, 90-93, k37, k37a
Missa Solemnis in D (Opus 123) 133, 140, k95

Songs

"Als die Geliebte sich trennen wollte" ("Empfindungen bei Lydiens Untreue", WoO 132) 163-165, k125
"Der freie Mann" ("Wer ist ein freier Mann?", WoO 117) 46, 61, k33
"Die Klage" (WoO 113) 62, 143
"Opferlied" (Opus 121b) 61, 140, k33, k100,
"Wenn jemand eine Reise thut" (Opus 52, No.1) 22